D0897940

Wave Propagation in Solids and Fluids

Forthcoming from Springer-Verlag:

Wave Propagation in Electromagnetic Media
by Julian L. Davis

Julian L. Davis

Wave Propagation in Solids and Fluids

With 58 Illustrations

Springer-Verlag
New York Berlin Heidelberg
London Paris Tokyo

Julian L. Davis
22 Libby Avenue
Pompton Plains, NJ 07444
U.S.A.

Library of Congress Cataloging-in-Publication Data
Davis, Julian L.
Wave propagation in solids and fluids.
Bibliography: p.
1. Solids—Mathematics. 2. Fluids—Mathematics.
3. Waves. 4. Differential equations, Partial.
5. Hamilton-Jacobi equations. I. Title.
QC176.8.W3D38 1988 531′.1133 88-4618

Typeset by J.W. Arrowsmith Ltd., Bristol, England.
Printed and bound by R.R. Donnelley & Sons, Harrisonburg, Virginia.
Printed in the United States of America.

9 8 7 6 5 4 3 2 1

ISBN 0-387-96739-7 Springer-Verlag New York Berlin Heidelberg
ISBN 3-540-96739-7 Springer-Verlag Berlin Heidelberg New York

Preface

The purpose of this volume is to present a clear and systematic account of the mathematical methods of wave phenomena in solids, gases, and water that will be readily accessible to physicists and engineers. The emphasis is on developing the necessary mathematical techniques, and on showing how these mathematical concepts can be effective in unifying the physics of wave propagation in a variety of physical settings: sound and shock waves in gases, water waves, and stress waves in solids. Nonlinear effects and asymptotic phenomena will be discussed.

Wave propagation in continuous media (solid, liquid, or gas) has as its foundation the three basic conservation laws of physics: conservation of mass, momentum, and energy, which will be described in various sections of the book in their proper physical setting. These conservation laws are expressed either in the Lagrangian or the Eulerian representation depending on whether the boundaries are relatively fixed or moving. In any case, these laws of physics allow us to derive the "field equations" which are expressed as systems of partial differential equations. For wave propagation phenomena these equations are said to be "hyperbolic" and, in general, nonlinear in the sense of being "quasilinear". We therefore attempt to determine the properties of a system of "quasilinear hyperbolic" partial differential equations which will allow us to calculate the displacement, velocity fields, etc. The technique of investigating these equations is given by the method of characteristics, which will be described in various portions of the book where appropriate. This method is essential in investigating large amplitude waves which are by nature nonlinear.

One of the unique features of this book is the treatment of the Hamilton–Jacobi theory in the setting of the variational calculus. It will be shown that this theory is a natural vehicle for relating classical mechanics and geometric optics; it guides us in developing the Schrödinger equation of quantum mechanics which will be taken up more fully in the second volume. The asymptotic expansion approach of J. Keller and his associates shows us that geometric optics is the infinite frequency approximation of an asymptotic expansion in powers of the wave number. The succeeding terms yield diffraction information. An asymptotic approach which handles nonlinear phenomena is also used in the chapter on water waves, following the methods of K.O. Friedrichs, J.J. Stoker, and their associates.

We envision a two-volume set. This first volume is concerned with wave propagation phenomena in nonconducting media. By a nonconducting medium we mean a continuum (solid, liquid, gas) that is not electrically or magnetically conducting. The second volume will be concerned with wave propagation in electromagnetically conducting media, and will cover such topics as wave propagation in electromagnetic media, plasmas, magnetohydrodynamics, and quantum mechanics including relativistic effects. Each volume is relatively self-contained and independent of the other.

The books are designed to be interdisciplinary in nature, in the spirit best expressed by the following quote from the Preface to *Methods of Mathematical Physics*, Volume 1, 1953, by Richard Courant:

> Since the seventeenth century, physical intuition has served as a vital source for mathematical problems and methods. Recent trends and fashions have, however, weakened the connection between mathematics and physics (and the engineering sciences); mathematicians turning away from the roots of mathematics in intuition, have concentrated on refinement and emphasized the postulational side of mathematics, and at times have overlooked the unity of their science with physics (and the engineering sciences). In many cases, physicists (and engineers) have ceased to appreciate the attitudes of mathematicians. This rift is unquestionably a serious threat to science as a whole; the broad stream of scientific development may split into smaller and smaller rivulets and dry out. It seems therefore important to direct our efforts toward reuniting divergent trends by clarifying the common features and interconnections of many distinct and diverse scientific facts. Only thus can the student attain some mastery of the material and the basis be prepared for further organic development of research.

These remarks of Professor Courant are as timely today as they were thirty-four years ago.

Julian L. Davis
January 15, 1988

Contents

CHAPTER 1

Oscillatory Phenomena

Introduction

The nature of wave propagation involves the oscillations or vibrations of a dependent variable such as gas pressure or lateral displacement of a string with respect to space and time. At a fixed point in the material the particle oscillates sinusoidally performing simple harmonic motion, if we neglect damping. On the other hand, a snapshot of the wave form at a fixed time shows a sinusoidal picture (again, for the case of no damping). These two types of vibrations, both in time and space, make the subject appear to be quite complicated for the beginner. Therefore, for pedagogical reasons we feel it is best to introduce the subject of wave propagation by an investigation of the harmonic motion of masses about their equilibrium points. Concerning this approach, we have an excellent example in the approach taken by Lord Rayleigh [32, Vol. I, p. 19]. He said, "The vibrations expressed by a circular function (sinusoidal function) of the time ... are so important in Acoustics that we cannot do better than devote a chapter to their consideration, before entering on the dynamical part of our subject (wave propagation)."

1.1. Harmonic Motion

We start with the simple example of a mass–spring combination as given by the following model: A spring is suspended vertically from a fixed point and a mass is attached to it at the lower end. At time $t=0$ the mass is displaced downward a small amount by a distance x_0 from its equilibrium position, and given an initial velocity in the vertical direction by an amount v_0. We let the x axis be directed vertically downward through the equilibrium position, which is the origin $x=0$ (the positive direction of x being below $x=0$). For the simplest case of $x_0 \neq 0$ and $v_0=0$, the mass will perform simple harmonic motion about the equilibrium position with an amplitude equal to x_0. The frequency will be discussed below. We assume the only external force acting on the mass is due to the compression or tension of the spring, and there are no external moments. (The motion is thus linear along the x axis.) Newton's second law of mechanics equates the change of linear momentum of the mass to the external force, yielding

$$m\ddot{x} = -kx, \qquad \ddot{x} \equiv \frac{d^2x}{dt^2},$$

where m is the mass and k is the spring constant in units of force/length. This equation may be rewritten as

$$\ddot{x} + \omega^2 x = 0, \tag{1.1}$$

where

$$\omega^2 = \frac{k}{m}. \tag{1.2}$$

Equation (1.1) is a second-order ordinary differential equation (ODE) with constant coefficients, and is the equation of motion for the simple harmonic oscillator described by the model. We will soon see from the nature of the solution that ω, given by (1.2), is the natural frequency of vibration of the mass in radians/sec. Since the equation is second order we require two conditions. For our problem we take them as *initial conditions* (ICs) on x and $\dot{x}$, namely,

$$x(0) = x_0, \qquad \dot{x}(0) = v_0, \tag{1.3}$$

x_0 is the initial displacement, and v_0 the initial velocity of the mass. We point out that this is not the only type of ICs we may have. For example, we may define the displacement x at two different times. However, we choose the ICs given by (1.3) thus defining an *initial value* (IV) problem. The general solution of (1.1) is

$$x(t) = A\cos(\omega t) + B\sin(\omega t), \tag{1.4}$$

where the constants A and B are to be determined from (1.3). Inserting (1.4) into (1.1) and using (1.3) yields

$$x(t) = x_0\cos(\omega t) + \left(\frac{v_0}{\omega}\right)\sin(\omega t). \tag{1.5}$$

It is easily seen that (1.5) satisfies the ODE (1.1) and the ICs (1.3). The right-hand side of (1.5) shows us that ω is the natural frequency of the mass as it undergoes simple harmonic motion. This is easily seen: Consider the case $v_0 = 0$; then $x = x_0\cos(\omega t)$. Clearly x_0 is the amplitude of the sinusoidal oscillations. If T is the period of oscillation then $\omega T = 2\pi$ so that $\omega = 2\pi/T$, which is the radial frequency.

We may put the solution given by (1.5) in the following form

$$x = c\cos(\omega t - \phi), \tag{1.6}$$

where c is the amplitude of x and ϕ is the *phase angle* which represents the phase of the vibration from $t = 0$. c and ϕ are now determined as known functions of the ICs x_0 and v_0. We expand the right-hand side of (1.6), use (1.5), equate coefficients of cosine and sine, and thereby obtain

$$c^2 = (x_0)^2 + \left(\frac{v_0}{\omega}\right)^2, \qquad \tan\phi = \frac{v_0}{\omega x_0}. \tag{1.7}$$

Clearly if the mass is initially at rest $v_0 = 0$, so that $c = x_0$ and the phase angle $\phi = 0$. The solution given by (1.5) or (1.6) is the simplest example of the *free vibrations* of a simple harmonic oscillator without damping. The ODE given by (1.1) is said to be homogeneous, since the right-hand side is zero.

Damped Free Vibrations

If the mass in the mass–spring model (given above) is considered to move in a viscous medium, then it is subject to an additional external force in the form of a *resistive force* proportional to the velocity of the mass. Figure 1.1 shows the model and the external forces acting on the mass (for simplicity x is shown horizontally). The resistive force is equal to $-b\dot{x}$ where the positive constant b represents damping. The ODE becomes

$$m\ddot{x} = -k\dot{x} - bx.$$

This is the equation of motion for a damped harmonic oscillator with no forcing function, i.e., with no external force other than the spring and resistive or damping force. The right-hand side represents the sum of the spring and the damping force. The reason for the "−" signs is that both these forces act as restoring forces, driving the mass to its equilibrium position. We may rewrite this ODE as

$$\ddot{x} + 2r\dot{x} + \omega^2 = 0, \qquad r = \frac{b}{2m}. \tag{1.8}$$

Clearly (1.8) is a generalization of (1.1) where the $2r\dot{x}$ term represents viscous damping. It is also a linear second-order homogeneous ODE with constant coefficients; also requiring two ICs. This tells us that there exist two unique linearly independent solutions, meaning: one solution does not depend on the other. From the nature of exponential functions, we assume

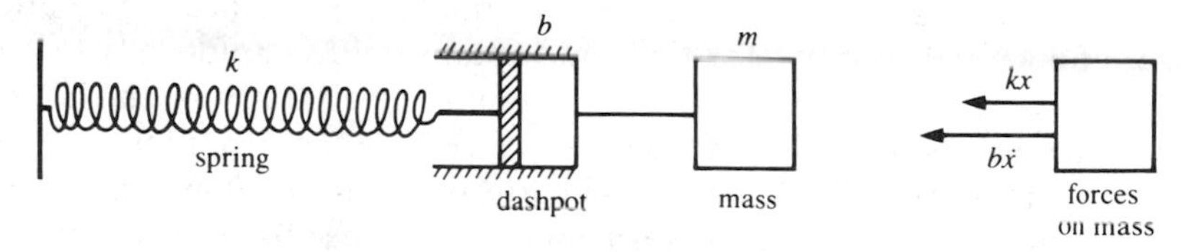

Fig. 1.1. Forced damped harmonic oscillator.

a solution of the form

$$x = \exp(\lambda t).$$

Substituting this expression into the ODE (1.8) we get the following algebraic equation for the constant λ:

$$\lambda^2 + 2r\lambda + \omega^2 = 0. \tag{1.9}$$

This quadratic equation for λ has three types of roots:

(1) real and unequal $\lambda_{1,2} = -r \pm \sqrt{r^2 - \omega^2}$, $r > \omega$;
(2) real and equal $\lambda_1 = \lambda_2 = -r$, $r = \omega$;
(3) complex conjugates $\lambda_{1,2} = -r \pm iq$, $q = \sqrt{\omega^2 - r^2}$, $r < \omega$.

Case (1) represents overdamping or nonoscillatory motion. Case (2) is critically damped. Case (3) represents damped oscillatory motion. The general solution of (1.8) can be written as

$$x = A\exp(\lambda_1 t) + B\exp(\lambda_2 t), \tag{1.10}$$

where A and B are arbitrary complex constants and λ_1 and λ_2 are given by the above three cases. Note that we want the real part of the right-hand side of (1.10). The solutions in real form corresponding to these cases are readily found to be

$$\begin{aligned} &(1) \quad x = c_1\exp(\lambda_1 t) + c_2\exp(\lambda_2 t), \qquad \lambda_1, \lambda_2 \text{ real},\\ &(2) \quad x = \exp(-rt)[c_1 t + c_2],\\ &(3) \quad x = \exp(-rt)[c_1 \cos qt + c_2 \sin qt]. \end{aligned} \tag{1.11}$$

The real constants c_1 and c_2 are determined from the two ICs.

A word about case (2): $\exp(-rt)$ is one solution. How do we obtain the other linearly independent solution for the case $r = \omega$? Recall that we have two ICs to satisfy. One method is to assume a solution of the form $x = f(t)\exp(-rt)$ and solve for function $f(t)$ by substituting back into (1.9) and setting $r = \omega$. This gives $\ddot{f} = 0$, which yields the result. Another more elegant method is to let r differ from ω by a small amount so that $\lambda_2 - \lambda_1$ is a very small quantity. Then, from the linearity of the ODE, the expression $(1/(\lambda_2 - \lambda_1))[\exp((\lambda_2 - \lambda_1)t)]$ is also a solution. In the limit as $\lambda_2 \to \lambda_1$, we have $(d/dt)\exp(-rt)$ as the required additional solution for this case. It is clear that for large t this solution damps out since t increases linearly, but the factor $\exp(-rt)$ decreases exponentially with t. Since the solution is zero for $t = 0$, is positive for $t > 0$, and damps out asymptotically to zero as $t \to \infty$, we have a maximum for the solution. It is easily seen that the maximum occurs for $t = 1/r$. Of course, we require the condition $r > 0$ for damping. This critically damped case corresponds to $q = 0$ and represents the transition from case (1) (overdamped or nonoscillatory) to case (3) (oscillatory).

If we set $r=0$ we get the undamped case which was already discussed. This is easily seen since $\lambda_{1,2}$ becomes $\pm i\omega$ and the solution is given in terms of sinusoidal functions (1.4). This can immediately be obtained from (1.11), case (3), by setting $r=0$.

For case (3), (1.11) may be put in the form

$$x = c\exp(-rt)\cos(qt-\phi), \tag{1.12}$$

where c and the phase angle ϕ are given by

$$c=\sqrt{(c_1)^2+(c_2)^2}, \qquad \tan\phi = \frac{c_2}{c_1}. \tag{1.13}$$

Equation (1.12) easily displays *damped harmonic oscillations.* The amplitude of the solution is the exponential decay expression $c\exp(-rt)$ and the cosine factor is a sinusoidal function with a radial frequency equal to $q=\sqrt{\omega^2-r^2}$. The multiplication of these two factors tells us that x varies with t as a damped sinusoidal function with a frequency q, the *damped natural frequency*, which is smaller than the undamped natural frequency ω of the undamped harmonic oscillator. Note that, for a fixed ω as r approaches ω from below, the damped frequency q becomes smaller and tends to zero as $r\to\omega$. This means that the critically damped case exhibits the following asymptotic phenomenon: It is the limit of the damped oscillations as the period of the damped oscillations becomes infinite. For r a little less than ω, q is very small or the period $T=2\pi/q$ is very large. For r a little greater thn ω, q and T are imaginary. In Fig. 1.2 the amplitude of x is plotted as a function of t for a given r. This is seen to be the two envelopes of the cosine curve for a given q and ϕ, meaning that the maxima and minima are tangent to the envelopes. In the engineering literature, the damping

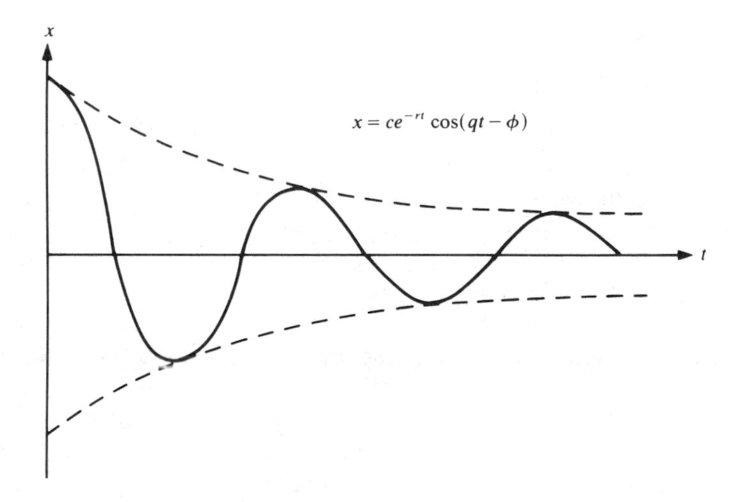

Fig. 1.2. Damped harmonic oscillations.

factor is frequently called the *logarithmic decrement* of the damped oscillation, meaning that the logarithm of the amplitude decreases at the rate r.

Satisfying the Initial Conditions

We assume ICs given by (1.3). We now show how to determine the constants c_1 and c_2 for the solutions given by (1.11).

Case (1): To satisfy the ICs we obtain the following algebraic equations for the constants:

$$c_1 + c_2 = x_0, \qquad \lambda_1 c_1 + \lambda_2 c_2 = \dot{x}_0.$$

Their unique solutions are

$$c_1 = \frac{\dot{x}_0 - \lambda_2 x_0}{\lambda_1 - \lambda_2}, \qquad -c_2 = \frac{\dot{x}_0 - \lambda_1 x_0}{\lambda_1 - \lambda_2}.$$

Case (2): Again using the ICs (1.3) we obtain

$$c_1 = r x_0 + x_0, \qquad c_2 = x_0.$$

Case (3): We insert the ICs into the solution given by (1.12). The equations determining the constants take the form

$$c \cos \phi = x_0, \qquad c(-r \cos \phi + q \sin \phi) = x_0.$$

The solutions for ϕ and c are

$$\phi = \arccos\left(\frac{x_0}{c}\right), \qquad c = \sqrt{[(x_0)^2 + (x_0 + r x_0)^2]}.$$

We have thus shown that, for each of the three cases, the general solution can be made to fit any arbitrary set of ICs.

Uniqueness of Solutions

How do we know that there is not at least another solution for any of the three cases that satisfies the same ICs? Well, there is a general theorem in the theory of differential equations that states that there is a unique solution for any given set of ICs, under rather general conditions. It is part of a classical existence and uniqueness theorem for ODEs. Rather than prove this theorem in its full generality, we give a proof of uniqueness for the special case of the damped harmonic oscillator. We are interested in proving that there is a unique solution to (1.8) for the ICs given by (1.3). The proof is by contradiction. Suppose there are two different solutions $u(t)$ and $v(t)$ to (1.8) both of which satisfy (1.3). Then it is clear that the difference $w(t) = u(t) - v(t)$ also satisfies the ODE (1.8). However, the ICs for $w(t)$ are $w(0) = 0$ and $\dot{w}(0) = 0$. These are called *homogeneous* ICs, since the right-hand sides are zero. For each of the three cases given by (1.11) an easy calculation gives $c_1 = c_2 = 0$, which immediately yields the fact that, for each case, $w(t)$ is zero for all $t > 0$. This gives the contradiction and completes the proof. There is another more elegant proof which involves integrating the ODE for $w(t)$. We multiply each term of the ODE by $2\dot{w}$ and recall that $2\dot{w}\ddot{w} = (d/dt)(\dot{w}^2)$ and $2w\dot{w} = (d/dt)w^2$. We thus obtain

$$\left(\frac{d}{dt}\right)(\dot{w}^2) + \left(\frac{d}{dt}\right)(\omega^2 w^2) + 4r\dot{w}^2 = 0.$$

If we integrate between $t = 0$ and $t = t$ and use the ICs $w(0) = \dot{w}(0) = 0$ we get

$$\dot{w}^2(t) + \omega^2 w^2(\bar{t}) + 2r \int_0^{\bar{t}} \left(\frac{dw}{dt}\right)^2 dt = 0.$$

The left-hand side of this equation is strictly positive for all $\bar{t} > 0$, unless $w(\bar{t})$ is identically zero. This proves our uniqueness theorem.

1.2. Forced Oscillations

The next stage in the discussion of the oscillations of a single particle, under the action of a spring restoring force and a viscous resistance, is an investigation of the *forced* oscillations that occur when a time-varying external force or *forcing function* is applied. In this analysis the forcing function is chosen as a sinusoidal function of t. The analysis is much less messy if, instead of using sines and cosines for the external force, we use complex variable notation. We recognize that sinusoidal functions of pt are the real parts of linear combinations of $\exp(\pm ipt)$, where p is the frequency of the external force. There is no loss in generality in assuming the external force to be of the form $c \exp(ipt)$, where c is the amplitude of the external force. A similar analysis holds for $\exp(-ipt)$ and suitable linear combinations will solve any given problem. Here we are invoking the *principle of superposition* which states that the solution to the sum of two or more external forces is equal to the sum of the solutions to each individual external force. After obtaining the solution for $x(t)$ in complex form, we must take the appropriate real part that satisfies the given applied external force.

Our ODE for a damped harmonic oscillator with a sinusoidal external force then becomes the following generalization of (1.8):

$$\ddot{x} + 2r\dot{x} + \omega^2 x = c \exp(ipt) = f(t). \tag{1.14}$$

Equation (1.14) is a nonhomogeneous linear second-order ODE with constant coefficients. The general solution of this equation is equal to the general solution of the homogeneous equation (called the *complementary solution*) plus *any* solution of the nonhomogeneous equation. Physically, this means that if we have forced oscillations due to an external force, and on it superimpose an arbitrary free oscillation (complementary solution), we obtain an oscillatory motion which satisfies the same nonhomogeneous equation as the original forced oscillations. If viscous damping is taken into account, the free motion will damp out exponentially as time goes on, and we will be left with the motion solely due to the external force. In physical language, this solution is called the *steady state* solution. Since the free motion, given by the complementary solution, involves the two ICs, as this motion damps out, the resulting motion does not depend on the ICs.

We now determine a particular integral of (1.14). Since any solution will do, we attempt to simplify the analysis by assuming a solution that has the same frequency p as the external force. To this end, we assume a solution of the form

$$x(t)=\bar{x}\exp(ipt), \tag{1.15}$$

where $\bar{x}$ is independent of t. Its dependence on p will now be determined. If we insert (1.15) into (1.14) we observe that we can factor out the exponential terms. Indeed, this is the reason for choosing the form given by (1.15). Solving for $\bar{x}$ gives

$$\bar{x}=\frac{c}{\omega^2-p^2+2irp}. \tag{1.16}$$

It is easy to verify that (1.15) and (1.16) satisfy the ODE (1.14). To express the meaning of this result clearly we must manipulate (1.16) into a convenient form. Specifically, we multiply the right-hand side of (1.16) by its complex conjugate divided by its complex conjugate. (Replace i by $-i$.) We obtain

$$\bar{x}=c\left[\frac{\omega^2-p^2-2irp}{(\omega^2-p^2)^2+4r^2p^2}\right]=c\alpha\exp(-i\beta), \tag{1.17}$$

where the positive *distortion factor* α and the *phase displacement* angle β are given by

$$\alpha=\frac{1}{\sqrt{[(\omega^2-p^2)^2+4r^2p^2]}}, \qquad \tan\beta=\frac{2rp}{\omega^2-p^2}. \tag{1.18}$$

Using this notation the solution given by (1.15) takes the form

$$x=c\alpha\exp(i(pt-\beta)). \tag{1.19}$$

Recall that the solution is obtained by taking the real part of the right-hand side. This means the following:

External force	$x(t)$
$c\cos(pt)$	$c\cos(pt-\beta)$
$c\sin(pt)$	$c\sin(pt-\beta)$

Resonance Curve

In order to grasp the physical significance of the solution we shall investigate the behavior of the distortion factor α as a function of the exciting frequency p, as given by (1.18), for a given set of "internal parameters" ω and r which define the oscillating particle. We observe that α is a positive function of p such that $\alpha(0)=1/\omega^2$ and $\alpha(\infty)=0$, meaning that α asymptotically tends to zero as p becomes infinite. In fact α vanishes to the order $1/p^2$. Therefore there must be a value of p which maximizes α. This value, which we call p_R, is called the *resonant frequency* of the system. It is obtained by setting $d\alpha/dp=0$ and solving for $p=p_R$. Performing the calculation yields

$$p_R=\omega^2-2r^2 \qquad \text{for} \quad \omega^2 \geq 2r^2. \tag{1.20}$$

Substituting this value of p into (1.18) yields the maximum value of the distortion factor α_R, or the distortion factor at resonance,

$$\alpha(p_R)=\alpha_R=\frac{1}{2r\sqrt{(\omega^2-r^2)}}. \tag{1.21}$$

As $r\to 0$ the value of α_R increase beyond all bounds. For $r=0$, that is, for the undamped oscillatory system, $\alpha(p_R)$ has an infinite discontinuity at the value $p=p_R=\omega$, and the spring will rupture in the neighborhood of p_R. This is a limiting case which will be given special consideration below.

The graph of the function $\alpha(p)$ is called the *resonance curve* of the system. The phenomenon of resonance is shown by the fact that each resonance curve has its maximum at $p=p_R$, which increases as the damping decreases, for a fixed ω. In Fig. 1.3 we have sketched a family of resonance curves,

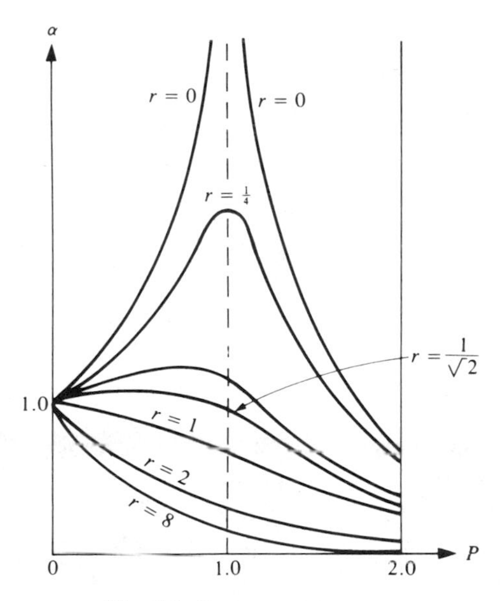

Fig. 1.3. Resonance curves.

normalized to $\omega_R = 1$. Each curve is specified by a value of the parameter r. We see that for small values of r well-marked resonance occurs near $p = 1$; and in the limiting case of $r = 0$ there would be an infinite discontinuity at $p = 1$ instead of a maximum value of α. As r increases, the maximum moves to the left and decreases in value. For the limiting case $p_R = 0$ we have $r = 1/\sqrt{2}$. This is the case of a statically applied external force which gives the response $\alpha(0) = 1$ for all p (no maximum occurs). In general, the resonance phenomenon ceases as soon as the following condition holds:

$$\omega^2 - 2r^2 \leq 0.$$

In the case of the equality sign, the resonance curve is the limiting case of the response to a static force, $\alpha(0) = 1/\omega^2$. Its tangent is horizontal there and after an initial almost horizontal course, it asymptotically tends to zero for large p.

We can now express the solution of the nonhomogeneous equation (1.14) by using (1.11) and (1.19). We obtain

$$x(t) = c\alpha \exp(i(pt - \beta)) + \exp(-rt)[c_1 \cos(qt) + c_2 \sin(qt)]. \quad (1.22)$$

Note that the solution of the free damped oscillatory system contains the ICs (which determine c_1 and c_2). However, the ICs must be used in (1.22) in order to determine the constants. Since for a given $r > 0$ these oscillations damp out asymptotically to zero, we are left with the steady state solution (the particular integral of the nonhomogeneous equation) which oscillates with the exciting frequency p.

It remains for us to discuss the undamped case $r = 0$. We recall that the solution fails when the exciting frequency is equal to the natural frequency, for then $\alpha(p_R) = \infty$. We therefore look for another solution to (1.14) for $p = \omega$. This solution is easily seen to be of the form

$$x = \bar{x}t \exp(i\omega t).$$

Substituting this expression into (1.14) for the case $r = 0$, $p = \omega$ gives

$$\bar{x} = \frac{1}{2i\omega}.$$

Thus when resonance occurs in an undamped system we have the solution

$$x = \left(\frac{t}{2i\omega}\right) \exp(i\omega t).$$

Using real notation, when $f(t) = \cos \omega t$, we have $x = (1/2\omega)t \sin \omega t$, and when $f(t) = \sin \omega t$, we have $x = -(1/2\omega)t \cos \omega t$.

We thus see that for the resonance condition in an undamped oscillator the solution is a sinusoidal function with an amplitude that increases linearly with time so that eventually the condition of small oscillations which led to the derivation of the ODE is violated and the solution becomes unbounded as time becomes infinite.

1.3. Combination of Wave Forms

We previously used the simple physical model of a coupled mass–spring dashpot to derive the ODE for an undamped and damped harmonic oscillator for free, and then forced, oscillations of a single particle. We now forget the ODE whose solutions are sinusoidal and damped sinusoidal motions, respectively; and concentrate on wave forms.† Neglecting damping, we think of $x(t)$ as the oscillatory motion of a particle (gas, liquid, or solid) centered at some point in space. This means that we are concerned with undamped wave forms at a given spatial point. We then investigate the wave forms that result from the combination of two or more pure sinusoidal waves.

The first question to answer is: What is the resulting wave form due to the coexistence of two or more harmonic oscillations of the same frequency and different amplitudes and phases? For the case of two harmonic motions given by $x_1 = a_1 \cos(\omega t - \phi_1)$ and $x_2 = a_2 \cos(\omega t - \phi_2)$, the combination is

$$x = x_1 + x_2 = a_1 \cos(\omega t - \phi_1) + a_2 \cos(\omega t - \phi_2) = \bar{x} \cos(\omega t - \phi), \quad (1.23)$$

where

$$\begin{aligned} \bar{x} &= (a_1)^2 + (a_2)^2 + 2a_1 a_2 \cos(\phi_1 - \phi_2), \\ \tan \phi &= \frac{a_1 \sin \phi_1 + a_2 \sin \phi_2}{a_1 \cos \phi_1 + a_2 \cos \phi_2}. \end{aligned} \quad (1.24)$$

We consider some particular cases.

If the two phases are equal, we have

$$x = (a_1 + a_2) \cos(\omega t - \phi),$$

so that the amplitude of the sum of the two waves is equal to the sum of the amplitudes of each wave.

If the phases differ by half a period, then

$$x = (a_1 - a_2) \cos(\omega t - \phi),$$

and the amplitude of the combined wave is equal to the difference of the amplitudes of each wave.

If, in addition, we have $a_1 = a_2$, then $x = 0$ for all t and we have the phenomenon of *interference* in the sense that the two sinusoidal waves of the same amplitude which are 90° out of phase cancel each other.

Next, we generalize to n simple harmonic waves of the same frequency and different amplitudes and phases. This is given by

$$x = \sum_{i=1}^{n} a_i \cos(\omega t - \phi_i) = \bar{x} \cos(\omega t - \phi), \quad (1.25)$$

---

† We shall call $x(t)$ a wave even though the dependent variable x is independent of space.

where

$$\bar{x} = \sum_{i=1}^{n} [(a_i \cos \phi_i)^2 + (a_i \sin \phi_i)^2],$$
$$\tan \phi = \frac{\sum_i \sin \phi_i}{\sum_i \cos \phi_i}. \tag{1.26}$$

It is clear that the case of two waves of this type is a special case of (1.24) and (1.25) for $n=2$.

We next consider the case of the coexistence of two sinusoidal waves of different amplitudes—frequencies and phases. The combined wave is given by

$$x_1 + x_2 = a_1 \cos(\omega_1 t - \phi_1) + a_2 \cos(\omega_2 t - \phi_2). \tag{1.27}$$

In general, we cannot put the right-hand side in the simple sinusoidal wave form $\bar{x} \cos(\omega t - \phi)$. Thus we see that the combined wave is *not* a simple sinusoidal function of t; but is a more complicated periodic wave form. This is a special case of a Fourier series representation of the combination of simple sinusoidal waves.

Consider the special case of (1.27) where $a_1 = a_2 = 1$, $\phi_1 = \phi_2 = 0$. Using a trigonometric identity, we then obtain

$$x_1 + x_2 = 2 \cos \tfrac{1}{2}(\omega_1 - \omega_2)t \cos \tfrac{1}{2}(\omega_1 + \omega_2)t. \tag{1.28}$$

The interpretation of (1.28) is: The sum of two cosine waves of the same amplitude and different frequencies that are in phase is equal to an amplitude modulated cosine wave oscillating at a frequency proportional to the sum of the frequencies. The modulating amplitude of the combined wave is a cosine function oscillating with a frequency proportional to the difference of the two frequencies. This amplitude (a low-frequency wave) may be considered as the envelope of the resulting cosine wave (high-frequency wave), in the sense that it consists of two curves tangent to the maxima and minima of the high-frequency wave. Figure 1.4 shows the two cosine waves

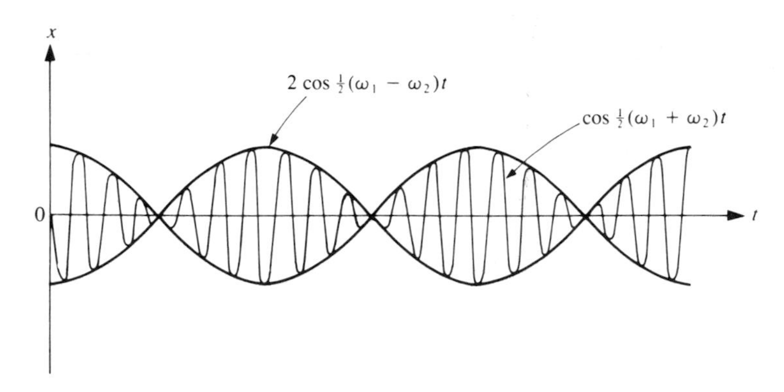

Fig. 1.4. Amplitude modulated wave as envelope of high frequency waves.

x_1 and x_2 and the resulting high-frequency wave inside its low-frequency envelope.

Suppose $\omega_2 - \omega_1 = \varepsilon$ where ε is small compared to ω_1 and $\omega_2 > \omega_1$. If we consider what happens over a time interval of a few periods, x_1 and x_2 which start off together as cosine waves maintain their synchronism since they continue to oscillate with almost equal frequencies. This means that they start off being equivalent to a cosine wave of the same frequency and twice the amplitude. This is shown by approximating (1.28) for small t, and very small ε. The approximations are $\cos \frac{1}{2}\varepsilon t \sim 1$ and $\cos \frac{1}{2}(\omega_1 + \omega_2)t \sim \cos \omega t$ for $\omega_1 = \omega$. However, as time goes on, this approximation becomes less and less valid so that x_1 and x_2 will not maintain this synchronism; the higher frequency wave will overtake the lower frequency one, thereby creating a variable difference in phase. The result is an amplitude modulated wave as shown by (1.28). *Note*: It is clear that by using a similar trigonometric identity, a combined wave form similar to (1.28) will hold for $x_1 = \sin \omega_1 t$, $x_2 = \sin \omega_2 t$.

This special combination of waves results in (1.28) which can be used to interpret the well-known phenomenon of *beat frequencies.* Suppose two tuning forks are simultaneously excited with the same amplitude but with frequencies that differ by one cycle/sec. The ear does not hear the individual waves but hears the beat frequency which is the low-frequency envelope. For example, a piano tuner uses this phenomenon to tune two notes to unison (same frequency) by fixing one note and tuning for a very small beat frequency. In the limit of zero beat frequency the two notes are exactly in unison. Lord Rayleigh points out that under favorable circumstances the ear can recognize beat frequencies as low as one in 30 seconds, meaning that the higher frequency note gains only two oscillations a minute over the lower. It is clear that the difference in two frequencies alone does not determine the interval between the two notes; that depends on the ratio of frequencies. This means, for example, that the beat frequencies are doubled when both notes are taken an octave higher.

We next investigate the generalization of (1.27) for the case of an infinite number of sinusoidal waves of different amplitudes and frequencies. We let the combined wave form be x, expressed by the following infinite series:

$$
\begin{aligned}
x &= \sum_{n=0}^{\infty} a_n \cos(\omega_n t - \phi_n) \\
&= A_0 + \sum_{n=1}^{\infty} \left[A_n \cos\left(\frac{2n\pi t}{T}\right) + B_n \sin\left(\frac{2n\pi t}{T}\right) \right],
\end{aligned} \tag{1.29}
$$

where T is the period, so that

$$
\omega_n = \frac{2n\pi}{T}.
$$

The series expression on the right-hand side is the *Fourier series representation* of x in terms of sine and cosine functions of time of different amplitudes and frequencies. The Fourier coefficients (A_n, B_n) are given by

$$A_0 = \frac{1}{T}\int_0^t x\,d\bar{t},$$

$$A_n = \frac{2}{T}\int_0^t x\cos\left(\frac{2n\pi\bar{t}}{T}\right)d\bar{t}, \tag{1.30}$$

$$B_n = \frac{2}{T}\int_0^t x\sin\left(\frac{2n\pi\bar{t}}{T}\right)d\bar{t}.$$

It is interesting to note that A_0 is the time average of x for one period. The Fourier coefficients are derived from the orthogonality relations for trigonometric functions, and are discussed in such works as [8, p. 50]. Conditions for a function to be expressible in terms of a Fourier series expansion including questions of convergence are given in any standard work on advanced calculus such as [23].

1.4. Oscillations in Two Dimensions

We now consider a particle that performs harmonic motion in the (x, y) plane. Let the dependent variables (u, v) be the displacement of the particle in the (x, y) directions, respectively. u and v are harmonic or sinusoidal functions of t of the form

$$u = \bar{u}\cos(\omega_1 t - \phi), \qquad v = \bar{v}\cos(\omega_2 t), \tag{1.31}$$

where $(\bar{u}, \bar{v})$ are the amplitudes of (u, v). We first take the case where the two frequencies are equal so that $\omega_1 = \omega_2 = \omega$. We can then eliminate t from (1.31) by manipulating the equations trigonometrically. We obtain

$$\left(\frac{u}{\bar{u}}\right)^2 + \left(\frac{v}{\bar{v}}\right)^2 - 2\left(\frac{\cos\phi}{\bar{u}\bar{v}}\right)uv = \sin^2\phi. \tag{1.32}$$

Equation (1.32) is the equation of a tilted ellipse in the (x, y) plane. If the phase angle $\phi = \pi/2$, so that the phases of u and v differ by one quarter of a period, then (1.32) reduces to

$$\left(\frac{u}{\bar{u}}\right)^2 + \left(\frac{v}{\bar{v}}\right)^2 = 1, \tag{1.33}$$

which is an ellipse whose semi-major and semi-minor axes $(\bar{u}, \bar{v})$ coincide with the coordinate axes.

If the amplitudes of u and v are equal, the ellipse becomes the circle

$$u^2 + v^2 = \bar{u}^2. \tag{1.34}$$

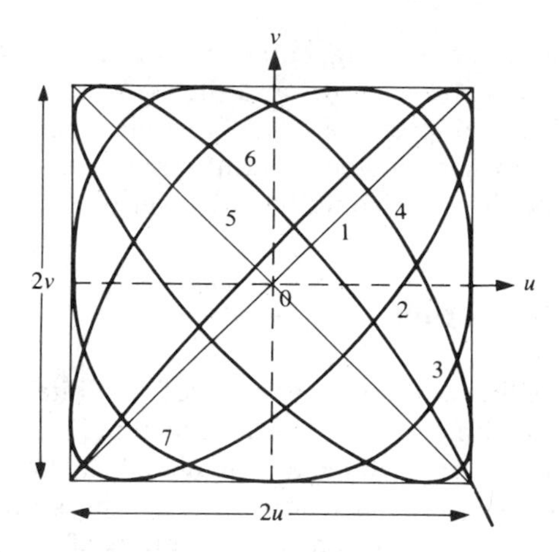

Fig. 1.5. Trajectory of family of ellipses with varying phases.

If u and v are in phase so that $\phi = 0$, the ellipse degenerates into the straight line

$$v = \left(\frac{\bar{v}}{\bar{u}}\right) u. \tag{1.35}$$

If u and v are 180° out of phase, we get the straight line

$$v = -\left(\frac{\bar{v}}{\bar{u}}\right) u. \tag{1.36}$$

Next we investigate the more realistic case where $\omega_1 \neq \omega_2$. Set $\omega_1 = \omega - \varepsilon$ and $\omega_2 = \omega$, where ε is a small constant. Then (1.31) becomes

$$u = u \cos(\omega t - \phi), \qquad v = v \cos \omega t, \qquad \phi = \varepsilon t. \tag{1.37}$$

Equation (1.37) shows us that the phase difference ϕ between u and v varies with t. For convenience we eliminated the constant in the phase angle so that u and v start in phase. For small time u and v remain almost in phase. Since ε is fixed, as t increases the phase difference between u and v continuously varies periodically. Therefore, instead of a single ellipse given by (1.32) we now have an infinite family of ellipses as t varies (a single ellipse for a given t).

We may now obtain a picture of how these ellipses vary with time. The extreme values of u and v are $\pm\bar{u}$ and $\pm\bar{v}$. Therefore the family of ellipses are inscribed in a rectangle whose sides are $2u$, $2v$. Figure 1.5 shows the trajectory of this family. For $t = 0$ we have $\phi = 0$. Equation (1.35) holds so that the ellipse is the straight line shown as curve (1) in the figure. As ϕ increases to $\pi/4$ we have curve (2) for $t = \pi/4\varepsilon$. Curve (3) occurs at $t = \pi/2\varepsilon$, which is given by (1.33), where one-quarter of a period has transpired. Curve (4) occurs at $t = 3\pi/4\varepsilon$, curve (5) at $t = \pi/\varepsilon$ where (1.36) holds (for one-half of a period). Continuing, curve (6) is at $t=5\pi/4\varepsilon$, curve (3) is at $t=3\pi/2\varepsilon$, curve (7) is at $t=7\pi/4\varepsilon$, and the cycle is completed with curve (1) at $t=2\pi/\varepsilon$. These ellipses are called *Lissajous figures.* They are easily demonstrated on a TV screen where u and v are sinusoidal voltages of the form (1.37); u is applied horizontally and v vertically.

1.5. Coupled Oscillations

We next investigate the oscillations of a system of finite masses and springs coupled together in a linear array. It was shown in [12, pp. 111 ff.] that the lateral vibrations of a string can be approximated by such a coupled system where each mass oscillates perpendicularly to the string at rest. In this discretization process the continuous distribution of matter describing the string is replaced by a finite number of equally spaced masses each connected by springs. A set of coupled ODEs (one for each mass) describes the system. Also, the same reference (pp. 128 ff.) describes a similar model for discretization of stress waves in a bar. The same mathematical description applies, but the oscillations of the masses are axial and the boundary conditions are different. In this section we shall investigate the mathematics of a coupled system of masses and springs (no damping considered) independent of a specific physical model.

We start with the simple case of two linearly coupled simple harmonic oscillators. Let the displacement of the ith mass from its equilibrium position be x_i $(i=1,2)$. Suppose the masses are uncoupled. Then the equations of motion are

$$m_i\ddot{x}_i + k_i x_i = 0.$$

Now let m_1 be attached to a spring on the left (of spring constant k_1) whose left end is fixed, and let m_2 be attached on the right to a spring (spring constant k_2) fixed at the right end. Next, let m_1 be attached to m_2 by a spring of spring constant k. Let m_1 be displaced by x_1 and m_2 by x_2. Figure 1.5 shows the equilibrium and displaced configurations of the masses and the forces acting on them. The additional spring force acting on m_1 is $k(x_2-x_1)$ and the additional spring force acting on m_2 is $k(x_1-x_2)$. Therefore, the equations of motion of the masses are

$$m_1\ddot{x}_1 = -k_1x_1 - k(x_2-x_1),$$

$$m_2\ddot{x}_2 = -k_2x_2 - k(x_1-x_2).$$

For convenience, we make the following change of notation:

$$\sqrt{m_1}\,x_1 = y_1, \qquad \sqrt{m_2}\,x_2 = y_2, \qquad \frac{k_1+k}{m_1} = (\omega_1)^2, \qquad \frac{k_2+k}{m_2} = (\omega_2)^2,$$

$$\frac{k}{\sqrt{m_1m_2}} = c.$$

Then the equations of motion become

$$\begin{aligned} \ddot{y}_1+(\omega_1)^2 y_1-c y_2=0, \\ \ddot{y}_2+(\omega_2)^2 y_2-c y_1=0. \end{aligned} \tag{1.38}$$

This is a pair of coupled linear homogeneous ODEs with constant coefficients. We assume a solution of the form $y_1=A\exp(i\omega t)$, $y_2=B\exp(i\omega t)$, where A, B, and ω are to be determined. Substituting these expressions into (1.38) gives the two following homogeneous algebraic equations for A and B:

$$\begin{aligned} (-\omega^2+\omega_1^2)A-cB=0, \\ -cA+(-\omega^2+\omega_2^2)B=0. \end{aligned} \tag{1.39}$$

In order to have nontrivial solutions for A and B the determinant of their coefficients must equal zero. This is easily seen by forming the ratio B/A in the first and second equations and then equating them. We obtain

$$\begin{vmatrix} (-\omega^2+\omega_1) & -c \\ -c & (-\omega^2+\omega_2^2) \end{vmatrix}=0. \tag{1.40}$$

Equation (1.40) is called the *secular equation* for ω. It is a quadratic equation for ω^2 whose roots are

$$\omega_\pm^2=\frac{\omega_1^2+\omega_2^2}{2}\pm\sqrt{\left[\frac{(\omega_1^2-\omega_2)^2}{4}+c^2\right]}. \tag{1.41}$$

This means we have two different possible frequencies of motion for the system. This is clear, since we should have two frequencies if the system were uncoupled (one frequency for one particle and the other for the other particle). Indeed, if we set $c=0$ in the right-hand side of (1.41), we get $\omega_+=\omega_1$, $\omega_-=\omega_2$, the natural frequencies of the uncoupled system. It is therefore natural to call c the *coupling coefficient.* If we expand the right-hand side of (1.41) in a Taylor series in c^2 and let c^2 be small so that we neglect terms of $O(c^4)$, we get the approximations

$$\begin{aligned} \omega_+^2 \sim \omega_1^2+\frac{c^2}{\omega_1^2-\omega_2^2}, \\ \omega_-^2 \sim \omega_2^2+\frac{c^2}{\omega_2^2-\omega_1^2}, \end{aligned} \tag{1.42}$$

again showing that $\omega_+\to\omega_1$, $\omega_-\to\omega_2$ as $c\to 0$. Equation (1.42) also shows that the frequencies are always spread apart by the coupling interaction: if $\omega_1>\omega_2$, then $\omega_+>\omega_1$, $\omega_-<\omega_2$, and conversely if the situation is reversed. It is easily shown that the following relations hold independent of the size of c:

$$\begin{gathered}\omega_+^2\omega_-^2=\omega_1^2\omega_2^2-c^2,\\ \omega_+^2+\omega_-^2=\omega_1^2+\omega_2^2,\\ (-\omega_+^2+\omega_1^2)(-\omega_-^2+\omega_2^2)=-c^2.\end{gathered}\tag{1.41'}$$

Having determined the two possible frequencies of vibration of the system, we next find the amplitudes A_+ and B_+ corresponding to ω_+, and A_- and B_- corresponding to ω_-. From the first of (1.39) we obtain

$$\frac{A_+}{B_+}=\frac{c}{d_+},\qquad \frac{A_-}{B_-}=\frac{c}{d_-},$$

where

$$d_+=-\omega_+^2+\omega_1^2,\qquad d_-=-\omega_-^2+\omega_1^2.$$

We see that the ratios of the amplitudes can be determined, but not the values themselves. However, if we normalize the amplitudes by setting $A_+^2+B_+^2=1$, $A_-^2+B_-^2=1$, we can indeed solve for the individual amplitudes. They become

$$\begin{aligned}A_+&=\frac{c}{\sqrt{(c^2+d_+)}}, & B_+&=\frac{d_+}{\sqrt{(c^2+d_+)}},\\ A_-&=\frac{c}{\sqrt{(c^2+d_-)}}, & B_-&=\frac{d_-}{\sqrt{(c^2+d_-)}}.\end{aligned}\tag{1.43}$$

We have the following situation: we have a possible solution $y_1=A_+\exp(i\omega_+t)$, $y_2=B_+\exp(i\omega_+t)$, where A_+ and B_+ are given by (1.43) subject to the normality condition on the amplitudes. Next we also have the solutions $y_1=A_-\exp(i\omega_-t)$, $y_2=B_-\exp(i\omega_-t)$, for the frequency ω_-. And now, on account of the linearity of the system of ODEs, we can make linear combinations of these solutions, obtaining

$$\begin{aligned}y_1&=A_+\exp(i\omega_+t)+A_-\exp(i\omega_-t),\\ y_2&=B_+\exp(i\omega_+t)+B_-\exp(i\omega_-t).\end{aligned}\tag{1.44}$$

This means that each coordinate has two periods in its motion, or is *doubly periodic.* Of course, there is a similar solution with $-i\omega_\pm t$ in the appropriate exponents. A unique problem is posed when we state the ICs on y_1 and y_2 which are: $y_1(0)=y_{10}$, $\dot{y}_1(0)=y_{11}$, $y_2(0)=y_{20}$, $\dot{y}_2(0)=y_{21}$; where the constants y_{10}, y_{11}, y_{20}, and y_{21} are prescribed.

It is of interest to investigate the physical nature of the solution given by (1.44). Let us assume that c is small so that the system is loosely coupled. Then one possible mode of oscillation has the frequency ω_+ which is only slightly greater than ω_1 (the natural frequency of m_1). Clearly both masses are oscillating. However, if we examine the coefficients A_+ and B_+ in this case, we find that B_+ is small compared with A_+. This means that the amplitude of y_2 is small compared with that of y_1. An easy calculation yields the approximation $B_+/A_+ \sim c/(\omega_1^2 - \omega_2^2)$. This is as if the first oscillator (m_1), vibrating with a frequency ω_+, which is approximately ω_1, and amplitude A_+, were *forcing* the second oscillator by virtue of the coupling, with a force cy_1 or $cA_+ \exp(i\omega_+ t)$, or approximately $cA_+ \exp(i\omega_1 t)$. This would produce a forcing function equal to $cA_+ \exp(i\omega_+ t)/(\omega_2^2 - \omega_+^2)$. Similarly, the second oscillator can vibrate almost by itself, with the frequency ω_- which almost equals ω_2; it also reacts back on the first and produces a small forcing function.

The phenomenon of beat frequencies, which was discussed above, can easily be demonstrated for this system. To relate to (1.28) we merely set $x_1 = y_1$, $x_2 = y_2$, $\omega_1 = \omega_+$, and $\omega_2 = \omega_+$. For our current system we may consider the physical model of a mass hung from a spiral spring and set to oscillate vertically. We identify y_1 with the vertical displacement and y_2 with an angle representing the torsional motion of the spring. Observing the oscillations of the mass we see that after a lapse of time the vertical motion will decrease, but there will be a torsional motion of considerable amplitude. As time goes on, these two forms of motion will alternatively take up large amplitudes. The coupling between the linear and torsional motion interchanges energy between the oscillating mass and the rotating spring in such a way as to illustrate the beat phenomenon when the frequencies of the two types of motion are close to each other.

We may continue in this vein by generalizing to a system of n coupled harmonic oscillators, and thereby study the resulting coupled set of ODEs which arise from the physical situation. However, we take a different tack. We shall use an approach due to the great French mathematical physicist, Lagrange, which is based on an energy method and a variational method involving the concept of virtual work. Consequently, the next topic will be a discussion of Lagrange's equations of motion, which will be used to study systems of N particles. Lagrange's equations along with Hamilton's equations, both based on variational principles, are the cornerstones of the branch of mathematical physics and engineering that deal with problems in dynamics. We can do no better at this point than to shift the subject to a study of Lagrange's equations, not only as a basis for vibration theory but also for a more general understanding of its role in mathematical physics.

1.6. Lagrange's Equations of Motion

Before we embark on a general discussion of Lagrange's equations of motion we shall introduce the subject, by starting with the concept of energy and illustrating the equations of motion, by a simple example. In a sense this is an opposite approach to what we had been doing, namely deriving the equations of motion from Newton's second law. For, from the equations of motion, a first integration yields the energy equation. A simple example: The single equation of motion for a simple harmonic oscillator is $m\ddot{x}+kx=0$. It is clear that $\ddot{x}=v\,dv/dx$ where $\dot{x}=v$. We thus have $v\,dv+kx\,dx=0$. A single integration yields $\frac{1}{2}mv^2+\frac{1}{2}kx^2=\frac{1}{2}mv_0^2+\frac{1}{2}kx_0^2=E$, where v_0 and x_0 are the ICs and the positive constant E is the total energy of the system. Let $\frac{1}{2}mv^2=T$ and $\frac{1}{2}kx^2=V$, where T is the kinetic energy and V the potential energy of the oscillator. The equation $T+V=T_0+V_0=E$ is the expression for the conservation of energy of the single particle system. Lagrange introduced the function L, the *Lagrangian* defined by

$$L=T-V. \tag{1.45}$$

Using this function he developed his equations of motion from the general principle of virtual work, which are the *Lagrange's equations of motion.* At this point we state Lagrange's equation for our single harmonic oscillator.

$$\frac{d}{dt}\frac{\partial L}{\partial \dot{x}}-\frac{\partial L}{\partial x}=0. \tag{1.46}$$

To illustrate the physical meaning of the terms on the left-hand side, we first calculate L for our single oscillator. We have $L=\frac{1}{2}mv^2-\frac{1}{2}kx^2$. We then obtain $\partial L/\partial\dot{x}=\partial T/\partial\dot{x}=mv=p$, the linear momentum of the mass. Also, we have $\partial L/\partial x=\partial V/\partial x=kx$, which is the spring force. We observe that, for our system, T depends only on the particle velocity v, and V only on the coordinate x. For the general case, T is a quadratic function of the generalized velocities where the coefficients may be at most functions of the generalized coordinates, and V is a quadratic function of the generalized coordinates. Generalized coordinates and velocities will be defined later. Equation (1.46) can be written as

$$\frac{dp}{dt}=-kx,$$

which states that the rate of change of the linear momentum is equal to the spring force.

The General Case

We now investigate Lagrange's equations of motion. To this end, we start with a system of n particles that are under no constraints. This means they are uncoupled so that each acts as a harmonic oscillator independent of the others. We first introduce the concept of *degrees of freedom.* The number of degrees of freedom of a particle is the number of *independent coordinates* necessary to specify its position in a given space. A particle undergoing translational and rotational motion in three space has six degrees of freedom, three for the motion of its center of gravity and the three Euler angles defining its rotational motion. For our purpose, since we do not consider the particle's rotation, each particle has three degrees of freedom. Instead of using $\mathbf{x}$ as our system of coordinates we introduce the vector $\mathbf{q} = (q_1, q_2, q_3)$ called the *generalized coordinates* of a free particle undergoing translational motion. The *generalized velocity* of the particle is $\dot{\mathbf{q}} = (\dot{q}_1, \dot{q}_2, \dot{q}_3)$. For N uncoupled particles undergoing translational motion we have $\mathbf{q} = (q_1, q_2, \ldots, q_{3N})$ and $\dot{\mathbf{q}} = (\dot{q}_1, \dot{q}_2, \ldots, \dot{q}_{3N})$. If we know $\mathbf{q}$ and $\dot{\mathbf{q}}$ as functions of t (the trajectory of the particles) we have completely specified its time history, and have thus solved the problem. Therefore each particle is completely characterized by its $(\mathbf{q}, \dot{\mathbf{q}})$. If a system of particles is acted on by one or more constraints, the number of degrees of freedom of the system is equal to the total number of coordinates $\mathbf{x}$ necessary to describe the system minus the number of constraints on the system. In the current analysis we shall not be concerned with constraints.

Hamilton's Principle

From an energy point of view, as illustrated for the single particle case, we assume that the mechanical properties of our system of N particles are completely determined by its kinetic energy T and potential energy V. At this stage we are concerned with conservative systems (nondissipative) where the conservation of energy holds. As mentioned above, we take T as a quadratic function of $\dot{\mathbf{q}}$ of the form

$$T = \sum_{i,j=1,1}^{n,n} P_{ij}(q_1, q_2, \ldots, q_n)\dot{q}_i\dot{q}_j, \tag{1.47}$$

where there are n generalized coordinates and velocities. In general, the coefficients P_{ij} are prescribed functions of $\mathbf{q}$ for nonlinear oscillations. For the linear case the P_{ij}'s are constants. We assume that V is a quadratic form in $\mathbf{q}$, of the form

$$V = \sum_{i,j=1}^{n,n} b_{ij}q_iq_j, \tag{1.48}$$

where the n^2 b_{ij}'s are constants. This means that the system is *conservative*, V does not depend on the $\dot{q}_i$'s so that no dissipative forces occur and the conservation of energy holds. The Lagrangian L becomes for the linear case

$$L(\mathbf{q}, \mathbf{q}) = T(\dot{\mathbf{q}}) - V(\mathbf{q}). \tag{1.49}$$

Hamilton's principle (applied to our system of N uncoupled particles) states the following: Between any two instants of time t_0 and t_1 the motion of the ensemble of particles defined by the trajectories $\mathbf{q} = \mathbf{q}(t)$ proceeds in such a way as to make

$$I = \int_{t_0}^{t_1} L(\mathbf{q}(t), \dot{\mathbf{q}}(t))\, dt, \tag{1.50}$$

a minimum. The integral I is taken between two fixed times t_0 and t_1. I is called a *functional* since it is a function of the integrand L which depends on the set of functions $(\mathbf{q}, \dot{\mathbf{q}})$. For each set of these functions we have the particle trajectories from t_0 to t_1, and a given value of I. We want that set of $(\mathbf{q}, \dot{\mathbf{q}})$ which minimizes I. The calculus of variations† tells us that $L(\mathbf{q}, \dot{\mathbf{q}})$ must satisfy a certain set of equations, which are the Lagrange equations of motion, whose solution gives the required trajectories that minimize I. They are

$$\frac{d}{dt}\frac{\partial L}{\partial \dot{q}_i} - \frac{\partial L}{\partial q_i} = 0, \qquad i = 1, 2, \ldots, n. \tag{1.51}$$

As shown in Chapter 9, if the functional $I = \int_{x_0}^{x_1} F(x, y, y')\,dx$, then there is a single Euler's equation given by $d/dx(F_{y'}) - F_y = 0$.‡ This is the simple case of finding a single trajectory $y = y(x)$ that makes I stationary in the (x, y) plane. Therefore, Lagrange's equations (1.51) is a generalization to $2n$ dependent variables $(\mathbf{q}, \dot{\mathbf{q}})$ of Euler's equation for $F = L$ and integration with respect to t instead of x. As seen, (1.51) is a set of n equations, one for each degree of freedom given by the ith generalized coordinate and velocity $(q_i, \dot{q}_i)$. Since the system is conservative so that V depends only on position, we may write (1.51) as

$$\frac{d}{dt}\frac{\partial T}{\partial \dot{q}_i} = -\frac{\partial V}{\partial q_i}, \qquad i = 1, 2, \ldots, n. \tag{1.52}$$

We recall that the term $\partial T/\partial \dot{q}_i$ is the linear momentum p_i corresponding to the ith coordinate q_i. The left-hand side then becomes dp_i/dt, so that the right-hand side is the ith *generalized force* corresponding to the ith generalized coordinate q_i which is a component of the gradient of the potential for a conservative system.

This sketchy survey of Lagrange's equations of motion is sufficient for our purpose—the study of oscillations of a system of particles. Further details including constraints, principle of virtual work, Hamilton's equations, etc. will be found in Chapter 9. We shall now use Lagrange's equations of motion in our investigations of oscillatory phenomena.

1.7. Formulation of the Problem of Small Oscillations for Conservative Systems

As given above, the potential energy V is assumed to be a function of position only (the generalized coordinates $\mathbf{q}$). Let $\mathbf{Q}$ be the generalized force

---

† Chapter 9 gives a detailed description of the calculus of variations in the setting of Lagrange's equations and Hamilton's equations. A complete description of the variational methods used in continuum mechanics is given in [13, Chap. 8].

‡ In this notation subscripts mean partial differentiation with respect to the subscript. For example, $u_x(x, y, z) \equiv \partial u/\partial x$, $u_{xy} \equiv \partial^2 u/\partial x\,\partial y$, etc.

due to the potential V. Then Lagrange's equations may be put in the following vector form for a conservative system:

$$\frac{d}{dt}\frac{\partial T}{\partial \mathbf{q}} = -\mathbf{Q}, \tag{1.53}$$

where

$$\mathbf{Q} = \mathbf{grad}\ V \qquad \text{or} \qquad Q_i = \frac{\partial V}{\partial q_i}. \tag{1.54}$$

The system is said to be *in equilibrium* when the generalized forces acting on the system vanish. The equilibrium conditions are

$$\mathbf{Q}_0 = (\mathbf{grad\ V})_0 = 0 \qquad \text{or} \qquad Q_{0i} = \left(\frac{\partial V}{\partial q_i}\right)_0 = 0. \tag{1.55}$$

The potential energy therefore has a *minimum* at the equilibrium configuration of the system $\mathbf{q}_0 = (q_{01}, q_{02}, \ldots, q_{0n})$. From the trivial case of the system being in equilibrium with zero initial velocity, then the system remains in equilibrium. An equilibrium configuration is said to be *stable* if a small disturbance of the system from equilibrium results only in small motion about the equilibrium or rest position, for example, a pendulum undergoing a small displacement from equilibrium. The equilibrium is *unstable* if an infinitesimal disturbance produces very large amplitude motion, for example, an egg standing on end. It is easily seen that when V is a minimum then the equilibrium is stable—for a conservative system. If V is a minimum at equilibrium, then any deviation from the equilibrium configuration will produce an increase in V. The conservation of energy tells us that $T + V = E$—the total constant energy. This increase in V will therefore produce a decrease in T or decrease the particle velocities. On the other hand, if V decreases as a result of some departure from equilibrium T, and hence the particle velocities will increase, thus yielding unstable equilibrium. The same conclusion may be arrived at by examining Fig. 1.6 which shows plots of V versus a particular q_i for the stable and unstable case. E_0 is the total energy at equilibrium.

Since we are concerned with small amplitude oscillations (within the framework of the linear theory), we shall be interested in the motion of our system within the neighborhood of its stable equilibrium configuration. Mathematically this means that, since the departures from equilibrium are small, all functions may be expanded in a Taylor series about the equilibrium, retaining only the lowest order terms. Let ξ_i be the deviation of q_i from q_{0i}, its equilibrium position. We have

$$\mathbf{q} = \mathbf{q}_0 + \boldsymbol{\xi} \qquad \text{or} \qquad q_i = q_{0i} + \xi_i, \qquad i = 1, 2, \ldots, n. \tag{1.56}$$

Expanding V about $\mathbf{q}_0$ gives $V(\mathbf{q})$ in terms of various spatial derivatives. Specifically, we get

$$V(\mathbf{q}) = V(\mathbf{q}_0) + \sum_i \left(\frac{\partial V}{\partial q_i}\right)_0 \xi_i + \tfrac{1}{2}\sum_{i,j}\left(\frac{\partial^2 V}{\partial q_i\, \partial q_j}\right)_0 \xi_i \xi_j + \cdots. \tag{1.57}$$

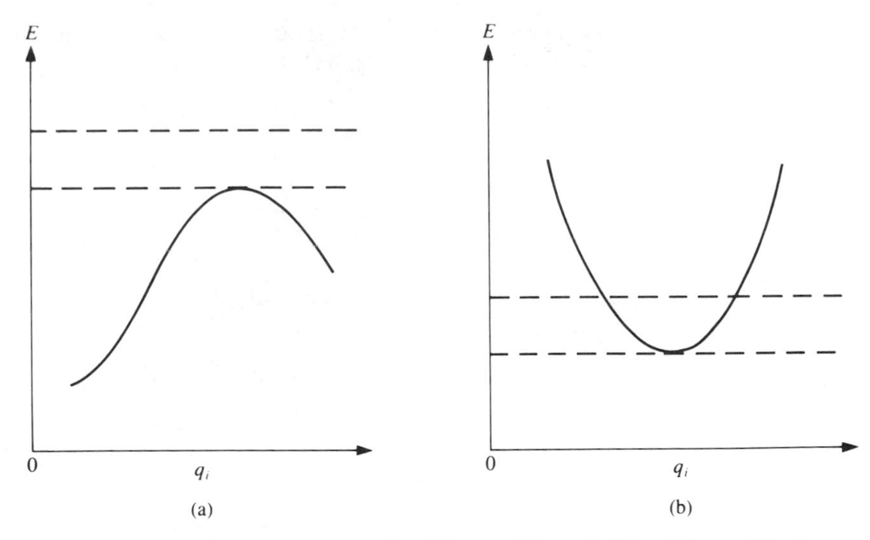

Fig. 1.6. Potential energy curves for (a) unstable and (b) stable equilibrium.

The linear terms in ξ_i vanish by invoking the equilibrium condition (1.55). By shifting the arbitrary zero of V to coincide with the equilibrium position we can make the first term on the right-hand side vanish. We therefore obtain

$$V = \frac{1}{2}\sum_{i,j}\left(\frac{\partial^2 V}{\partial q_i\,\partial q_j}\right)_0 \xi_i\xi_j = \frac{1}{2}\sum_{i,j} V_{ij}\xi_i\xi_j, \qquad V_{ij} \equiv \frac{\partial^2 V}{\partial x_i\,\partial x_j}, \tag{1.58}$$

where the n^2 constants $V_{,ij}$ denote the second spatial derivatives of V at equilibrium. It is clear that the V_{ij}'s are symmetrical, i.e., $V_{ij} = V_{ji}$. We see that V is a quadratic form in the displacement vector ξ from the equilibrium configuration $\mathbf{q}_0$.

Performing a similar expansion on the kinetic energy we obtain

$$T = \frac{1}{2}\sum_{i,j} m_{ij}\dot{q}_i\dot{q}_j = \frac{1}{2}\sum_{i,j} m_{ij}\dot{\xi}_i\dot{\xi}_j.$$

The coefficients m_{ij} are in general functions of $\mathbf{q}$. However, if they are expanded in a Taylor series about the equilibrium configuration, only the constant terms persist, since the terms such as $(\partial m_{ij}/\partial q_k)_0\xi_k$ must not appear, otherwise we would not have a pure quadratic form in the $\dot{\xi}_i$'s for T. Denoting the constant values of the m_{ij}'s at equilibrium by T_{ij},† we write the kinetic energy as

$$T = \frac{1}{2}\sum_{i,j} T_{ij}\dot{\xi}_i\dot{\xi}_j. \tag{1.59}$$

† Note that the subscripts in T_{ij} do not mean differentiation in this case.

It is clear that the constants T_{ij}'s are symmetric since an interchange of i and j does not change the right-hand side. The Lagrangian L now becomes

$$L = \frac{1}{2} \sum_{i,j} (T_{ij} \dot{\xi}_i \dot{\xi}_j - V_{ij} \xi_i \xi_j). \tag{1.60}$$

Taking the ξ_i's as the generalized coordinates, and using (1.60), Lagrange's equations (1.51) become

$$\sum_j T_{ij} \ddot{\xi}_j + \sum_j V_{ij} \xi_j = 0, \qquad i = 1, 2, \ldots, n, \tag{1.61}$$

where use has been made of the symmetry property of the T_{ij}'s and V_{ij}'s. In general, each of the n equations (1.61) will involve all of the n coordinates ξ_i so that (1.61) represents a set of simultaneous second-order coupled homogeneous ODEs which must be solved to obtain the small amplitude motion in the neighborhood of equilibrium.

1.8. The Eigenvalue Equation

Since the system (1.61) has constant coefficients we may, in the usual manner, assume an oscillatory solution of the form

$$\xi_i = \bar{\xi}_i \exp(i\omega t), \qquad i = 1, 2, \ldots, n, \tag{1.62}$$

where the $\bar{\xi}_i$'s are the amplitudes of each of the n modes of vibration and ω is the natural frequency. Inserting (1.62) into (1.61) yields

$$\sum_j (-\omega^2 T_{ij} + V_{ij}) \bar{\xi}_j = 0. \tag{1.63}$$

Equations (1.63) (called the *eigenvalue equations*) constitute a set of n linear homogeneous algebraic equations for the $\bar{\xi}_i$'s. By the usual algebraic argument (1.63) can have a nontrivial solution for the $\bar{\xi}_i$'s only if the determinant of their coefficients vanish. We have

$$\begin{vmatrix} -\omega^2 T_{11} + V_{11} & -\omega^2 T_{12} + V_{i2} & \cdots & -\omega^2 T_{1n} + V_{1n} \\ -\omega^2 T_{21} + V_{21} & -\omega^2 T_{22} + V_{22} & \cdots & -\omega^2 T_{2n} + V_{2n} \\ \vdots & \vdots & & \vdots \\ -\omega^2 T_{n1} + V_{n1} & -\omega^2 T_{n2} + V_{n2} & \cdots & -\omega^2 T_{nn} + V_{nn} \end{vmatrix} = 0. \tag{1.64}$$

The determinantal condition (1.64) is the *secular equation* for ω. This has a familiar ring since we had the same type of condition given by (1.40) for a two degree of freedom system. When expanded in the usual way, (1.64) is actually an nth degree algebraic equation for ω^2. Therefore, the n roots ω^2 of (1.64) give the n *modes* of oscillation, one for each frequency. This means that for each of the roots ω^2, (1.63) may be solved for the $n-1$ amplitudes $\bar{\xi}_i$ in terms of an arbitrary $\bar{\xi}_i$.

We now put the system (1.63) into the following vector form:

$$\mathbf{V}\boldsymbol{\xi} = \lambda \mathbf{T}\boldsymbol{\xi}, \tag{1.65}$$

where $\mathbf{V}$ and $\mathbf{T}$ are $n \times n$ matrices; the ijth component of $\mathbf{V}$ is V_{ij}, and that of $\mathbf{T}$ is T_{ij}, and $\omega^2 = \lambda$. $\boldsymbol{\xi}$ is an *eigenvector* and λ is the corresponding *eigenvalue.* This means that to each eigenvalue λ_k which is a root of the secular equation (1.64) there corresponds an eigenvector $\boldsymbol{\xi}_k$ $(k = 1, 2, \ldots, n)$. Let $\boldsymbol{\xi}_k = (\xi_{1k}, \xi_{2k}, \ldots, \xi_{nk})$. Then (1.65) can be written in the expanded form

$$\sum_j V_{ij}\xi_{jk} = \lambda_k \sum_j T_{ij}\xi_{jk}. \tag{1.65$'$}$$

We see that the meaning of the matrix equation (1.65) is: The matrix $\mathbf{V}$ operates on the eigenvector $\boldsymbol{\xi}_k$ to produce a multiple of the result of the matrix $\mathbf{T}$ acting on $\boldsymbol{\xi}_k$. The multiplying factor on the right-hand side is of course the kth eigenvalue λ_k.

Both $\mathbf{T}$ and $\mathbf{V}$ are symmetric matrices. A *symmetric matrix* is one that is equal to its transpose. The *transpose* of a matrix is defined as a matrix obtained by interchanging the rows and columns of the original matrix. This means that if the ijth element of a matrix $\mathbf{M}$ is m_{ij} and if $\mathbf{M}^*$ is defined as the transpose of $\mathbf{M}$ then the ijth element of $\mathbf{M}^*$ is equal to m_{ji} (the indices are interchanged). If $\mathbf{M}$ is symmetric then $\mathbf{M} = \mathbf{M}^*$ or $m_{ij} = m_{ji}^*$. For the moment we asume that the eigenvector $\boldsymbol{\xi}$ is complex, and shall represent $\boldsymbol{\xi}$ by the column matrix whose elements are complex. We shall let $\boldsymbol{\xi}^*$ be the transpose of the conjugate of $\boldsymbol{\xi}$. $\boldsymbol{\xi}^*$ is a row vector whose elements are the complex conjugates of the corresponding elements of $\boldsymbol{\xi}$. An additional fact we need about matrices is: The transpose of the product of two matrices is equal to the product of their transposes in reverse order (matrices do not necessarily commute). Specifically, if $\mathbf{A}$ and $\mathbf{B}$ are $n \times n$ matrices then $(\mathbf{AB})^* = \mathbf{B}^*\mathbf{A}^*$. This is easily seen by invoking the definition of matrix multiplication. Let $\mathbf{C} = \mathbf{AB}$. Let c_{ij} be the ijth element of $\mathbf{C}$, etc. Then

$$c_{ij} = a_{ik}b_{kj}, \qquad i, j, k = 1, 2, \ldots, n,$$

which is a set of n^2 equations for the elements c_{ij}. We use the *tensor summation convention.* This means that we sum over the double index k from 1 to n in the right-hand side. Taking the transpose of this equation, we obtain

$$c_{ij}^* = c_{ji} = a_{jk}b_{ki} = b_{ki}a_{jk} = b_{ik}^* a_{kj}^*,$$

since the products of the pairs of elements commute. This proves our statement that $\mathbf{C}^* = \mathbf{B}^*\mathbf{A}^*$.

We are now in a position to prove the following statement: Since $\mathbf{V}$ and $\mathbf{T}$ are real symmetric matrices, it follows that the eigenvalues λ are real. We assume that the secular equation is such that the eigenvalues are distinct. To this end we premultiply both sides of (1.65) by ξ^*, first set $\lambda = \lambda_k$, then set $\lambda = \lambda_k^*$ (the complex conjugate of λ_k) and subtract the second from the first equation. Since the left-hand side of the resulting difference equation vanishes, we obtain

$$0 = (\lambda_k - \lambda_k^*)\xi^* \mathbf{T} \xi. \tag{1.66}$$

We shall now show that the quadratic form given by $\boldsymbol{\xi}^*\mathbf{T}\boldsymbol{\xi}$ is real. To prove this assertion we set $\boldsymbol{\xi} = \boldsymbol{\alpha} + i\boldsymbol{\beta}$. The transpose of the complex conjugate of $\boldsymbol{\xi}$ is $\boldsymbol{\xi}^* = (\boldsymbol{\alpha}^* - i\boldsymbol{\beta}^*)$. We then get

$$\boldsymbol{\xi}^*\mathbf{T}\boldsymbol{\xi} = (\boldsymbol{\alpha}^* - i\boldsymbol{\beta}^*)\mathbf{T}(\boldsymbol{\alpha} + i\boldsymbol{\beta}) = \boldsymbol{\alpha}^*\mathbf{T}\boldsymbol{\alpha} + \boldsymbol{\beta}^*\mathbf{T}\boldsymbol{\beta} + i(\boldsymbol{\alpha}^*\mathbf{T}\boldsymbol{\beta} - \boldsymbol{\beta}^*\mathbf{T}\boldsymbol{\alpha}).$$

The imaginary part of the right-hand side is

$$\boldsymbol{\alpha}^*\mathbf{T}\boldsymbol{\beta} - \boldsymbol{\beta}^*\mathbf{T}\boldsymbol{\alpha} = 0.$$

This is shown as follows: Since every quadratic form is symmetric and $\mathbf{T}$ is symmetric, we have

$$\boldsymbol{\beta}^*\mathbf{T}\boldsymbol{\alpha} = (\boldsymbol{\beta}^*\mathbf{T}^*\boldsymbol{\alpha})^* = \boldsymbol{\alpha}^*\mathbf{T}^*\boldsymbol{\beta} = \boldsymbol{\alpha}^*\mathbf{T}\boldsymbol{\beta}.$$

We have thus proved that $\boldsymbol{\xi}^*\mathbf{T}\boldsymbol{\xi}$ is real, from which it follows that $\lambda_k^* = \lambda_k$, so that the eigenvalues are real. Equation (1.65) tells us that the ratio of the eigenvector components of $\boldsymbol{\xi}$ must all be real. Of course, there is still some indeterminateness since a particular one of the components of $\boldsymbol{\xi}$ can be chosen at will. Premultiplying (1.65) by $\boldsymbol{\xi}^*$ we solve for each of the λ_k's and obtain

$$\lambda_k = \frac{\boldsymbol{\xi}^*\mathbf{V}\boldsymbol{\xi}}{\boldsymbol{\xi}^*\mathbf{T}\boldsymbol{\xi}}. \tag{1.67}$$

Equation (1.67) tells us that the λ_k's are given as the ratio of two quadratic forms, the numerator involving the potential energy and the denominator involving the kinetic energy. In fact, the denominator is equal to twice the kinetic energy of the velocity vector $\dot{\boldsymbol{\xi}}$. Since the eigenvalues are real and positive, the numerator and denominator are both *positive definite* quadratic forms.† Incidentally, we have a mathematical proof of the demonstration given above that the potential energy is a minimum for equilibrium. Were the potential not a strict minimum the numerator might be negative, giving rise to imaginary frequencies which would give rise to an unbounded exponential increase of the amplitude of $\boldsymbol{\xi}$ with time, thus leading to instability.

† Let $Q = \mathbf{x}^*\mathbf{A}\mathbf{x}$ for an n-dimensional column vector $\mathbf{x}$, a row vector $\mathbf{x}^*$, and an $n \times n$ matrix $\mathbf{A}$. The right-hand side of this equation is a quadratic expression of the form $x_i a_{ij} x_j$ (summed over i, j from 1 to n) where a_{ij} is the ijth component of $\mathbf{A}$. The left-hand side is a number (since the quadratic form is a scalar). The quadratic form is positive definite if $Q > 0$ except for the roots $\mathbf{x}$ which make $Q = 0$.

1.9. Similarity Transformation and Normal Coordinates

We return to the coupled system of ODEs given by (1.61). These equations become in matrix form

$$\mathbf{T}\ddot{\boldsymbol{\xi}} = -\mathbf{V}\boldsymbol{\xi}. \tag{1.68}$$

Let $\mathbf{T}^{-1}$ be the *inverse*† of the matrix **T**. Premultiplying (1.68) by $\mathbf{T}^{-1}$ yields

$$\ddot{\boldsymbol{\xi}} = -\mathbf{T}^{-1}\mathbf{V}\boldsymbol{\xi}.$$

With a slight change of notation, we put this equation in the form

$$\ddot{\mathbf{u}}(t) = \mathbf{A}\mathbf{u}(t), \tag{1.69}$$

where the matrix $\mathbf{A} = \mathbf{T}^{-1}\mathbf{V}$ and the n-dimensional column vector $\mathbf{u} = \boldsymbol{\xi}$. Clearly (1.69) is a matrix equation which represents a coupled system of harmonic oscillators whose solution is given by $\mathbf{u} = \mathbf{u}(t)$. It is coupled because there are nonzero off-diagonal elements of **A**. On the other hand, if **A** were a *diagonal matrix* (a matrix with only nonzero diagonal elements), then the system of oscillators would be *uncoupled.* In any case, **A** is a symmetric matrix. Suppose we have an n-dimensional eigenvector $\mathbf{v}(t)$ and a diagonal matrix **D** such that

$$\ddot{\mathbf{v}}(t) = \mathbf{D}\mathbf{v}(t), \tag{1.70}$$

where the ijth element of **D** is d_{ii} for $i = j$ and 0 for $i \neq j$. It turns out that $d_{ii} = \lambda_i$ the ith eigenvalue. It is easily seen that (1.70) is a second-order system of *uncoupled* ODEs. We seek a transformation from the eigenvector **u** to the eigenvector **v** such that **v** satisfies (1.70). This vague statement can be made precise by the following assertion: Since **A** is a symmetric matrix we can find an *orthogonal transformation* represented by the *orthogonal matrix*‡ **R** such that

$$\mathbf{R}^{-1}\mathbf{A}\mathbf{R} = \mathbf{D}. \tag{1.71}$$

---

† The inverse of an $n \times n$ matrix **M** is another $n \times n$ matrix $\mathbf{M}^{-1}$ such that $\mathbf{M}\mathbf{M}^{-1} = \mathbf{M}^{-1}\mathbf{M} = \mathbf{I}$ where **I** is the $n \times n$ *identity matrix* defined by the fact that its ijth element is the Kronecker delta $\delta_{ij} = 1$ for $i = j$, 0 for $i \neq j$. In order for the inverse of a matrix to exist, it must be *nonsingular* meaning that the determinant of the matrix must not vanish.

‡ An orthogonal matrix **R** is defined as a nonsingular matrix whose transpose is equal to its inverse: $\mathbf{R}^* = \mathbf{R}^{-1}$. Example: In three dimensions let $\mathbf{x} = (x_1, x_2, x_3)$ be a Cartesian coordinate system. Then the orthogonal transformation $\mathbf{R}\mathbf{x} = \mathbf{y}$ rotates the **x** into the **y** coordinate system and the elements of R are the direction cosines which determine the rotation. In general, if r_{ij} is the ijth element of **R**, then we have the following orthogonal condition: $r_{ik}r_{kj} = \delta_{ij}$ (summed over k from 1 to n). This means that the scalar product of any two different rows or columns is zero, and the scalar product of a row or column with itself is equal to unity (the condition of normality); therefore **R** is an *orthonormal* matrix.

Equation (1.71) expresses a *similarity transformation.* We shall now justify (1.71). Let $\mathbf{r}_i$ be the ith row vector, and $\mathbf{r}_j$ the jth column vector of $\mathbf{R}$; and similar notation for $\mathbf{A}$. Since $\mathbf{R}^{-1} = \mathbf{R}^*$, the ith column of $\mathbf{R}$ equals the ith row of $\mathbf{R}^* = \mathbf{r}_i^*$. Recall that $\mathbf{R}$ is an orthonormal matrix so that the scalar products $\mathbf{r}_i \cdot \mathbf{r}_j = \delta_{ij}$. We now manipulate the matrix $\mathbf{AR}$ in (1.71).

$$\mathbf{AR} = \begin{pmatrix} \mathbf{a}_1 \\ \mathbf{a}_2 \\ \vdots \\ \mathbf{a}_n \end{pmatrix} (\mathbf{r}_1 \quad \mathbf{r}_2 \quad \ldots \quad \mathbf{r}_n) = \begin{pmatrix} \mathbf{a}_1^*\mathbf{r}_1 & \mathbf{a}_1^*\mathbf{r}_2 & \cdots & \mathbf{a}_1^*\mathbf{r}_n \\ \mathbf{a}_1^*\mathbf{r}_1 & \mathbf{a}_2^*\mathbf{r}_2 & \cdots & \mathbf{a}_2^*\mathbf{r}_n \\ \vdots & \vdots & & \vdots \\ \mathbf{a}_n^*\mathbf{r}_1 & \mathbf{a}_n^*\mathbf{r}_2 & \cdots & \mathbf{a}_n^*\mathbf{r}_n \end{pmatrix}.$$

If we now premultiply the right-hand matrix by $\mathbf{R}^*$, and use the orthonormal property of $\mathbf{R}$ we see that the result is a diagonal matrix.

To get the correspondence between the eigenvectors $\mathbf{u}$ and $\mathbf{v}$ we obtain $\mathbf{v}$ by the orthogonal transformation given by $\mathbf{R}$ by letting

$$\mathbf{u} = \mathbf{R}\mathbf{v} \qquad \text{or} \qquad \mathbf{v} = \mathbf{R}^{-1}\mathbf{u}. \tag{1.72}$$

Then the system of coupled ODEs (1.69) becomes

$$\mathbf{R}\ddot{\mathbf{v}} = \mathbf{AR}\mathbf{v}.$$

Multiplying this equation by $\mathbf{R}^{-1}$ yields

$$\ddot{\mathbf{v}} = \mathbf{R}^{-1}\mathbf{AR}\mathbf{v} = \mathbf{D}\mathbf{v},$$

by using (1.71). We have thus retrieved (1.70), the uncoupled set of ODEs for $\mathbf{v}(t)$.

To solve the system (1.69) in the usual way we assume a solution of the form

$$\mathbf{u}(t) = \mathbf{U}\exp(i\omega t) \qquad \text{or} \qquad u_j = U_j \exp(i\omega t), \tag{1.73}$$

where $\mathbf{U}$ is a constant vector whose jth component is U_j. Inserting (1.73) into (1.69) yields the following matrix equation:

$$(\mathbf{A} + \lambda\mathbf{I})\mathbf{U} = 0, \qquad \omega^2 = \lambda, \tag{1.74}$$

where $\mathbf{I}$ is the identity matrix. Given $\mathbf{A}$, (1.74) is a system of n homogeneous algebraic equations for the components of $\mathbf{U}$. We therefore have the condition

$$\det(\mathbf{A} + \lambda\mathbf{I}) = 0, \tag{1.75}$$

where $\det(\mathbf{M})$ means the determinant of the matrix M. Equation (1.75) is the secular equation which is an nth degree algebraic equation for the roots λ.

Similarly, set

$$\mathbf{v}(t) = \mathbf{V}\exp(i\omega t),$$

where $\mathbf{V}$ is an n component constant. Substituting this expression into (1.70) gives

$$(\mathbf{D}+\lambda\mathbf{I})\mathbf{V}=0. \tag{1.76}$$

In order for $\mathbf{V}$ to be not zero we must obtain the secular equation

$$\det(\mathbf{D}+\lambda\mathbf{I})=0, \tag{1.77}$$

for the eigenvalues in terms of $\mathbf{D}$ instead of $\mathbf{A}$ as for (1.75).

Now $\mathbf{V}$ is obtained from $\mathbf{U}$ by the orthogonal transformation $\mathbf{U}=\mathbf{RV}$. We can therefore apply the similarity transformation (1.71) to (1.76) and obtain

$$\det(\mathbf{R}^{-1}\mathbf{AR}+\lambda\mathbf{I})=\det(\mathbf{A}+\lambda\mathbf{I}), \tag{1.78}$$

since $\mathbf{R}$ is orthogonal. Equation (1.78) tells us that the roots of the secular equation *are invariant with respect to a similarity transformation.* Since the roots of (1.75) are the same as those of (1.77) it is simpler to use (1.77) to find the eigenvalues, which gives the λ_i's as the elements of $\mathbf{D}$.

The above analysis leads to the conclusion that a similarity transformation given by (1.71) applied to the coupled oscillatory system (1.69) yields the uncoupled oscillatory system given by (1.70). Clearly this means that each oscillator acts independently of the others. This process of using the similarity transformation to uncouple an oscillating system is said to be a transformation into *normal coordinates.* For the special case of a three degree of freedom system, $\mathbf{R}$ plays the role of rotating the coordinate system for the coupled oscillators to another coordinate system in which the oscillators are uncoupled. The analogy in elasticity theory is a principle axis transformation, where the symmetric stress tensor is transformed to a diagonal matrix by a similarity transformation, which rotates the original stress axes to the principle axes in which the stress has no shear components. A similar argument holds for the strain tensor. See [13, Chap. 1] for further details.

We now return to physical notation. We go back to (1.58) which tells us that V is a quadratic form in $\boldsymbol{\xi}$. More precisely, it is a *bilinear form*, meaning that in general the mixed terms $V_{ij}\neq 0$ for $i\neq j$. Equation (1.59) shows us that a similar statement holds for T, which is a bilinear form in $\dot{\boldsymbol{\xi}}$. Suppose $\mathbf{R}$ is an orthogonal matrix that converts the bilinear forms for V and T into pure quadratic form in which the coefficients of the mixed terms analogous to $\xi_i\xi_j$ and $\dot{\xi}_i\dot{\xi}_j$ vanish for $i\neq j$ in another coordinate system. We want an $\mathbf{R}$ that obeys the similarity transformation (1.71) on $\mathbf{A}$ which is now interpreted as the matrix of coefficients V_{ij} and T_{ij}. Specifically, we perform an orthogonal transformation from the coordinate system given by the eigenvectors $\boldsymbol{\xi}$ to another given by the eigenvectors $\boldsymbol{\eta}$ such that

$$\boldsymbol{\xi}=\mathbf{R}\boldsymbol{\eta}. \tag{1.79}$$

Equation (1.58) in matrix form becomes

$$V=\tfrac{1}{2}\boldsymbol{\xi}^*\mathbf{V}\boldsymbol{\xi}. \tag{1.80}$$

The left-hand side is the positive value of the scalar V and the right-hand side is the bilinear representation of V in terms of the eigenvectors. It is therefore positive definite. Inserting the new eigenvectors $\boldsymbol{\eta}$ defined by the transformation (1.79) into (1.80) yields

$$V = \tfrac{1}{2}\boldsymbol{\eta}^*\mathbf{D}\boldsymbol{\eta}, \qquad (1.81)$$

where $\mathbf{D}$ is given by

$$\mathbf{D} = \mathbf{R}^*\mathbf{V}\mathbf{R}. \qquad (1.82)$$

Equation (1.82) is the similarity transformation that transforms $\mathbf{V}$ into the diagonal matrix $\mathbf{D}$ whose iith element is the eigenvector $\lambda_i = \omega_i^2$. In extended form (1.81) appears as

$$V = \tfrac{1}{2}\sum_{i=1}^{n} \omega_i^2 n_i, \qquad (1.83)$$

$$\dot{\boldsymbol{\xi}}_k^*\mathbf{T}\dot{\boldsymbol{\xi}}_l = \mathbf{I} \qquad \text{or} \qquad T_{ij}\dot{\xi}_{il}\dot{\xi}_{jk} = \delta_{lk} \qquad \text{(sum over } i, j). \qquad (1.84)$$

We now use (1.84) to show that the kinetic energy has an even simpler quadratic form in the transformed velocity components $\dot{\eta}_i$. Since $\dot{\boldsymbol{\xi}}$ obeys the same orthogonal transformation (1.79), (1.59) becomes

$$T = \tfrac{1}{2}\dot{\boldsymbol{\xi}}^*\mathbf{T}\dot{\boldsymbol{\xi}} = \tfrac{1}{2}\dot{\boldsymbol{\eta}}^*\mathbf{R}^*\mathbf{T}\mathbf{R}\dot{\boldsymbol{\eta}} = \tfrac{1}{2}\dot{\boldsymbol{\eta}}^*\dot{\boldsymbol{\eta}} \qquad \text{or} \qquad T = \tfrac{1}{2}\dot{\eta}_i\dot{\eta}_i. \qquad (1.85)$$

where we have used the orthogonality condition $\mathbf{R}^*\mathbf{R} = \mathbf{I}$. Equation (1.85) tells us that the kinetic energy is a positive definite pure quadratic in the velocity components $\dot{\eta}_i$ (in the transformed coordinate system given by the eigenvectors $\boldsymbol{\eta}$).

A method of diagonalizing a symmetric matrix by use of the similarity transformation is given in [13, Chap. 1], wherein it was shown that the rotation matrix $\mathbf{R}$ is made up of column vectors that are the eigenvectors $\boldsymbol{\eta}_k$ in the transformed coordinate system. These eigenvectors corresponding to two different normal modes are orthogonal to each other. Furthermore, if we normalize the $\boldsymbol{\eta}_k$ then they form an orthonormal set which we use to construct $\mathbf{R}$. Now if we consider the left-hand side of (1.71) and let the matrix $\mathbf{AR} = \mathbf{C}$ we observe that all the elements of the first column of $\mathbf{C}$ involve the first column of $\mathbf{R}$ which is $\boldsymbol{\eta}_1$, the second column of $\mathbf{C}$ involves $\boldsymbol{\eta}_2$, etc. This means that if we then premultiply $\mathbf{C}$ by $\mathbf{R}^*$ and use the symmetry property of $\mathbf{A}$, we then obtain the diagonal matrix $\mathbf{D}$ by using the orthogonal property $\mathbf{R}^*\mathbf{R} = \mathbf{I}$.

The Lagrangian in the transformed coordinate system becomes

$$L = \tfrac{1}{2}\sum_j (\dot{\eta}_j^2 - \omega_i^2\eta_j^2). \qquad (1.86)$$

Lagrange's equations of motion for the η_j's then become

$$\ddot{\eta}_j + \omega_j^2\eta_j = 0, \qquad j = 1, 2, \ldots, n. \qquad (1.87)$$

The solution of this uncoupled system is

$$\eta_j = \bar{\eta}_j \exp(i\omega_j t), \tag{1.88}$$

where the natural frequencies ω_j satisfy the secular equation (1.77) for the n roots $\omega_j = \lambda_j^2$. Equation (1.88) tells us that each of the transformed coordinates η_j is a periodic undamped function of t involving only *one* of the natural frequencies ω_j. It is therefore customary to call the η_j's the *normal coordinates* of our system of n oscillating particles, and these component oscillations are said to be the *normal modes of vibration.* Each *mode* is characterized by an amplitude $\bar{\eta}_j$ and frequency ω_j. All the particles in each mode oscillate with the same frequency and phase, the amplitudes being determined by the components of the $\bar{\eta}_j$'s. (Particles may be 180° out of phase if the η_j's have opposite sign.) The complete solution to a given initial value problem can then be constructed by taking that linear combination of normal modes that satisfies the prescribed ICs, using the method of Fourier analysis. Harmonics of the fundamental frequencies are absent because of the linearity due to the assumption of small amplitude oscillations.

To review, equations (1.83) and (1.85) tell us that, in the transformed coordinate system $(\eta_1, \eta_2, \ldots, \eta_n)$, V is the sum of the squares of the η_i's and T is the sum of the squares of the $\dot{\eta}_i$'s (there are no cross product terms). This process of using **R** in the similarity transformation (1.71) on **V** and **T** results in *diagonalizing* these forms into the pure quadratic forms. **R** is made up of the columns of the set of orthogonal eigenvectors $\boldsymbol{\eta}_k$. The Lagrangian in the transformed coordinate system leads to Lagrange's uncoupled equations of motion which yield the normal modes of vibration.

As pointed out previously, the similarity transformation is used in transforming the stress and strain tensors into principle axes. Thus, this transformation is a *principal axis transformation* which transforms stress and strain matrices to diagonal matrices. In mechanics this transformation on the inertia tensor transforms the moment of inertia to the pure quadratic form (the sum of squares). A theorem of algebra roughly stated tells us that, given a symmetric matrix **M** we can find an orthogonal matrix **R** such that the similarity transformation $\mathbf{M}' = \mathbf{R}^*\mathbf{M}\mathbf{R}$ transforms **M** to a diagonal matrix $\mathbf{M}'$, where **R** consists of the eigenvectors and $\mathbf{M}'$ the eigenvalues. In our case the similarity transformation involving the same **R** diagonalizes both **V** and **T**.

To illustrate the above theory we take as an example the linear oscillations of two masses coupled together by three springs as described in Section 1.5. The equations of motion with respect to (y_1, y_2) are (1.38). It is easily seen that if we take the following expressions for V and T and form L we obtain Lagrange's equations of motion in the form given by (1.38):

$$V = \tfrac{1}{2}(\omega_1^2 y_1 - 2cy_1 y_2 + \omega_2^2 y_2),$$

$$T = \tfrac{1}{2}(\dot{y}_i^2 + \dot{y}_2^2).$$

Note that T is already diagonalized for this example. The matrix representations of these quadratic forms are easily seen to be

$$V = \tfrac{1}{2}\mathbf{y}^*\mathbf{V}\mathbf{y}, \qquad T = \tfrac{1}{2}\dot{\mathbf{y}}^* I \dot{\mathbf{y}},$$

where

$$\mathbf{V} = \begin{pmatrix} \omega_1^2 & -c \\ -c & \omega_2^2 \end{pmatrix}.$$

The secular equation is

$$\det(\mathbf{V} - \lambda^2 \mathbf{T}) = 0,$$

which is the same as (1.40). The roots are given by (1.41) and the eigenvectors y_1 and y_2 are given by

$$\boldsymbol{\eta}_+ = \begin{pmatrix} c/\sqrt{(c^2 + \omega_1^2 - \omega_+^2)} \\ (\omega_1^2 - \omega_+^2)/\sqrt{(c^2 + \omega_1^2 - \omega_+^2)} \end{pmatrix}, \qquad \boldsymbol{\eta}_- = \begin{pmatrix} c/\sqrt{(c^2 + \omega_1^2 - \omega_-^2)} \\ (\omega_1^2 - \omega_-^2)/\sqrt{(c^2 + \omega_1^2 - \omega_-^2)} \end{pmatrix}.$$

where the eigenvalues are ω_+^2 and ω_-^2. These expressions for the orthonormal eigenvectors are essentially (1.43). It is easily seen that $\boldsymbol{\eta}_+$ and $\boldsymbol{\eta}_-$ are orthonormal. The orthogonality condition is $\boldsymbol{\eta}_+ \cdot \boldsymbol{\eta}_- = 0$, and the normality conditions are $\boldsymbol{\eta}_+ \cdot \boldsymbol{\eta}_+ = 1$, $\boldsymbol{\eta}_- \cdot \boldsymbol{\eta}_- = 1$. To justify these conditions use (1.41).

CHAPTER 2

The Physics of Wave Propagation

Introduction

This chapter presents a physical foundation for the chapter on partial differential equation (PDEs) of the hyperbolic type and characteristic theory which is rather abstract. It also gives the reader an introduction to the rich variety of wave forms that occur in nature. The physics of water waves is clearly different from that of stress waves in solids, for example. However, the mathematical structure of these seemingly different wave forms yields a unifying theory which encompasses the different physical properties. This mathematical theory is based on three fundamental conservation laws of physics: conservation of mass or continuity, conservation of momentum, and conservation of energy. These will be discussed in this chapter, the mathematical formulation being treated from the point of view of physics.

The propagation of waves in liquids, gases, and solids is, at least in a qualitative sense, familiar to all of us. We have all seen water waves roll in and break on a beach and heard sound waves in air. We have also observed the beautiful interference patterns produced by the interaction of two sets of expanding circular surface waves generated by dropping two stones a distance apart in a quiet pool.

Getting more quantitative, we have studied the traveling and standing waves in a stretched string in physics courses and in applied mathematics courses, where analytical methods such as Laplace transforms were used to solve the linear wave equation associated with the vibrating string. The vibrating string is an excellent prototype for the study of transverse waves. A taut finite string with fixed ends, initially plucked at some interior point and suddenly let go, is a model for a vibrating harp, or piano string if a point on the string suffers an impulsive load which represents the striking of a padded hammer. Mathematically, this presents us with an initial value (IV) problem, the so-called "Cauchy Problem" for the solution of the one-dimensional wave equation. The classical solution was given by D'Alembert. The solution can also be obtained by the method of characteristics which has the advantage of allowing us to solve a wider class of problems which includes large amplitude or nonlinear wave phenomena. If the same string is stroked normal to its length by a bow, we have a model for the vibrations of a stringed instrument. In this case the linear motion of the bow normal to the string sets the string into forced vibrations due to the friction between the bow and the string. In this case we must solve a more complicated PDE, i.e., the wave equation with a damping term and an external forcing function. The vibrations of the string are transmitted through the bridge to the body of the musical instrument. The body acts as a resonator and, in a very complicated and not really understood way, produces sound waves that travel in air and are picked up by a receiver such as the ear. The various modes of vibration of the body can be approximately attacked by applying the theory of vibrating shells. This is beyond the scope of this book. There are many variations on the vibrating string problem. For example, we may impose an arbitrary initial displacement and velocity of the string and then have one or both of the ends perform prescribed lateral oscillations. This gives a combined initial and boundary value problem. As long as the string undergoes small lateral oscillations the wave equation is linear so that any of the operational methods such as Laplace transforms or D'Alembert's method can be used to generate solutions. We can also superimpose a forcing function, which means we must solve the nonhomogeneous wave equation.

Since a musical instrument generates sound waves in air, it appears that the next class of wave propagation problems to discuss would be this problem of sound wave propagation. This is a problem in the investigation of wave propagation in a one-dimensional unsteady compressible gas. To this end, we consider the model of a semi-infinite tube of air with a vibrating diaphragm at the front end. This is an excellent model for describing the generation of longitudinal sound waves. The physics of sound wave propagation in gases involves an equation of state in which the pressure is a prescribed function of the gas density. Specifically, it was experimentally found that an adiabatic equation of state (no heat loss or gain to the surroundings) is a fairly accurate representation of the pressure-density relationship for wave propagation in a gas, as long as shock waves are not considered. In a sense, the theory of propagation of sound waves using an adiabatic gas is an approximation, since (as we shall show in the text) a sound wave is a limiting form of a shock wave where the shock strength is infinitely weak. Incidentally, sound waves cannot be polarized since they are longitudinal, in contrast to transverse waves which can be polarized.

The phenomenon of stress wave propagation in solids offers us a rich variety of waves. There are longitudinal waves, the so-called "P waves" which propagate with a characteristic wave speed. They are irrotational waves and represent a change in volume. There are the "S waves" or shear waves which are transverse waves that are rotational in nature and represent no volume change ("equivoluminal waves"). They propagate with a different wave speed, which is smaller than that of the longitudinal waves. All these wave forms, and more, are useful to the materials, aeronautical, and civil engineer interested in the dynamic response of materials to unsteady loads. In addition, "Love waves" occur in layered media and are important in seismology. A.E.H. Love showed that they are essentially horizontally polarized shear waves trapped in a superficial layer of material and propagated by multiple total reflections. Lord Rayleigh discovered a type of surface wave, "Rayleigh waves". These waves satisfy the boundary condition on the surface of zero traction. They propagate along the surface with a characteristic wave speed determined from the boundary conditions, and the amplitude decays exponentially into the material. They are also important in seismology, see [14].

In studying the propagation of waves in water, we must clearly take into account the fact that a liquid propagates waves in a manner somewhat different from a gas and very different from a solid. This is clearly due to the difference in physical properties of the liquid. For example, the equation of state for an elastic solid undergoing very small displacements is given by Hooke's law, which is the constitutive equation expressing the assumption that the stress tensor is a linear function of the strain tensor. In the more general type of elastic solid the constitutive equation is given by a nonlinear relation between the stress and strain tensors. On the other hand, the constitutive equation for a gas or liquid is an equation of state that relates pressure to density. The equation of state for water is somewhat like the adiabatic equation of state for a gas. However, the ratio of specific heats for water is 7, rather than the lower value of 1.4 for air. According to Sir Horace Lamb [25] in his classic work *Hydrodynamics*: "One of the most interesting and successful applications of hydrodynamical theory is to the small oscillations, under gravity, of a liquid having a free surface." He goes on to say that these oscillations may combine to form progressive waves traveling (to within the linear approximation) with no change of form over the surface. This line of investigation leads to the theory of the tides which involves gravitational oscillations. The free surface (no tractions) of a body of water in equilibrium is in a gravitational field and is essentially a planar surface. If under the action of some external force, such as the gravitational pull of the moon, a point on the surface is moved from its equilibrium position, oscillatory motion will occur and spread out over the surface in the form of waves. These are called "gravity waves"† since they are due to the action of a gravitational field. They appear mainly on the surface, but also affect the interior of the liquid and damp out with depth. Shallow water waves of large amplitude involve a nonlinear analysis, not susceptible to the classical methods of linear analysis. We may apply the method of Riemann, using characteristic theory, to this class of problems.

---

† It is clear that these gravity waves are different from the gravity waves that arise in the general theory of relativity.

2.1. The Conservation Laws of Physics

The above description of wave propagation in different media shows physical properties that depend on the medium. As mentioned, these properties are characterized by constitutive equations or equations of state. They arise from the conservation of energy. This brings us to the laws of physics that unify the seemingly different properties of the great variety of waves in the different media. These laws are the three conservation laws: *conservation of mass*, *momentum*, and *energy*. Before we sketch these laws let us point out that we are dealing with a continuum, which is defined as a continuous distribution of matter, in the sense that the properties of a differential volume of the material are the same as those of the material in the large, so that the molecular structure is not considered. There are two ways of describing the trajectory of a typical particle of our continuum: The Lagrange and Euler representations. (Historically speaking, they are actually both due to Euler.) The *Lagrange representation* assigns coordinates to each particle so that the dependent variables such as stress, displacement, strain, etc. are differentiable functions of the particle (Lagrange) coordinates and time. This representation is useful in describing the dynamics of solids since there is small particle motion. The dynamics of the situation is that the solution of a given problem may be expressed in terms of families of plots of particle trajectories (particle coordinates versus time); the parameter defining a given curve is the laboratory fixed spatial coordinates. Another way of looking at the situation is: A problem is solved if we can find a continuous transformation from the Lagrange or particle coordinate system to the Euler or spatial coordinate system (described below). This means that the problem is solved if we can find the Euler coordinates as differentiable functions of the Lagrange coordinates and time, so that the dynamical and thermodynamic variables are expressed as functions of the Lagrangian coordinates. The *Euler representation* is useful for gases and liquids. Physically speaking, in this representation we station ourselves at a given point in space. Each such point is in a laboratory fixed space. The set of all such points is called the *Euler coordinates*. To clarify the situation: A material called "the body" is characterized by a coordinate system fixed in the body. This coordinate system is called the Lagrangian set of coordinates. The body is embedded in the laboratory fixed coordinate system characterized by the Euler coordinates. For convenience we may fix the origin of the Lagrange coordinate system at the center of gravity and let it coincide with the origin of the Euler coordinate system. This representation is useful in fluid mechanics investigations. We observe the particles of fluid flowing by our station. Each particle is described by its Lagrange coordinates and has associated with it all the dynamical and thermodynamic variables (particle velocity, pressure, temperature, etc.). These dependent variables are differentiable functions of the Euler coordinates and time. The Euler coordinates are defined as the spatial or laboratory fixed coordinates of a particle.

A solution of a given problem may be described in terms of a family of plots of Euler coordinates versus time, where the parameter describing each curve consists of the Lagrangian coordinates. Or, a solution is obtained by finding the transformation from the Euler to the Lagrange coordinates, so that the dynamical and thermodynamic variables are obtained as functions of the Eulerian coordinates.

We now come back to the conservations laws. These three laws of physics complctely describe a continuum. They are essentially the "field equations" whose solutions yield the dynamical fields: particle displacements, velocities, stress, and strain, and the thermodynamic fields such as temperature and entropy. The conservation laws may be expressed by the Lagrangian or Eulerian representations, depending on whether we are concerned with solids or fluids. The conservation of mass tells us that the mass associated with a given particle of the material, called "the material particle" (expressed mathematically by a differential volume element) is invariant with respect to time, meaning that, as time goes on the mass does not change. The conservation of momentum expresses Newton's second law of mechanics which tells us that the time rate of change of linear momentum of the material particle is equal to the resultant of the external forces acting on the particle. An analogous conservation law holds for the conservation of angular momentum where, for a rotating body, the time rate of change of the angular momentum is equal to the resultant of the external moments on the body. These two conservation laws are valid for all continuous media, be they solids, liquids, or gases. Clearly this means that we do not have enough field equations to solve for the dynamical and thermodynamic fields. The additional equation that is needed is obtained from the conservation of energy. This is an energy balance equation. It invokes the first and second laws of thermodynamics. Usually we can get away with not using the second law explicitly in problems not involving shock waves. In shock wave phenomena, the second law is used. It supplies an additional equation which states that the entropy increases across a shock wave. This allows us to prove that a shock wave in a gas is a compression wave. Neglecting the phenomenon of shock wave propagation, the three conservations laws supply the three sets of field equations, in general, in vector form. If we do not neglect the inertial terms (density times particle acceleration) in the conservation of momentum we have a time-varying or unsteady state which leads to the propagation of waves.

2.2. The Nature of Wave Propagation

At this stage it is probably appropriate to say a few words about the nature of wave propagation. What is a wave and how is it propagated in a given medium? Consider a dependent variable v, which may stand for gas pressure, lateral displacement of a stretched string, etc. In solving the conservation or field equations we obtain $v = v(\mathbf{x}, t)$ which means v is a given function of the radius vector $\mathbf{x}$ (in some space) and time. The following simple example illustrates the main features of traveling waves: Suppose v stands for the pressure in a gas tube, the front end of which has a diaphragm that is oscillating axially with a fixed frequency and with small oscillations so that sound waves are propagated in the tube away from the front end. We have a one-dimensional problem so that $\mathbf{x} = x$, the axis of the tube, and the oscillating diaphragm is at $x = 0$. v propagates as a sound wave in the following sense: Suppose we station ourselves at some point x in the tube. If at $t = 0$ the diaphragm is set into oscillations, the elementary particle of gas (of length dx) in the neighborhood of $x = 0$ $(x > 0)$ will suddenly suffer alternate axial compressions and expansions in a periodic manner (simple harmonic motion) with the same frequency and amplitude of the diaphragm, thus sending sound signals into the gas. This means that dt later the next neighboring particle will be set into the same harmonic motion, and dt later, the next particle will be set into similar motion, etc., so that sound signal emanating from the diaphragm will propagate into the tube. The rate of propagation will be constant for small amplitude vibration of the diaphragm if the gas is homogeneous. Let c be the speed of the signal propagation. c is called the speed of the sound wave or wave speed. Nothing will happen at x until the sound signal reaches that point, in the time $t = x/c$. For all $t > x/t$ the particle of gas at x will perform simple harmonic motion with an amplitude and frequency of the diaphragm. It is clear that there are two types of motion: the simple harmonic motion of each gas particle after it responds to the signal from the diaphragm, and the motion down the tube of the signal which propagates from particle to particle. c is clearly much larger than the variable speed of the vibrating particles. Suppose the diaphragm oscillates sinusoidally with an amplitude v_0 and frequency ω. Then at $x = 0$ we have $v(0, t) = v_0 \sin \omega t$. At x, v becomes $v(x, t) = v_0 \sin(\omega t - (x/2\pi\lambda))$, where λ is the wavelength, ω is the radial frequency, $c = 2\pi\lambda\omega$, and $t > x/2\pi c$. $v(x, t)$ satisfies the one-dimensional wave equation. The equation for $v(x, t)$ shows it to be a progressing wave. The most general solution of the one-dimensional wave equation is $v(x, t) = F(x - ct) + G(x + ct)$ where F and G are arbitrary functions of their arguments. If the argument $x - ct = \text{const.}$ then F is constant. Suppose the constant for $x - ct$ is adjusted to make F a maximum value. If we plot $x - ct = \text{const.}$ in the (t, x) plane, then along that straight line curve (called a "characteristic") F propagates with its maximum value. As we increase x, t must increase at the same rate in order for F to maintain its maximum. ($dx/dt = c$ is the reciprocal slope of the characteristic curve.) For this reason $F(x - ct)$ is called a "progressing wave". Similarly, positive x must decrease as t increases in order for G to remain constant. Therefore, $G(x + ct)$ is called a "regressing wave".

2.3. Discretization

We may discretize wave propagation phenomena such as the transverse waves produced by vibrating strings and the longitudinal sound waves produced in a gas tube. This means that we replace the string as a continuous distribution of matter by a system of discrete particles coupled together by springs. This discrete model approximates the continuum by a finite distribution of matter. The mathematical description of such discrete systems involves a finite array of coupled ordinary differential equations (ODEs), each of which describes the vibrational motion (simple harmonic motion) of a mass representing a particle of the physical model. This aspect was discussed in Chapter 1 where the oscillatory motion of a system of n coupled particles was investigated, and a method was developed and used for uncoupling the system into normal vibrational modes. If we increase the number of particles in a finite region (for example, increasing the number of masses of a string of finite length) by a limiting process, the system of ODEs will, in the limit, converge to a single PDE, the wave equation.

For example, the standing waves produced by a plucked harp string fixed at both ends may be approximated by a discrete system composed of a series of masses and springs coupled together linearly and having the end springs fixed at their outer ends. Suppose an interior mass is initially displaced laterally (normal to the axis of the mass-spring system) and then suddenly let go. Then that mass will begin to oscillate laterally and keep oscillating with a fixed amplitude and frequency as long as we neglect friction. This means that the excited mass will undergo lateral simple harmonic motion. At a short time after the initial lateral displacement of the excited mass (the "response time") the two neighboring masses will be set into lateral harmonic motion; a response time later the next set of neighboring masses that were motionless will then be set into harmonic motion, etc. This means that signals were simultaneously sent by the initially excited mass to its two neighbors. These masses will begin to undergo the same type of harmonic motion. After a response time the same type of signals initiating harmonic motion of the next set of neighbors will begin, etc. Therefore two traveling wavefronts are set up, which move in opposite directions. The wavefronts are the loci of points of the positions of the masses that were initially excited into motion. These wavefronts will move to the left and right (respectively) with the same speed, thus giving a discrete representation of two traveling waves. After simultaneous reflections from the fixed ends, standing waves are set up. This discrete model of transverse wave propagation can be made continuous by fitting an infinite number of masses and springs into a finite length. By a limiting process, which will be shown in the text, we may transform this model to a continuous model of a vibrating string. Mathematically, a finite set of ODEs representing the coupled harmonic motion of a finite number of masses is transformed into a single PDE, which is the one-dimensional wave equation whose solution gives standing waves for the lateral wave motion of the harp string.

A similar process of discretization can be done for sound waves in a gas tube. The simplest model is one where a semi-infinite axial array of coupled masses and springs is constructed. The leading mass (at the front end) acts as an oscillating diaphragm, oscillating axially with a fixed frequency and a small amplitude. The same set of coupled ODEs is derived; but the boundary and initial conditions are different from the harp model. The solution is interpreted differently—in terms of longitudinal waves for this case. Also, the physical constants such as the wave speed are different; for the string the square of the wave speed equals the ratio of the string tension to the linear density. The sound speed in a gas is equal to the ratio of specific heats times the ratio of the undisturbed pressure to density.

2.4. Sinusoidal Wave Propagation

We mentioned above that a function of the form $F(x-ct)$ is a progressing wave, i.e., moving in the direction of positive x with a wave speed c, and $G(x+ct)$ is a regressing wave moving in the direction of negative x with the same wave speed. We now give a physical model that illustrates how a progressing transverse wave is formed and how it propagates down a string. To this end, we consider a semi-infinite stretched string represented by the positive x axis in the (x, y) plane. The front end of the string at $x=0$ is attached to a mass-spring combination that oscillates in the y direction as shown in Fig. 2.1(a). Suppose at $t=0$ the mass is pulled down a distance A and then released from rest. It will then perform a simple harmonic motion with an amplitude A and radial frequency ω in radians/sec while the frequency f is in cycles/sec. The relationship between the two is $\omega=2\pi f$. Since the end of the string is attached to the mass, the mass will act as a source of a sinusoidal transverse wave that travels to the right along the string in the manner indicated by the family of curves in Fig. 2.1(b). These curves show successive "snapshots" of the sinusoidal shape of the string during one half cycle after the motion has been established. This means that at a given time the "wave profile" is a sine wave in x displaced by an amount depending on the time and the wave speed. Let T be the period. The respective curves show the wave form at $t=0$, $T/8$, $T/4$, $3T/8$, and $T/2$. The wavelength λ is the distance between adjacent maxima or minima of a given wave. Each time the mass (at $x=0$) makes a complete cycle (corresponding to the period T), going from $y=A$, to 0 to A to 0, and back to A, the wave travels in the x direction a distance λ, so that the constant wave speed c can be written as

$$c=\frac{\lambda}{T}=f\lambda=\frac{\omega\lambda}{2\pi}. \tag{2.1}$$

One way of describing the motion of the wave as it travels down the string is to again refer to Fig. 2.1(b). Suppose we take a wire and bend it in the shape of the sine curve of the wave form and place it on the curve

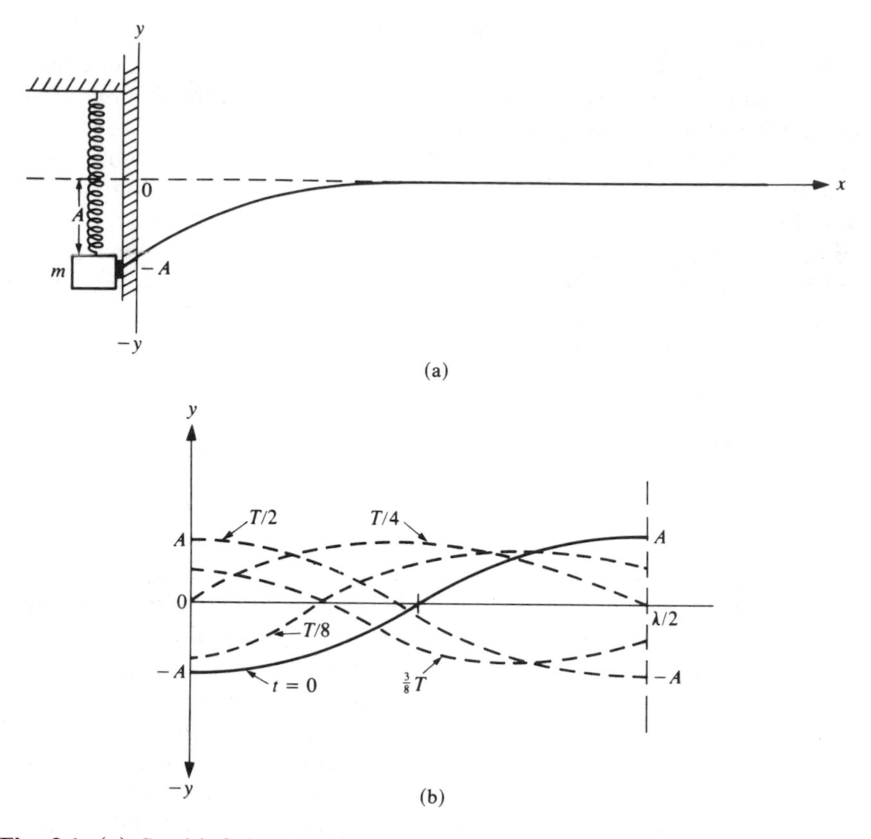

Fig. 2.1. (a) Semi-infinite string pulled down by a vertical mass spring at $t=0$. (b) Sinusoidal wave profiles at $t=0$, $T/8$, $T/4$, $3T/8$, $T/2$ over a $\lambda/2$ interval.

corresponding to $t=0$. The equation of this curve is then

$$y=-A\sin\left(\frac{2\pi x}{\lambda}\right), \qquad t=0.$$

We then move the wire to the right with a speed c so that the axis of the sine curve is coincident with the x axis. At $t=T/8$, $T/4$, $3T/8$, and $T/2$ the wire takes the positions shown by the corresponding curves. This demonstration clearly tells us that the sine wave does not change its shape during its translational motion down the x axis. The equations of the wave form corresponding to the appropriate curves in Fig. 2.1(b) are

$$y=A\sin\left(\frac{\pi}{4}-\frac{2\pi x}{\lambda}\right), \qquad t=\frac{T}{8},$$

$$y=A\sin\left(\frac{\pi}{2}-\frac{2\pi x}{\lambda}\right), \qquad t=\frac{T}{4},$$

$$y = A\sin\left(\frac{3\pi}{4} - \frac{2\pi x}{\lambda}\right), \qquad t = \frac{3T}{8},$$

$$y = A\sin\left(\pi - \frac{2\pi x}{\lambda}\right), \qquad t = \frac{T}{2}.$$

The boundary condition at $x = 0$ is given by the lateral harmonic motion of the mass, so that

$$y = A\sin\left(\frac{2\pi t}{T}\right), \qquad x = 0.$$

We now write y as the function of (x, t) that satisfies the above values of t for any x and the boundary condition at $x = 0$ for all $t > 0$.

$$y(x, t) = A\sin\left(\frac{2\pi t}{T} - \frac{2\pi x}{\lambda}\right). \tag{2.2}$$

The right-hand side of (2.2) is a specific case of the general form $F(x - ct)$ which is a progressing traveling wave. Note that the argument of the right-hand side of (2.2) is given by

$$\frac{2\pi t}{T} - \frac{2\pi x}{\lambda} = -\left(\frac{2\pi}{\lambda}\right)(x - ct), \qquad \text{using} \quad (2.1).$$

For the more general case, let $u(x, t) = F(x - ct)$ and $v(x, t) = G(x + ct)$ so that u is a progressing wave and v is a regressing wave. If we form u_{xx}, u_{tt}, and v_{xx}, v_{tt} we see that both u and v satisfy the following PDE:

$$c^2 w_{xx} - w_{tt} = 0. \tag{2.3}$$

This pivotal PDE is the *wave equation* in one dimension. It is a linear PDE with constant coefficients, and $w = u + v$ is also a solution. Therefore the general solution of the wave equation (2.3) is

$$w(x, t) = F(x - ct) + G(x + ct). \tag{2.4}$$

Key facts about (2.3) are that all solutions are of the general form (2.4), and no function that is not of the form (2.4) is a solution of (2.3). We shall prove these facts when we investigate the one-dimensional wave equation in detail.

2.5. Derivation of the Wave Equation

We now derive the wave equation (2.3) for the vibrating string undergoing small amplitude oscillations, and we shall show that the wave speed c is given by

$$c = \sqrt{\frac{T}{\rho}}, \tag{2.5}$$

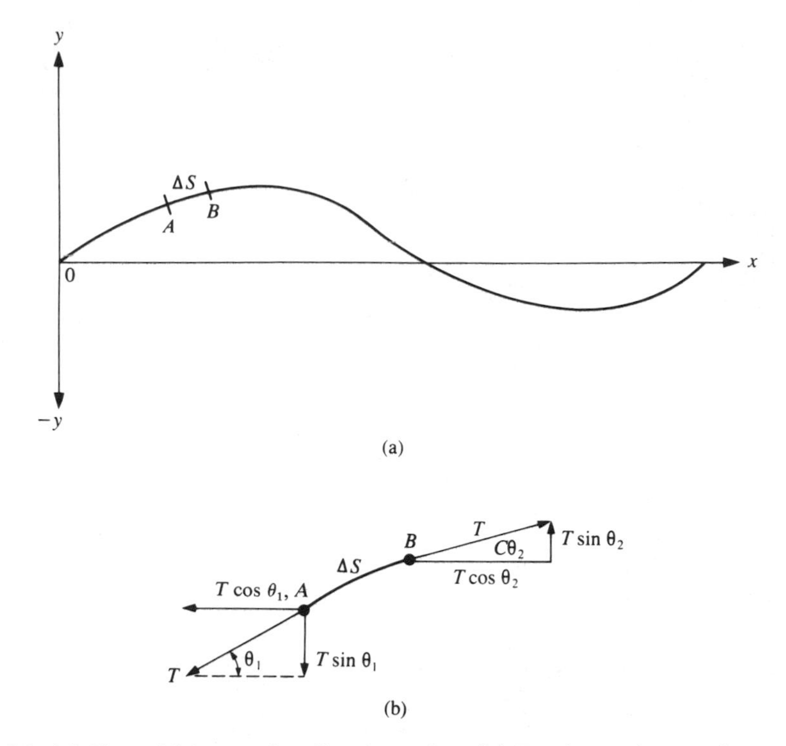

Fig. 2.2. (a) Sinusoidal wave for vibrating string. (b) Tension acting on element of string.

where T is the tension† on the string and ρ is the linear density (mass per unit length of string). Figure 2.2(a) shows the sinusoidal wave form of the string at a given t. Figure 2.2(b) shows a "blow up" of a typical element s of the string and the tensions acting on the ends of the element. Assume the string has a unit cross-sectional area. $y(x, t)$ is the lateral displacement of the string from its equilibrium position. Let $\mathbf{F} = \mathbf{i}X + \mathbf{j}Y$ be the resultant of the force acting on the ends of the element, where $(\mathbf{i}, \mathbf{j})$ are unit vectors in the (x, y) directions, (X, Y) are the corresponding components of $\mathbf{F}$, and $\mathbf{F} = \mathbf{F}_1 + \mathbf{F}_2$. It is easily seen that $X_1 + X_2 = 0$ for small angles. We need only consider the y component of $\mathbf{F}$ which is

$$Y = Y_2 + Y_1 = T(\sin\theta_2 - \sin\theta_1) \sim T(\tan\theta_2 - \tan\theta_1),$$

again invoking the assumption of small angles. But $\tan\theta_i$ $(i = 1, 2)$ is the slope of the curve at the points 1 and 2. Let these points have the coordinates (x, y), $(x + \Delta x, y + \Delta y)$, respectively. Then, as $\Delta x \to dx$ and $\Delta y \to dy$, in the

† Tension is defined as the tensile stress acting on the string, or the force per unit cross-sectional area of string causing the string to stretch.

limit, the approximation for Y becomes

$$Y = \lim \left[T\left(\frac{1}{\Delta x}\right)(y_x(x+\Delta x, y+\Delta y) - y_x(x, y))\Delta x \right] = Ty_{xx}\,dx.$$

Using Newton's second law we equate this expression for Y to the mass times the particle acceleration y_{tt} of the element and thereby write the equation of motion in the form

$$\left(\frac{T}{\rho}\right) y_{xx} = y_{tt}. \tag{2.6}$$

If we use (2.5) we see that (2.6) is the same as the PDE given by (2.3) where w is identified with y the lateral displacement of the string. Therefore (2.6) is the wave equation for the vibrating string, where the wave speed c is given by (2.5).

2.6. The Superposition Principle, Interference Phenomena

Thus far we have investigated the passage of a *single* wave disturbance—a sinusoidal pulse (due to a single cycle of the oscillating mass at the end of the string) or a continuous wave train (due to many cycles of the mass). Now, what happens when two or more waves pass simultaneously along the string? Within the assumption of small amplitude lateral string displacement, the effect of the sum of the displacements is equal to the sum of the effect of each individual wave. Mathematically, if y_i is the ith wave form then

$$y = \sum_{i=1}^{n} y_i,$$

where we sum over the n waves. Physically, we may state this *superposition principle* as follows: When two or more waves move simultaneously through a region of space, each wave proceeds independently, as if the other were not present. The resulting effect (displacement of the string, pressure of a sound wave, etc.) at any point and time is the vector sum of the effects of the individual waves. This principle is valid for mechanical waves on strings, springs, and liquid surfaces, and for sound waves in fluids, and stress waves in solids, providing the displacements are small so that the linear theory of wave propagation holds. The principle also holds for electromagnetic waves, in which case "displacement" refers to electric or magnetic intensity vector.

The mathematical representation of the superpositon refers to solutions of the linear wave equation. The general form of the linear wave equation is

$$c^2 \nabla^2 w = w_{tt}, \tag{2.7}$$

where ∇^2 is the Laplacian operator in three dimensions given by $\nabla^2 = \partial^2/\partial x^2 + \partial^2/\partial y^2 + \partial^2/\partial z^2$. In general, the density of the medium may be

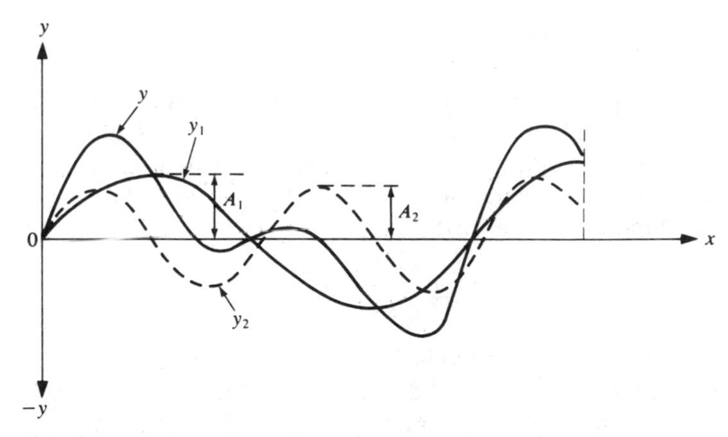

Fig. 2.3. Superposition of two waves moving in the $+x$ direction.

spatially dependent so that c is a known function of space. Equation (2.7) is then a linear PDE with a variable coefficient that depends on (x, y, z). The superposition theorem applied to this equation states: Suppose w_1 and w_2 are two solutions of (2.7). Then $w = c_1 w_1 + c_2 w_2$ (where c_1 and c_2 are constants) is also a solution. This means that w is a *linear combination* of w_1 and w_2. In general, if there are n wave forms represented by n solutions of (2.7), then a linear combination of these solutions is also a solution. This generalizes the principle of superposition by stating that we can combine n separate wave forms by finding the n constants c_i that satisfy the specific requirements of the problem.

Suppose we have two progressing waves of amplitudes A_1 and A_2 and frequencies ω_1 and ω_2 and wave speed c, given by

$$y_1 = A_1 \sin\left(\omega_1\left(\frac{x}{c} - t\right)\right), \qquad y_2 = A_2 \sin\left(\omega_2\left(\frac{x}{c} - t\right)\right).$$

At a particular t the profiles of these two sinusoidal waves combine to give a resultant wave form specified by $y = y_1 + y_2$ which is periodic but non-sinusoidal. Figure 2.3 shows plots of y_1, y_2 and y for the special case $t = 0$, $A_2 = \frac{3}{4}A_1$, $\omega_2 = 2\omega_1$. Since the speeds of the wave trains y_1 and y_2 are the same, the resultant wave train will have the same speed as the components.

Standing Waves

We now consider the important case of two sinusoidal waves of equal amplitude and the same frequency traveling *in opposite directions* along the string. This is a case frequently met in practice. Let the progressing and regressing waves, y_1 and y_2, respectively, be given by

$$y_1 = A \sin \omega\left(t - \frac{x}{c}\right), \qquad y_2 = A \sin \omega\left(t + \frac{x}{c}\right),$$

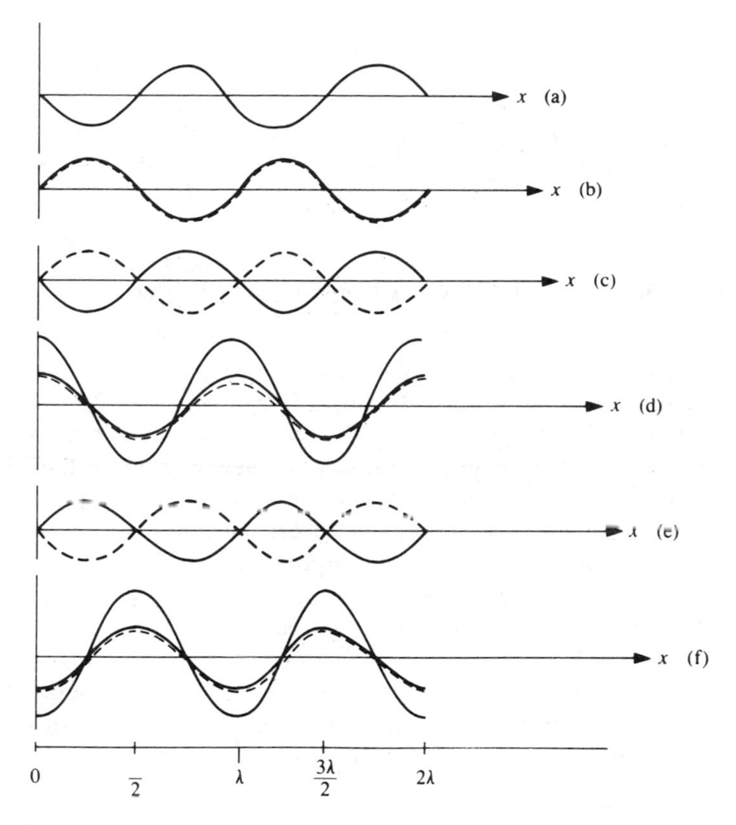

Fig. 2.4. Standing wave produced by interaction of progressing and regressing wave.

and let the resultant wave be given by $y = y_1 + y_2$. Then, by using a trigonometric identity, we obtain

$$y = y_1 + y_2 = 2A\left[\cos\frac{\omega x}{c}\right]\sin\omega t. \tag{2.8}$$

The wave pattern given by (2.8) is called a *standing wave*.† Note that the amplitude is twice that of y_1 or y_2. We can look at the right-hand side in two ways:

(1) the sinusoidal function of t is spatially modulated sinusoidally;
(2) the profile which is a sinusoidal function of x is sinusoidally modulated with respect to t.

The motion associated with this standing wave is illustrated in Fig. 2.4. Figure 2.4(a), (b) shows a profile of a progressive wave y_1 and a regressive

† In Chapter 5 standing wave solutions are discussed in detail from the point of view of water waves.

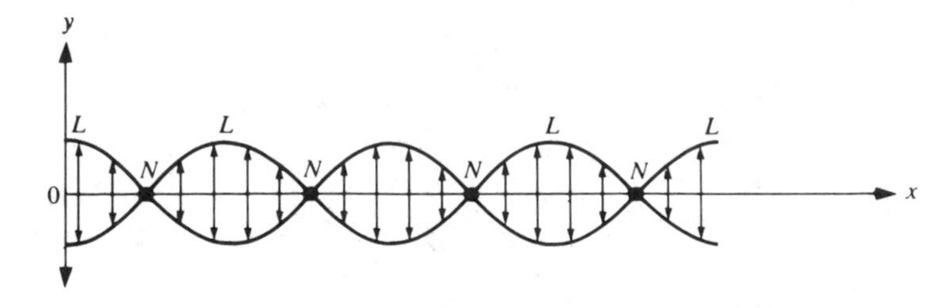

Fig. 2.5. Standing wave showing loops, nodes and particle motion.

wave y_2 traveling at $t=0$. The resultant wave form y is shown four times during a complete cycle in Fig. 2.4 (c, d, e, f) at times $t=0$, $T/4$, $T/2$, and $\frac{3}{4}T$, respectively, where the period $T=2\pi/\omega$. The particles in the standing wave execute simple harmonic motion of frequency ω about their equilibrium positions, but the amplitude of their motion is sinusoidal in x. Referring again to Fig. 2.4, we see that at certain points on the string the particles never move (are always in the equilibrium position). These values of x are called *nodes*. The points midway between the nodes, at which the amplitude of vibration is a maximum, are called *antinodes* or *loops*. Note that the wavelength λ of y_1 and y_2 which produce the standing wave y is equal to the distance between *alternate* nodes or alternate antinodes. Figure 2.5 illustrates a standing wave pattern showing loops L and nodes N, and the particle velocities (given by the arrows).

Reflection

Let us now investigate what happens when a transverse sinusoidal pulse advancing to the right along the string of length L arrives at the boundary which is a rigid support at the end of the string. The boundary condition at $x=L$ is: no motion of the string. Mathematically, this means

$$y(L, t)=0, \qquad t>0.$$

It is easily seen that, in order to satisfy this boundary condition, the incident pulse (the progressing wave) must be 180° out of phase with the reflected pulse (regressive wave). This is shown in Fig. 2.6. Note that the pulse shape remains the same.

Continuous transverse sinusoidal wave forms are also reflected from the fixed ends of a string of finite length, thus setting up standing waves. The model is: Given a string of length L, fixed at $x=L$, and having a sinusoidal continuous wave train set up at $x=0$. After a certain time the end $x=0$ is also fixed, and standing waves are thereby set up due to the interaction of the progressing and regressing waves. It is clear that the ends $x=0$, L are nodes. This means that physically the standing waves are composed of an integral number of half-wave lengths. The standing wave of the longest

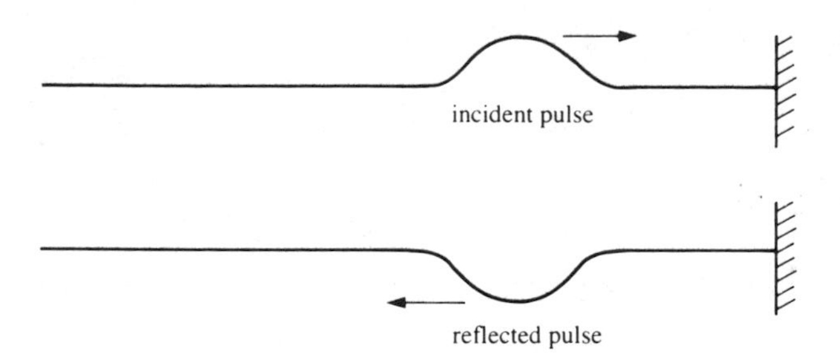

Fig. 2.6. Reflection of a transverse pulse at a rigid wall.

wavelength is a half-wave whose wavelength is $\lambda_1 = 2L$, called the *fundamental mode of oscillation.* The next *mode* is given by $\lambda_2 = 2L/2$ which is two half-waves, . . . , and the nth mode is given by $\lambda_n = 2L/n$, $n = 1, 2, 3, \ldots$. This gives the permissible values for the wavelengths that permit standing waves to be propagated.

2.7. Concluding Remarks

As mentioned in the Introduction, the purpose of this chapter is to present a physical foundation for the more abstract mathematics to be subsequently presented. After giving a qualitative description of various types of waves, and the conservation laws, the physics of interacting waves was presented vis-à-vis the vibrating string. The treatment of wave propagation in water, sound waves in gases, and stress waves in solids, as well as a more detailed treatment of the vibrating string, will be reserved for separate chapters. It is felt that this chapter is sufficient to give the reader a background on which we can construct the more mathematical aspects of wave propagation.

CHAPTER 3

Partial Differential Equations of Wave Propagation

Introduction

Even though this book is solely concerned with wave propagation phenomena, which is mathematically described by a certain type of partial differential equation (PDE), we can do no better than briefly discuss the PDEs that the discipline of mathematical physics uses to describe the rich variety of natural phenomena.

In general, a set of PDEs is a system of equations relating the dependent variables and their various partial derivatives. More specifically, a PDE for a function $u(x, y, \ldots)$ with partial derivatives u_x, u_y, $u_{xx}, \ldots$ is a relation of the form

$$F(x, y, \ldots, u, u_x, u_y, u_{xx}, \ldots) = 0, \tag{3.1}$$

where F is a given function of the variables $x, y, \ldots, u, u_x, u_y, u_{xx}, \ldots$ (where only a finite number of derivatives occur). It is obvious that the function $u(x, y, \ldots)$ is a solution of (3.1) if in some region of space of its independent variables $\mathbf{x} = (x, y, \ldots)$, the function and its derivatives satisfy the equation identically. In expressing natural phenomena by the methods of mathematical physics, these equations can generally be expressed as second-order PDEs. In general, they arise from the three conservation laws of physics, described in the previous chapter. These PDEs are called the *field equations*, and their solution gives the u field in space and time (pressure, temperature, etc.). An example is the PDE for the scalar $u(x, y)$ of two independent variables (x, y) given by

$$au_{xx} + 2bu_{xy} + cu_{yy} + d = 0, \tag{3.2}$$

where the coefficients $a, \ldots, d$ are known functions of (x, y, u, u_x, u_y). u may stand for temperature, potential, gas pressure, etc., and may also be extended to vectors and systems of equations. Equation (3.2) is said to be *quasilinear*, meaning that the second-order (highest) partial derivatives occur linearly so that the coefficients are at most known functions of the first-order derivatives of u. This is a special type of nonlinearity. For example, the equation $u_{xx}^2 + u_y u_{yy} = 0$ is nonlinear but *is not quasilinear*, since the highest derivative term u_{xx} does not occur linearly. If the coefficients depend at most on (x, y) then (3.2) is *linear*. It is clear that (3.2) can be extended to three-dimensional space by inserting the derivatives u_{xz}, u_{yz}, u_{zz}. Equation (3.2) can be considered the prototype for two-dimensional problems in mathematical physics.

It will be shown in this chapter that (3.2) can be classified into three categories: (1) elliptic, (2) parabolic, and (3) hyperbolic. These terms will be shown to have special meaning that depends on the nature of the roots of an associated quadratic equation. These three categories will be shown to depend only on the *principal part* of the PDE. The principal part of a PDE involves only the highest order partial derivatives. These categories give the mathematical representation of the following three types of phenomena that occur in nature:

(1) Gravitational attraction where a massive body in space instantly sets up a gravitational field that is felt everywhere in space. For example, if a star suddenly explodes so that its mass distribution is changed, that effect is *instantaneously felt* at every field point in space. A *gravitational potential* exists that satisfies Laplace's equation which is a linear PDE of the type (3.2) for two dimensions, $u_{xx} + u_{yy} = 0$, and is elliptic.
(2) The phenomenon of diffusion and heat transfer, which is a diffusion of heat, involves dissipation of thermal energy. A source of heat is instantaneously felt at any field point in space but is damped or smeared out with distance. The one-dimensional unsteady Fourier heat conduction equation $u_{xx} - u_t = 0$ is an example of (3.2); it is parabolic.†
(3) Wave propagation phenomena (the subject of this book). The essential characteristic of this type of phenomenon is that a source of wave propagation at a given point in space has an effect at any field point that involves a *time delay* depending on the distance of the field point from the source. A wave travels with characteristic wave speed that depends on the properties of the medium. No need to mention that the fastest speed is that of an electromagnetic wave propagating in a vacuum.

The mathematical representation that fits the PDE (3.2) is the one-dimensional wave equation $c^2 u_{xx} - u_{tt} = 0$, which is hyperbolic. If we impose damping on the traveling wave, that does not destroy the hyperbolicity of the PDE since damping is expressed by a first-order derivative u_t, and therefore does not involve the principle part of the PDE.

It is clear that the second-order PDE given by (3.2) can be expressed by an *equivalent first-order system*. Using the standard notation of PDEs, we

† The phenomenon of diffusion is associated with a stochastic or random process such as the random walk on Brownian motion.

set $u_x = p$, $u_y = q$. Then the single equation (3.2) can transformed into the equivalent first-order system

$$ap_x + bp_y + bq_x + cq_y + d = 0, \tag{3.3a}$$

$$p_y - q_x = 0, \tag{3.3b}$$

which are two PDEs for the functions $(p(x, y), q(x, y))$. Then $u(x, y)$ can be determined by integrating $u_x - p = 0$, $u_y - q = 0$. For three dimensions, we add the notation $r = u_{zx}$, $s = u_{yz}$, and $t = u_{zz}$, and arrive at five PDEs for the variables p, q, r, s, t.

The plan of this chapter is to start with the simplest first-order system of PDEs. By way of motivation, we will first show how the one-dimensional wave equation can be put into an equivalent first-order system of two PDEs. We then concentrate on a single quasilinear first-order PDE and introduce the notion of directions derivative which is then used to construct the *theory of characteristics*—the backbone of the technique used in solving the PDEs of the hyperbolic type to determine wave phenomena. We then extend the method of characteristics to first-order systems. It is seen that this method is a natural one for the solution of quasilinear systems, which cannot be attacked by the classical operational methods such as Laplace transforms, since these methods are restricted to linear PDEs. We then proceed to the second-order PDE given by (3.2) and classify this equation into the three categories mentioned above. The method of characteristics will then be developed for this second-order system and the correspondence with the first-order system will be shown. It will be shown that the pivotal problem in wave phenomena is the *Cauchy initial value* (IV) *problem.* The wave equation is an example of this problem. However, characteristic theory can be used to solve the Cauchy problem involving the nonlinear wave equation. The Cauchy problem will be reserved for a detailed treatment in a subsequent chapter.

3.1. Wave Equation as an Equivalent First-Order System

To motivate our investigations of the PDEs of wave propagation we shall start with the wave equation in one dimension with constant wave speed c.

$$c^2 u_{xx} - u_{tt} = 0. \tag{3.4}$$

Recall that, since this is a second-order PDE in x and t, we need two initial conditions (ICs) and two boundary conditions (BCs) for a well-defined solution. For this investigation we need not concern ourselves with these conditions since we shall be interested only in the nature of the general solution. (Note that the time t takes the place of y.) We now let

$$\varepsilon = u_x, \qquad v = u_t. \tag{3.5}$$

To get a physical orientation: u is the particle displacement, v is the particle velocity, and ε is the one-dimensional linear strain. Using (3.5), (3.4) can be put into the equivalent first-order system of two PDEs.

$$c^2\varepsilon_x - v_t = 0, \tag{3.6a}$$

$$v_x - \varepsilon_t = 0. \tag{3.6b}$$

As an example in *nonlinear elasticity* c, instead of being constant, is a prescribed function of ε. It turns out that c^2 is proportional to the slope of the stress–strain curve. When we discuss a system of first-order PDEs the system (3.6) will be treated in detail by the method of characteristics. It is easily seen that, for the case $c = \text{const.}$, the general solution of (3.6) is $\varepsilon = f(x+t)$, $v = g(x+c^2t)$, where f and g are arbitrary functions of their arguments.

A more symmetric notation for the system (3.6) is: let $\phi = c\varepsilon$. Then (3.6) becomes

$$\begin{aligned} c\phi_x - v_t &= 0, \\ cv_x - \phi_t &= 0, \end{aligned} \tag{3.7}$$

for $c = \text{const.}$ It is easily seen that both ϕ and v have a general solution of the form $F(x-ct)$, where F is an arbitrary of the argument $x-ct$. This shows traveling progressive waves with a wave speed $c>0$. ϕ has physical significance. As pointed out above, for nonlinear elasticity, $c = c(\varepsilon)$. Then ϕ is defined by $\phi(\varepsilon) = \int_{\varepsilon_0}^{\varepsilon} c(\bar{\varepsilon})\, d\bar{\varepsilon}$, where the integration is taken from an initial state ε_0 to a variable ε. For $c = \text{const.}$ the integration reduces to the original definition of ϕ. It was shown in [12, p. 152] in the section on "Impact Loading" in connection with the nonlinear reference on an elastic-plastic bar, that there is a physical reason to call ϕ the *impact velocity*. The first-order system given by (3.6) or (3.7) is classified as hyperbolic—v, ϕ, and ε exhibit wave properties, as evidenced by the fact that x and t appear as real linear combinations. It will be shown in a detailed study of characteristic theory that any real linear combination of x and t is a necessary consequence that the PDE be hyperbolic (with constant coefficients) so that traveling waves are propagated.

It is instructive to now give an example of a first-order system that does not exhibit wave properties. Let V be a solution of the two-dimensional Laplace equation $V_{xx} + V_{yy} = 0$. V is said to be a *harmonic function*. Then, the necessary and sufficient condition that V be harmonic (satisfy the above Laplace equation) is that the *Cauchy–Riemann equations* be satisfied for the functions u and v, which are also harmonic. These are the solutions to the first-order system

$$u_x - v_y = 0, \qquad v_x + u_y = 0. \tag{3.8}$$

It is easily shown that both u and v are also harmonic functions. We can see that solutions are of the form $u = F(x+iy)$, $v = G(x+iy)$, where F and

G are arbitrary functions of $z = x + iy$ so that u and v are the real and imaginary parts, respectively, of an analytic function of a complex variable $w = f(z)$. This means $w = u + iv$. Since the argument z in the arbitrary functions F and G is complex, there are no real linear combinations of x and y, which shows that u and v do not have wave properties. Indeed, the Cauchy-Riemann equations are of the elliptic type and have no real characteristics. It will be shown in our classification of PDEs that the elliptic equation has imaginary characteristics, while the hyperbolic equation has real characteristics, which is reflected in the fact, as mentioned above, that all solutions of hyperbolic PDEs in two dimensions must have the arguments of the general solution in the form of $ax + by$ where a and b are real constants (for the linear case with constant coefficients). The obvious generalization to three dimensions tells us that the corresponding hyperbolic system with constant coefficients involves arguments of the solutions as real linear combinations of x, y, z.

3.2. Method of Characteristics for a Single First-Order Quasilinear Partial Differential Equation

We begin our detailed study of the PDEs of wave propagation by developing the method of characteristics for a single first-order quasilinear PDE in the (x, y) plane.

$$au_x + bu_y = cu + d, \tag{3.9}$$

where the coefficients a, b, c, d are prescribed functions of (x, y, u). The left-hand side of (3.9) is a linear combination of the derivatives u_x and u_y for any fixed value of (x, y, u). It represents the derivative of u in the direction given by the vector whose components are (a, b). For this reason du/dx is called the *directional derivative*. This will be made clear by the following analysis: We first expand du/dx, obtaining

$$\frac{du}{dx} = u_x + u_y\left(\frac{dy}{dx}\right). \tag{3.10}$$

We now divide (3.9) by $a(x, y, u)$ (assuming that $a \neq 0$ in the domain of (x, y) and equate the right-hand side of (3.10) to the left-hand side of (3.9). We thereby obtain

$$\frac{du}{dx} = u_x + u_y\left(\frac{dy}{dx}\right) = u_x + \left(\frac{b}{a}\right)u_y.$$

In order for this equation to be valid in the domain of the solution we must have

$$\frac{dy}{dx} = \frac{b(x, y, u)}{a(x, y, u)}. \tag{3.11}$$

Equation (3.11) is an ordinary differential equation (ODE) for $y(x)$. Equation (3.10), coupled with (3.11), makes it clear why we call du/dx the directional derivative. It is the total derivative of u with respect to x in the direction given by the slope dy/dx of a one-parameter family of curves, which is obtained by integrating (3.11), called the *characteristic curves.*

For the *linear case* we know the right-hand side of (3.11) as a function of (x, y). Therefore (3.11) is an ODE (albeit in general nonlinear) whose solution is $y = y(x, \alpha)$, where α is the constant of integration (hence, a *one-parameter family* of curves), and that family is independent of u.

For the nonlinear case, the right-hand side of (3.11) is a function of u as well as x and y so that we have a more complicated situation in which *the family of curves depends on the solution u.* Using (3.10), (3.9) becomes

$$\frac{du}{dx} = \left(\frac{c}{a}\right)u + \frac{d}{a}. \tag{3.12}$$

The system of ODEs given by (3.11) and (3.12) is called the *characteristic equations* corresponding to (3.9). Equation (3.12) is the characteristic ODE for the directional derivative in terms of u, a, and b. The system of characteristic equations allow us to construct a graphical solution to (3.9) for this quasilinear case, even though an analytical solution, in general, is not available.

We point out the following:

(1) For the linear case, (3.11) and (3.12) are uncoupled in the sense that the solution of (3.11), for the one-parameter family of characteristic curves, is independent of (3.12), and thus independent of u.

(2) For the more complicated situation where the PDE is quasilinear (3.11) and (3.12) are coupled in the sense that, in order to integrate (3.11) to obtain the family of characteristic curves, we must simultaneously integrate (3.12) to obtain $u(x, y)$.

Thus, for the quasilinear case the characteristic equations (3.11) and (3.12) represent a system of simultaneous or coupled ODEs.

Geometric Considerations

We first consider the case $c = 0$ in (3.9). Setting $u_x = p$, $u_y = q$, (3.9) is specialized to give

$$ap + bq - d = 0. \tag{3.13}$$

We now interpret (3.13) geometrically for the quasilinear case by first considering the (x, y, u) space. Suppose a solution to (3.13) exists in some domain in this space. We then have an *integral surface* in this domain. An integral surface is defined by the property that every point on this surface is a solution to (3.13).

Let P: (x, y, u) be any point on the integral surface. We put (3.13) in the following vector form:

$$(a, b, d) \cdot (p, q, -1) = 0, \tag{3.14}$$

which tells us that the scalar product of two vectors equals zero. Since P is on the integral surface, $(a, \ldots, q)$ are known functions of (x, y, u) at P. Equation (3.14) tells us that the known normal to the integral surface at P, given by $(p, q, -1)$, is normal to the direction given by (a, b, d). This clearly means that (a, b, d) defines the tangent plane of the surface at P. Suppose we have a family of integral surface going through P. Then, in the neighborhood of P, all these integral surfaces are tangent planes through P and they form an *axial pencil of planes* through a straight line, since p and q appear linearly in the quasilinear PDE. This line is called the *Monge axis* and the family of planes is called the *Monge pencil.* The directions of the Monge axes everywhere on the integral surface form a direction field in our domain. The integral curves of this direction field are determined by integrating the characteristic ODE (3.11).

We thus have the following situation: Every surface generated by a one-parameter family of characteristic curves, obtained by integrating (3.11), is an integral surface $u(x, y)$ of the PDE. Conversely, every integral surface $u(x, y)$ is generated by a one-parameter family of characteristic curves.

This last statement is easily proved. On every integral surface $u(x, y)$ of (3.13) a one-parameter family of curves given by $y = y(x, \alpha)$ arises by integrating the characteristic equation (3.11). Along such a curve the PDE (3.13) becomes

$$\frac{du}{dx} = \frac{d(x, y, u)}{a(x, y, u)},$$

upon using the definition of the directional derivative du/dx. Thus our one-parameter family of curves satisfies the characteristic equations (3.11) and (3.12) and, hence, is a one-parameter family of characteristic curves.

Since solutions of the system of characteristic equations are *uniquely* determined by the initial values of x, y, u, we easily obtain the following theorem:

Every characteristic curve which has one point in common with an integral surface lies entirely on that surface.

Moreover, every integral surface is generated by a one-parameter family of characteristic curves for our first-order PDE. We will see below when we study the second-order PDE that two families of characteristic curves exist for a second-order hyperbolic system.

Cauchy Initial Value Problem

Now suppose that $u(x, y)$ is assigned initial values (IV) $u = f(x, g(x))$ along some curve Γ defined by $y = g(x)$. Γ is called the IV curve. The *Cauchy IV*

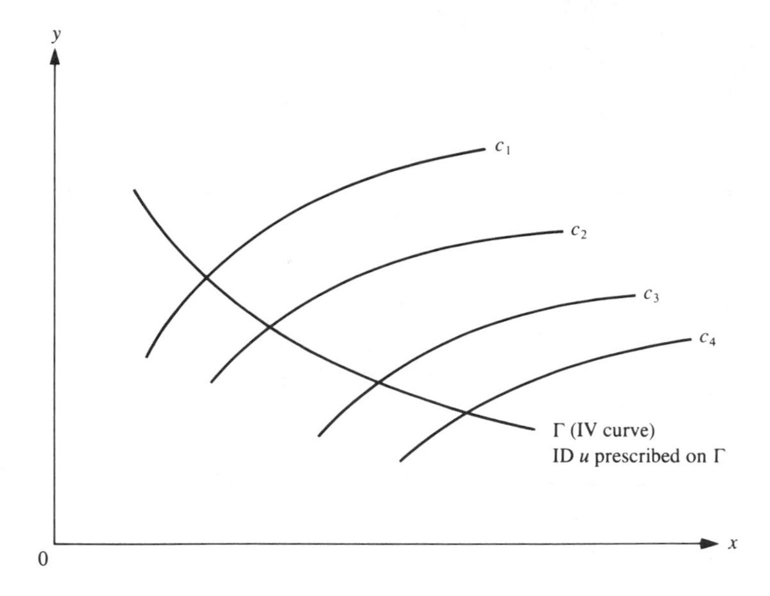

Fig. 3.1. Characteristic curves crossing an initial value curve.

Problem for a first-order PDE is formulated as follows: Given $u = f(x, g(x))$ on the IV curve Γ: $y = g(x)$, determine $u = u(x, y)$ that satisfies the following:

$$\begin{aligned} au_x + bu_y &= cu + d, \\ u = f(x, g(x)) \quad &\text{on} \quad \Gamma\text{: } y = g(x), \end{aligned} \tag{3.15}$$

Clearly Γ must at no point have the direction of a characteristic curve, since there u is determined as a solution of an ODE. The characteristic equations suggest that if u is prescribed along Γ and we know the one-parameter family of characteristic curves C, then we can determine a unique solution $u(x, y)$ in the domain covered by that characteristic family. This is shown in Fig. 3.1 where a family of characteristics C_α cross the IV curve Γ. The subscript α defines a particular characteristic. At any field point P not on Γ there exists a characteristic going through P which, when drawn back to cross Γ, picks up the IV of u. By integration along that characteristic we can calculate u_P. We point out again that it is only for the linear case that we can determine the one-parameter family of characteristic curves independent of the solution $u(x, y)$. For the quasilinear case the characteristics depend on the solution for u and can be determined locally by building up a characteristic net starting in the neighborhood of the IV curve. Each field point depends on the solution developed at the neighboring field points.

Parametric Representation

We now reformulate the method of characteristics in a *parametric form* in the following sense: We introduce the parameter t to generate a characteristic

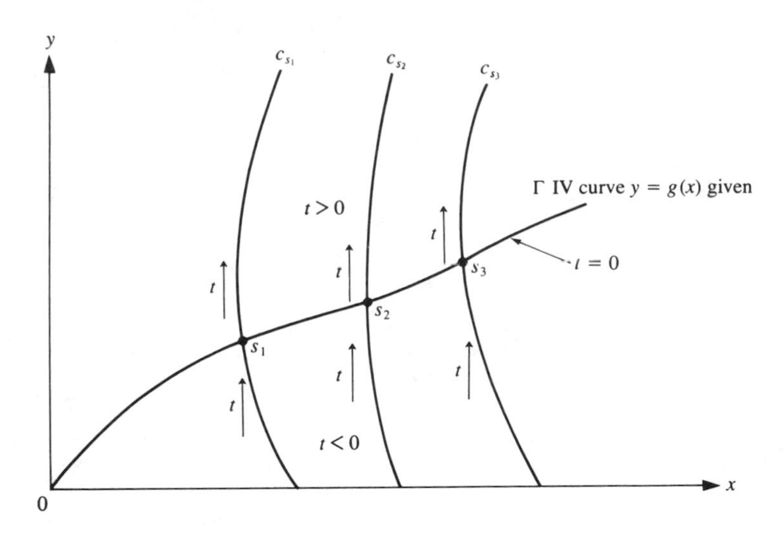

Fig. 3.2. Characteristic curves C_s crossing an initial value curve Γ in parametric form.

curve C and the parameter s to generate the IV curve Γ. This parametric representation has the advantage of symmetrizing the formulation, since x and y would then play equal roles. We may think of t as denoting the arc length of a particular characteristic and s, the arc length of Γ. We can modify Fig. 3.1 by introducing the coordinate system such that $s = 0$ at the intersection of Γ with the origin. As s increases we generate Γ. The family of characteristics C_i cross Γ. Each characteristic is defined by the value of s it picks up at its intersection with Γ. At this intersection $t = 0$, and as t varies we generate a characteristic having a particular s as its defining parameter. This is shown in Fig. 3.2 where three characteristics denoted by C_1, C_2, C_3 cross the IV curve at s_1, s_2, s_3, respectively. In general, we can extend the characteristic family for $t < 0$; but if t is denoted as time, then we do not know the past history of u and we are thereby only interested in the domain for $t > 0$.

Suppose we solve a Cauchy IV Problem. Then we have the following transformation equations:

$$x = x(s, t), \qquad y = y(s, t), \qquad u = u(x(s, t), y(s, t)).$$

The inverse transformation

$$s = s(x, y), \qquad t = t(x, y), \qquad u = u(s(x, y), t(x, y)),$$

allows us to go from the (s, t) to the (x, y) plane. Such an inverse exists and is unique if and only if the determinant of the Jacobian of the mapping $\det[J(x_s, x_t / y_s, y_t)] \neq 0$.

The IV for u is

$$u = u(s, 0) = f(s) \quad \text{on} \quad \Gamma\colon x = x(s, 0), \qquad y = y(s, 0),$$

where $f(s)$ is given. Each C_i is defined by

$$C_i\colon x = x(s_i, t), \qquad y = y(s_i, t). \tag{3.16}$$

Each field point P: (x, y) is defined by a particular characteristic coming from Γ, or by a particular value of (s, t).

We now obtain the characteristic equations in terms of (s, t) by first expanding $du(s, t)/dt$. We get

$$\frac{du(s,t)}{dt} = \left(\frac{dx}{dt}\right)u_x + \left(\frac{dy}{dt}\right)u_y = au_x + bu_y = cu + d,$$

upon using (3.9). We immediately obtain the characteristic ODEs in the (s, t) plane given by (3.17) and (3.18).

$$\frac{dx}{dt} = a(x(s,t), y(s,t), u(s,t)), \qquad \frac{dy}{dt} = b(x(s,t), y(s,t)u(s,t)), \tag{3.17}$$

$$\frac{du}{dt} = cu + d. \tag{3.18}$$

Integrating the system (3.17) yields the characteristics in the form $x = x(s, t, u(s, t))$, $y = y(s, t, u(s, t))$. For the quasilinear case (3.17) is coupled with the integration of (3.18) which gives $u = u(x(s), y(s, t))$. Referring to (3.16), we see that s_i is the integration constant that defines a characteristic. It is clear that (3.17) and (3.18) are related to (3.11) and (3.12), since $dy/dt/dx/dt = dy/dx$ and $du/dt/dx/dt = du/dx$.

Example 1. The first example to illustrate the method is to obtain the solution of the linear PDE obtained from (3.9) by setting $a = x$, $b = y$, $c = \text{const.}$, $d = 0$, for the IV $u = f(x)$ on $y = 1$. We therefore want the solution $u(x, y)$ to the following Cauchy IV Problem:

$$xu_x + yu_y = cu, \qquad \text{IC } u = f(s) \qquad \text{given on} \qquad \Gamma\colon t = 0, \quad x = s, \quad y = 1.$$

The characteristic equations (3.17), (3.18) become

$$\frac{dx}{dt} = x, \qquad \frac{dy}{dt} = y, \qquad \frac{du}{dt} = cu.$$

Integrating and using the IC yields

$$x = se^t, \qquad y = e^t, \qquad u = f(s)e^{ct}.$$

We have thus obtained $u(s, t)$. To get $u(x, y)$ we invert this transformation and obtain

$$u = f\left(\frac{x}{y}\right) y^c,$$

from $s = x/y$, $e^{ct} = y^c$.

Now, what does this problem look like in the (s, t) plane? We have

$$u_x = u_s s_x + u_t t_x, \qquad u_y = u_s s_y + u_t t_y.$$

The above transformation gives $s_x = y^{-1} = e^{-t}$, $s_y = -x/y^2 = -se^{-t}$, $t_x = 0$, $t_y = e^{-t}$. We then obtain $u_t = cu$. Integrating and using the IC gives $u = f(s)e^{ct}$.

Example 2. A more complicated linear problem is: Find the solution to the following Cauchy IV Problem:

$$yu_x - xu_y = cu, \qquad u = f(s) \quad \text{on} \quad \Gamma\colon x(s, 0) = 2, \quad y(s, 0) = ms.$$

The characteristic equations become

$$\frac{dx}{dt} = y, \qquad \frac{dy}{dt} = -x, \qquad \frac{du}{dt} = cu.$$

From the first two we obtain

$$\frac{d^2x}{dt^2} = -x, \qquad \frac{d^2y}{dt^2} = -y,$$

whose solutions are

$$x = A\cos t + B\sin t, \qquad y = B\cos t - A\sin t \quad \text{or} \quad x^2 + y^2 = A^2 + B^2 = \alpha^2.$$

This, of course, can easily be obtained in nonparametric form from the characteristic equations $dx/y = -dy/x$. We solve for the coefficients A and B by using the IC $u(s, 0) = f(s)$ on $x(s, 0) = s$, $y(s, 0) = ms$, and obtain

$$x(s, t) = s\cos t + (ms)\sin t, \qquad y(s, t) = ms\cos t - \sin t$$

or

$$x^2 + y^2 = \alpha^2 = (1 + m^2)s^2.$$

By integrating the third characteristic equation we obtain the solution for $u(s, t)$. Using the above transformation equations gives $u(x, y)$. We get

$$u(x, y) = f(s)e^{ct},$$

$$s = \frac{x}{\cos t + m\sin t} = \frac{y}{ms\cos t - s\sin t}, \qquad \tan t = \frac{mx - y}{x + my}.$$

Example 3. We now give an example of a quasilinear PDE by obtaining the solution to the Cauchy IV Problem:

$$uu_x + u_y = 1, \qquad u(s, 0) = Us \quad \text{on} \quad \Gamma\colon x(s, 0) = s, \quad y(s, 0) = 0,$$

where U is a given constant. The characteristics C_s are $x = x(s, t)$, $y = y(s, t)$. They are obtained from the three coupled characteristic equations

$$\frac{dx}{dt} = u, \qquad \frac{dy}{dt} = 1, \qquad \frac{du}{dt} = 1.$$

The third equation gives

$$u(s, t) = t + Us.$$

Clearly this satisfies the IC $u(s, 0) = Us$. To find $s = s(x, y)$, $t = t(x, y)$ we need to integrate the first two characteristic equations and use the above solution for $u(s, t)$. We get

$$x = \frac{t^2}{2} + Ust + s, \qquad y = t.$$

Using these results we obtain the solution for $u(x, y)$, which is

$$u = y + \left[\frac{U}{2(Uy+1)}\right](2x - y^2).$$

Since the solution for u is simpler in the (s, t) plane (namely, $u = t + Us$) we transform the Cauchy Problem, as in Example 1, to (s, t) coordinates. Using the above transformations and omitting details, we obtain the Cauchy Problem in the (s, t) plane

$$(u - t - Us)u_s + (Ut+1)u_t = Ut + 1, \qquad u(s, 0) = Us.$$

The characteristic equations become

$$\frac{ds}{u - t - Us} = \frac{dt}{Ut + t} = \frac{du}{Ut + 1},$$

which gives $u - t - Us = 0$ (satisfying the IC), yielding $ds/dt = 0$. It follows that the characteristics in the (s, t) plane consist of a family of straight lines $s = \text{const.}$ along which $u = t + Us$. The integral surface in (s, t, u) space is therefore a plane which depends on the given value of U.

We now extend the method of characteristics to the solution of a first-order system of quasilinear PDEs. To do this we work with a single second-order PDE and transform it to an equivalent first-order system.

3.3. Second-Order Quasilinear Partial Differential Equation

We now investigate the single second-order quasilinear PDE given by (3.2). To develop the method of characteristics for this equation, we work with the equivalent system of two first-order PDEs given by (3.3a, b) (recall that $p = u_x$, $q = u_y$). We now write this system according to the following notation:

$$L_1 = ap_x + bp_y + bq_x + cq_y = 0, \tag{3.19}$$

$$L_2 = p_y - q_x = 0. \tag{3.20}$$

In this notation we see that L_1 and L_2 are forms that are linear in (p_x, p_y, q_x, q_y) where the coefficients (a, b, c) depend on (x, y, u, p, q) for the quasilinear case and (x, y) for the linear case. Equation (3.20) arises from the continuity of the derivatives. We now introduce the *directional derivatives* dp/dx and dq/dx. We attempt to find conditions for which these directional derivatives are *in the same direction*, which is called the *characteristic direction*—tangent to the characteristic curves. These conditions will allow us to determine the characteristic direction. It turns out that the characteristic direction arises from the solution of a quadratic equation whose roots give the slope of the characteristic curves. It is the character of these roots that allow us to classify the second-order PDE into three types. For the hyperbolic case, which yields wave propagation phenomena, there are two characteristic directions which give two families of characteristic curves. (But we are getting ahead of ourselves!)

To carry out this program we seek the proper linear combination of the forms L_1 and L_2 (since the four derivatives $p_x \ldots q_y$ are embedded in these expressions). To this end, we introduce a parameter λ such that $L = L_1 + \lambda L_2$. We shall determine that value of λ which puts dp/dx and dq/dx in the same direction. In general (for the quasilinear case), λ depends on the field point P: (x, y) and the values of u, p, and q at P. The *crux of the method* is the following: In order for dp/dx and dq/dx to be in the characteristic direction (which we have yet to determine) the following rule must be observed:

$$\text{coef.}(p_y) : \text{coef.}(p_x) = \text{coef.}(q_y) : \text{coef.}(q_z) = z, \qquad \text{where} \quad z = \frac{dy}{dx}. \tag{3.21}$$

As mentioned above, the set of ratios given by (3.21) is the key to the method of characteristics applied to our system. A little reflection will show that these ratios must be valid in order to have that combination of the derivatives $p_x \ldots q_y$ which put dp/dx and dq/dx in the same direction. Note that the rule given by (3.21) is essentially used for any system of first-order quasilinear PDEs. This means that the rule can be applied to the fluid dynamics of two-dimensional steady flow, one-dimensional unsteady compressible flow, as well as certain wave propagation problems in solid continua. In general, it can be applied to any two-dimensional wave propagation problem which is governed by a quasilinear system. It can be extended to three-dimensional quasilinear systems, see, for example, [6] and [11, Vol. II].

Using (3.19) and (3.20), we get for L,

$$L = ap_x + (b - \lambda)q_x + (b + \lambda)p_y + cq_y + d = 0. \tag{3.22}$$

Applying the rule (3.21) to (3.22) gives

$$\frac{b+\lambda}{a} = \frac{c}{b-\lambda} = \frac{dy}{dx} = z. \tag{3.23}$$

From (3.23) we get the following quadratic equation for z (as promised above):

$$az^2 - 2bz + c = 0. \tag{3.24}$$

Table 3.1. Classification of quasilinear PDEs in two dimensions.

Discriminant	Type of PDE	Nature of roots	Example
$D=b^2-ac$			
>0	hyperbolic	real unequal	wave equation $b=0$, $ac<0$
$=0$	parabolic	real equal	unsteady heat equation $b=c=0$, $a\neq 0$
<0	elliptic	complex conjugates	Laplace equation $b=0$, $ac>0$

Equation (3.24) is called the *characteristic equation* and is a pivotal result. It tells us that the roots z of this quadratic equation depend on the coefficients (a, b, c) involved in the PDE (3.2). Moreover, since these coefficients are involved only with the second-order derivatives, z depends only on the *principle part* of the second-order PDE. By definition $z=dy/dx$, therefore the roots yield the characteristic directions given by the slopes z. The roots z are thus the *characteristic ODES* associated with the first-order system (3.19) and (3.20) or the original second-order PDE (3.2). Clearly these roots depend on the discriminant $D=b^2-ac$. The nature of D allows us to categorize our system (3.19) and (3.20) into three separate types (as mentioned previously). There are three possibilities for D:

(1) $D>0$, giving real and unequal roots.
(2) $D=0$, giving real and equal roots.
(3) $D<0$, giving complex conjugate roots.

Table 3.1 summarizes the classification of quasilinear PDEs according to the nature of the roots z of (3.24). Note that for $D>0$ the PDE is called *hyperbolic*, for $D=0$, *parabolic*, and for $D<0$, *elliptic*. The only case for which we can have wave propagation is the hyperbolic case. Since the original PDE is second order, the hyperbolic case yields *two families* of characteristic curves. The reason is that the associated ODE for z, (3.24), is quadratic. In general, for an nth order PDE or n first-order PDEs the ODE for z is nth degree so that there are n roots. This means that such a hyperbolic system has n families of characteristics.

3.4. Method of Characteristics for Second-Order Partial Differential Equations

There is another way of developing the method of characteristic for a second-order quasilinear PDE, and that is by investigating directly the solutions of this PDE (3.2) without transforming it into a first-order system. The basic idea of this approach is illustrated by considering the Cauchy IV

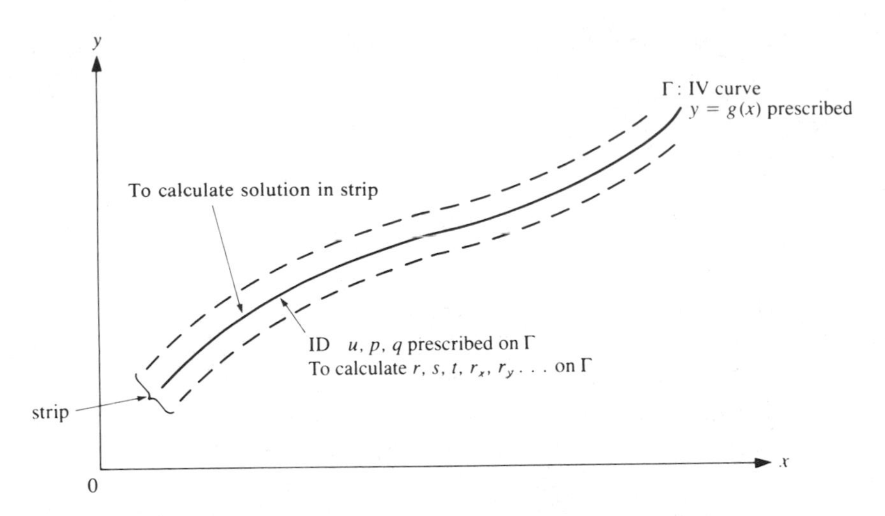

Fig. 3.3. Initial value curve showing initial data and strip.

Problem for (3.2). This means that we prescribe an IV curve Γ in the (x, y) plane on which are given *initial data*: (u, p, q). Since the PDE is second order we must prescribe both u and the first partial derivatives p and q. From p and q we can calculate the derivative of u normal to the IV curve. The idea is to obtain a solution to (3.2) in the neighborhood of Γ (in a strip surrounding Γ) which satisfies the ICs on Γ. In order to do this, we must be able to calculate all the higher-order derivatives on Γ using the initial data. We extend the notation to second-order derivatives (standard in PDEs).

$$u_x = p, \qquad u_y = q, \qquad u_{xx} = r, \qquad u_{xy} = s, \qquad u_{yy} = t. \tag{3.25}$$

This means that, given the initial data (u, p, q) on the IV curve, we must calculate $(r, s, t, r_x, r_y, \ldots)$ on Γ. We then use these derivatives to obtain the solution in the strip surrounding Γ by a Taylor series expansion. Note that this is an extension to PDEs of the Taylor series expansion method about a point for ODEs. Of course, the number of derivatives we use for the expansion depends on the accuracy we want. Having obtained the solution in the strip we then, by the same process of a Taylor series expansion, extend the domain of the solution. Figure 3.3 shows Γ, the initial data, and the strip.

Thus our first task is to derive algebraic equations for the values of r, s, t on Γ, and then use these values to calculate the higher-order derivatives on Γ. Using (3.25), (3.2) becomes

$$ar + 2bs + ct = -d. \tag{3.26}$$

We have three unknowns, therefore we need two additional equations. They are obtained by expanding $dp/dx \equiv p'$ and $dq/dx \equiv q'$. We get

$$r + zs = p', \tag{3.27}$$

$$s + zt = q'. \tag{3.28}$$

Since z, p', and q' are known on Γ (3.26)-(3.28) are the three algebraic equations to determine r, s, t on Γ. The solution for r, s, t is

$$r = -\frac{\Delta_1}{\Delta_4}, \qquad s = \frac{\Delta_2}{\Delta_4}, \qquad t = -\frac{\Delta_3}{\Delta_4}, \tag{3.29}$$

where the determinants are

$$\Delta_1 = \begin{vmatrix} 2b & c & d \\ z & 0 & -p' \\ 1 & z & -q' \end{vmatrix}, \qquad \Delta_2 = \begin{vmatrix} a & c & d \\ 1 & 0 & -p' \\ 0 & z & -q' \end{vmatrix},$$

$$\Delta_3 = \begin{vmatrix} a & 2b & d \\ 1 & z & -p' \\ 0 & 1 & -q' \end{vmatrix}, \qquad \Delta_4 = \begin{vmatrix} a & 2b & c \\ 1 & z & 0 \\ 0 & 1 & z \end{vmatrix}. \tag{3.30}$$

As long as $\Delta_4 \neq 0$, r, s, t are finite and can be uniquely determined from (3.29).

Now an interesting question arises. What happens when $\Delta_4 = 0$? For this case, we cannot solve (3.29) for r, s, t on Γ. This means we cannot expand in a Taylor series and thereby obtain the solution in the strip surrounding Γ. Therefore such a curve is not an admissible IV curve since we cannot use the ICs to obtain solutions off the curve. Setting $\Delta_4 = 0$ yields

$$az^2 - 2bz + c = 0,$$

which is (3.24). Since the roots of (3.24) give the characteristic directions, we obtain the important result: The IV curve Γ shall never be in a characteristic direction z, since the curves for which initial data do not allow us to obtain solutions are the families of characteristic curves. We see that this second approach vis-à-vis the second-order PDE gives new insight into the meaning of characteristic curves as those curves along which we cannot solve the Cauchy IV Problem.

Suppose that in addition to $\Delta_4 = 0$ we have $\Delta_2 = 0$. Then $\Delta_1 = \Delta_3 = 0$, so that r, s, t are indeterminant. This yields

$$azp' + cq' + zd = 0. \tag{3.31}$$

Equation (3.31) can also be obtained by multiplying (3.27) by az, (3.28) by c, adding the result, and using (3.26) and (3.24). Equation (3.31), combined with (3.24), tells us that if (3.26) is hyperbolic, then there are two families of characteristic curves arising from the two real and unequal roots z of (3.24). Along each characteristic there is a relationship between the directional derivatives p' and q' given by (3.31). The characteristic ODEs (3.24)

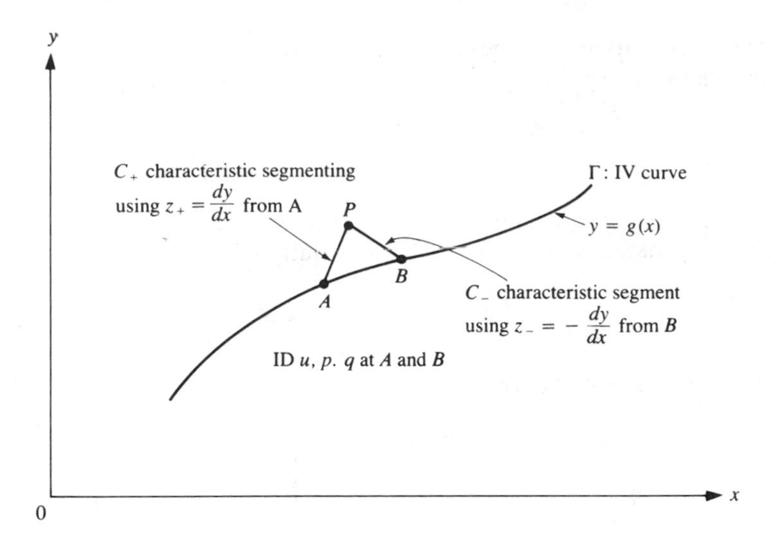

Fig. 3.4. Field point P obtained from intersection of C_+ from A and C_- from B.

and (3.31) are the pivotal equations for the method of characteristics. They allow us to solve the quasilinear hyperbolic second-order PDE graphically.

We now show how these characteristic ODEs are used to construct the solution field in the neighborhood of an IV curve Γ (in the strip) for a typical Cauchy IV Problem. A *characteristic net* can thereby be constructed. The net is composed of two families of characteristic line segments. Since the PDE (3.26) is quasilinear, the line segments forming the net must be determined from the solution at neighboring field points. This will be shown in the graphical description for the construction of the solution field. But first we show how to construct the solution at a typical field point P in the strip surrounding Γ, which we define by the curve $y = g(x)$. Figure 3.4 shows two neighboring points A and B on Γ. We use the notation C_+ as the characteristic line segment (of very small length) whose slope is $z_+ = dy/dx$ and C_-, the characteristic line segment of slope $z_- = -dy/dz$. These characteristic line elements emanate from points such as A, B where data are given to calculate the appropriate slopes. The initial data at A and B are x_A, y_A, u_A, p_A, q_A, and x_B, y_B, u_B, p_B, q_B, respectively. We then calculate the coefficients (a, b, c) at A and B since they are known functions of (x, y, u, p, q), which are the given data at these points. Next, we use (3.24) to calculate the roots $z_{+,A}$ at A and $z_{-,B}$ at B. Using the definition of z we get

$$\begin{aligned} y_P - y_A &= z_{+,A}(x_P - x_A), \\ y_P - y_B &= z_{-,B}(x_P - x_B). \end{aligned} \tag{3.32}$$

From (3.32) we solve for the coordinates (x_P, y_P) of the field point P. Thus we see that the position of P is determined from the intersection of a C_+

characteristic from A and a C_- characteristic from B. For this quasilinear case, we see that if any of the initial data for u, p, or q are changed at A or B then the position of P is changed so that the position of the field points in the strip depends on these initial data. For the linear case, the situation is much more simple in that the positions of the field points in the strip only depend on the curve $y = g(x)$ (are independent of the initial data). We next solve for p_P and q_P by using (3.31) in differential form, by approximating dp and dq by finite differences. We get

$$a_A z_{x,A}(p_P - p_A) + c_A(q_P - q_A) + d_A(x_P - x_A) = 0 \quad \text{along} \quad C_+\text{: } AP,$$

$$a_B z_{-,B}(p_P - p_B) + c_B(q_P - q_B) + d_B(x_P - x_B) = 0 \quad \text{along} \quad C_-\text{: } BP. \tag{3.33}$$

Having calculated the coordinates of P we then use (3.33) to calculate (p_P, q_P) in terms of the known quantities at A and B. Next we calculate u_P by integrating the differential

$$du = p\,dx + q\,dy.$$

One method of calculating u_P is to integrate this differential along the C_+ segment AP using the average values of p_A, p_P, q_A, q_P; integrate along the C_- segment BP using the average of p_B, p_P, q_B, q_P; and thereby obtain the average value of u_P from the resulting equations, which are

$$u_P - u_A = \tfrac{1}{2}(p_A + p_P)(x_P - x_A) + \tfrac{1}{2}(q_A + q_P)(y_P - y_A) \quad \text{along } AP,$$

$$u_P - u_B = \tfrac{1}{2}(p_B + p_P)(x_P - x_B) + \tfrac{1}{2}(q_B + q_P)(y_P - y_B) \quad \text{along } BP.$$

We solve these equations for u_P and thereby obtain u_P as the average value of the two equations.

$$u_P = \tfrac{1}{2}(u_A + u_B) + \tfrac{1}{4}(p_A + p_P)(x_P - x_A) + \tfrac{1}{4}(p_B + p_P)(x_P - x_A)$$
$$+ \tfrac{1}{4}(q_A + q_P)(y_P - y_A) + \tfrac{1}{4}(q_B + q_P)(y_P - y_B). \tag{3.34}$$

Equation (3.34) thus gives us the solution at the field point p in terms of the IV at A and B. The process is repeated, using, for example, data at A and its left-hand neighboring point on Γ, constructing the C_+ from that point, the C_- segment from A, and determining the point of intersection of these two segments to define another field point, etc. If these calculations are repeated for a finite given distribution of points on Γ we can then build up a solution field at a finite number of points in the strip which involves one finite difference step in y. If y corresponds to time then we have determined the field points in the strip for one time interval after the initial time (starting from Γ). Having then calculated as many field points in the strip as we feel necessary (depending on the accuracy we desire for a given problem), we then use these results as "new initial conditions" and continue the procedure to construct solutions for the second time step. This is then continued for the nth time step so that an approximation to the solution field is thereby obtained. Referring again to Fig. 3.4 we now let A and B

be two neighboring *field points* at which we have obtained the solution at a particular time. Then point P (which corresponds to the next time step) can be determined as the intersection of the appropriate characteristic segments and u_P can be determined from (3.34). It is clear that this way of applying characteristic theory is actually a numerical method. It is an approximation method whose accuracy depends on the number of data points chosen, the closeness of two neighboring data points, the method of integration, etc. The curvilinear families of characteristics are replaced by an approximate characteristic net made up of characteristic line segments constructed from neighboring points. For more details on this finite difference method, including conditions for the stability of solutions, the reader is referred to [12].

3.5. Propagation of Discontinuities

In the theory of PDEs it can be shown that PDEs of the elliptic and parabolic type have analytic solutions even when the boundary or initial conditions are discontinuous. However, hyperbolic PDEs are different in the sense that *discontinuities are propagated along characteristics* into the solution domain. This statement must be modified: it is only true for weak discontinuities. For strong or large amplitude discontinuities, strong shock waves are produced which propagate along curves that are different from the characteristic curves, which are curves along which *infinitely weak shock waves*, such as sound waves, propagate. In this section we shall deal with the propagation of weak discontinuities. A proof that these discontinuities are propagated only along characteristic curves will be given in Chapter 7, Section 7.1 on propagation of discontinuities. The solution u_P can be calculated in terms of the initial data on the interval $[A, B]$. This statement will be verified for the linear case when we discuss D'Alembert's solution to the Cauchy IV Problem (for the one-dimensional wave equation). For the linear PDE the characteristics AP and BP, which are curvilinear in general, can be calculated independently of u, p, q. For the quasilinear case these characteristics can only be determined by building up solutions by a stepwise process in y as described above. Because of the distinctness of the slopes $z_\pm$, *no two characteristics of the same type intersect.* It follows that the solution for u at every field point P inside the curvilinear triangle ABP (refer to Fig. 3.5) is determined by the ICs on the closed interval $[A, B]$ on Γ. We introduce the following standard definitions in hyperbolic PDEs: The *domain of dependence*† of a field point P in the triangle ABP is defined as the segment $[A, B]$ of the IV curve Γ on which the solution at P depends. The *range of influence* of the segment $[A, B]$ is defined as the region consisting of all points in the triangle ABP.

† Refer also to Chapter 7, Section 7.1 on domain of dependence and range of influences.

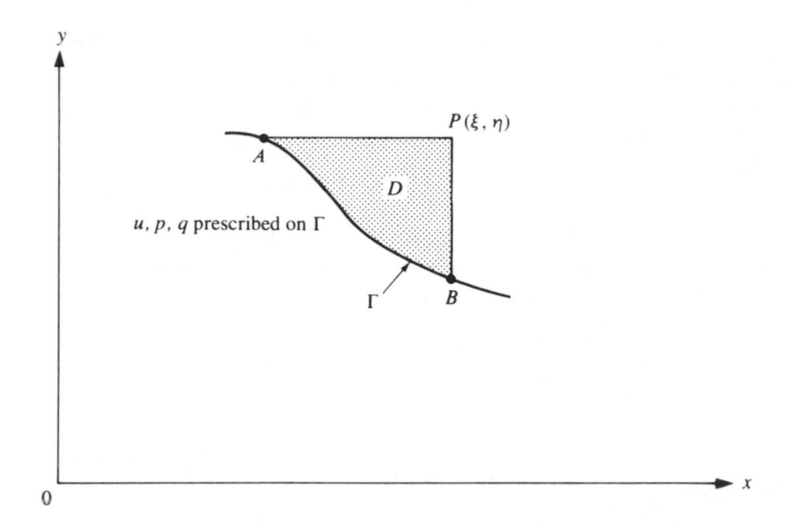

Fig. 3.5. Initial value curve Γ in characteristic plane showing domain of dependence *D*.

A characteristic can separate two different solutions. An important feature of these second-order PDEs is that these two solutions together with their first-order partial derivatives can be continuous across the dividing characteristic. However, we shall that the higher derivatives are discontinuous across the characteristic. This is seen as follows: The three equations (3.26), (3.27), and (3.28) allow us to solve for the three independent derivatives r, s, t *as long as they do not lie on a characteristic.* We now assume r, s, t are *not* linearly independent. By manipulating these equations we can eliminate r and t, for instance, and thereby obtain

$$s[az^2 - 2bz + c] - [ap'z + cq' + dz] = 0. \tag{3.35}$$

Once a value of s is assigned, then r and t can be calculated from any pair of the equations (3.26), (3.27), or (3.28). Assume $s \neq 0$. Then (3.35) tells us that the brackets on the left-hand side vanish, thus giving the characteristic ODEs (3.24) and (3.31). Indeed, this is another way of deriving these equations. In fact, this is another way of demonstrating that r, s, t *are not linearly independent on a characteristic.* Incidentally, the same analysis applies to r or t, i.e., we can eliminate s and r or s and t to allow for arbitrary values of t or r. Now, suppose we attempt to obtain a solution for u in the neighborhood of one side of a characteristic by a Taylor series expansion in terms of $u, p, \ldots, t$ and higher-order derivatives on the characteristic. As shown, on the characteristic, one of the terms r, s, t is arbitrary and thus can be prescribed. If we now attempt a similar Taylor series expansion on the other side of the same characteristic, we are free to choose another value of either r, s, or t. This tells us that we can obtain two different solutions on either side of the characteristic; which means that as we approach the characteristic from either side we get a jump discontinuity in the solution. This violates the uniqueness condition for the solution of our PDE. The conclusion is: If initial data u, p, q are given on a characteristic then the solution at any field point is not unique. Therefore, a *characteristic cannot be used as an* IV *curve.*

As an example of this nonuniqueness of solutions when using a characteristic as an IV curve, we consider the one-dimensional wave equation in dimensionless coordinates (y plays the role of time).

$$u_{xx} - u_{yy} = 0.$$

The characteristic directions are

$$C_+: z_+ = \frac{dy}{dx} = 1, \qquad C_-: z_- = \frac{dy}{dx} = -1.$$

The differential relations dp, dq derived from (3.31) yield the following results:

$$\text{on } C_+: \quad y - x = \text{const.}, \quad p - q = \text{const.},$$

$$\text{on } C_-: \quad y + x = \text{const.}, \quad p + q = \text{const.}$$

Suppose we erroneously supply the initial data $u = 2$, $p = -1$, $q = 1$ on a C_+ characteristic $y - x = 0$. Then it is easily seen that u_1 and u_2 are two different solutions that satisfy these ICs on $y - x = 0$, where

$$u_1 = 1 + \sin(y - x) + \cos(y - x), \qquad u_2 = 2 + (y - x)^2 + (y - x).$$

Thus we have two different solutions that satisfy the same ICs on a characteristic. However, if we calculate r, s, t on $y - x = 0$ from u_1 and u_2, we see that u_1 gives $r = -1$, $s = 1$, $t = -1$, and u_2 gives $r = 2$, $s = -2$, $t = 2$. In both cases we have $r - s = 0$ (which, of course, is the wave equation), but they are different values. The example illustrates the nonuniqueness of the second derivatives on a characteristic. This fact is yet another illustration that we cannot get a unique solution in a strip surrounding a characteristic on which we prescribe initial data, either by a Taylor series expansion or any other method, for that matter.

3.6. Canonical Form for Second-Order Partial Differential Equations with Constant Coefficients

A canonical form of an equation is defined as its simplest or standard form. We learned previously that second-order PDEs can be classified into three types according to the nature of the roots of an associated quadratic equation (the characteristic equation). This quadratic equation guides us in the transformation of the PDE into its canonical form.

Characteristic Coordinates

We introduce a transformation into new coordinates called *characteristic coordinates.* They are related to the roots of the characteristic equation. We start with (3.2) and consider the case where the coefficients a, b, c are constants and d is a linear function of u, u_x, u_y plus a known function of x, y for the nonhomogeneous case. We learned that only the principal part of (3.2) is involved in the classification of this PDE. The principal part of (3.2) is

$$au_{xx}+2bu_{xy}+cu_{yy}=L[u], \tag{3.36}$$

where the linear operator $L=a\,\partial^2/\partial x^2+2b\,\partial^2/\partial x\,\partial y+c\,\partial^2/\partial y^2$. The characteristic equation (3.24) can be written in the following factored form:

$$\left(\frac{dy}{dx}-z_+\right)\left(\frac{dy}{dx}-z_-\right)=0. \tag{3.37}$$

The solution of this characteristic equation consists of the two families of straight lines

$$y-z_+x=\alpha=\text{const.},\qquad y-z_-x=\beta=\text{const.}, \tag{3.38}$$

where, of course, the character of the roots $z_\pm$ of (3.24) classify the PDE into the three types: hyperbolic, parabolic, or elliptic. We now use these roots z to transform the (x, y) coordinates to the *characteristic coordinates* (α, β) defined by

$$\alpha=y-z_+x,\qquad \beta=y-z_-x. \tag{3.39}$$

Equation (3.39) can be written in implicit form as

$$\phi(x,y)=\alpha,\qquad \psi(x,y)=\beta. \tag{3.39$'$}$$

The first equation of (3.39′) is a one-parameter (α) family of solutions of the characteristic equation $dy/dx=z_+$, and the second equation is a one-parameter (β) family of solutions of $dy/dx=z_-$. This means: If β is kept constant and α is varied, a family of straight lines $y-z_+x=\alpha$ is generated. Conversely, if α is kept constant and β is varied, a family of straight lines $y-z_-x=\beta$ is generated. The two-parameter families of straight lines (of slopes z_+, z_-) in the (x, y) plane are thereby mapped into lines parallel to the coordinate axes in the (α, β) or characteristic plane.

In order to transform the PDE (3.2) to characteristic coordinates we must transform the derivatives $u_x, \ldots, u_{yy}$ into $u_\alpha, \ldots, u_{\beta\beta}$ by expanding the first and second derivatives and using (3.39). For example, $u_x=u_\alpha\alpha_x+u_\beta\beta_x$, etc. Thus, by using (3.39), (3.2) transforms into

$$Au_{\alpha\alpha}+2Bu_{\alpha\beta}+Cu_{\beta\beta}+D=0, \tag{3.40}$$

where A, B, C are known functions of a, b, c, z_+, z_-, and D is a linear combination of u, u_α, u_β. Omitting the calculational details we obtain

$$
\begin{aligned}
A &= a(\phi_x)^2 + 2b\phi_x\phi_y + c(\phi_y)^2 = Q(\phi_x, \phi_y) = a - z_+ 2b + z_+^2 c, \\
2B &= a\phi_x\psi_x + 2b(\phi_x\psi_y + \psi_x\phi_y) + c\phi_y\psi_y = a - 2b(z_+ + z_-) + z_+ z_-, \qquad (3.41)\\
C &= a(\psi_x)^2 + 2b\psi_x\psi_y + c(\psi_y)^2 = Q(\psi_x, \psi_y) = a - 2bz_- + z^2.
\end{aligned}
$$

An important observation about the form of A and C is that A is the same quadratic form Q in (ϕ_x, ϕ_y) that C is in (ψ_x, ψ_y). (B is a mixed quadratic form in $(\phi_x, \ldots, \psi_y)$.)

The equations (3.39′) yield the following system:

$$
\begin{aligned}
&\phi_x + \left(\frac{dy}{dx}\right)\phi_y = 0 \quad \text{or} \quad \frac{dy}{dx} = z_+ = -\frac{\phi_x}{\phi_y} \quad \text{on } \alpha = \text{const.}, \\
&\psi_x + \left(\frac{dy}{dx}\right)\psi_y = 0 \quad \text{or} \quad \frac{dy}{dx} = z_- = -\frac{\psi_x}{\psi_y} \quad \text{on } \beta = \text{const.}
\end{aligned}
\qquad (3.42)
$$

Assume the hyperbolic case, so that the roots $z_\pm$ of (3.24) are real and unequal. The system (3.42) is of the form $dy/dx = -f_x/f_y$, where $f = \phi$ for $dy/dx = z_+$, $f = \psi$ for $dy/dx = z_-$. The partial derivatives $\phi_x, \ldots, \psi_y$ are real. Inserting this expression into (3.24) yields

$$
a(f_x)^2 + 2bf_xf_y + c(f_y)^2 = Q(f_x, f_y) = 0. \qquad (3.43)
$$

Comparing (3.43) with (3.41) tells us that

$$
A = Q(\phi_x, \phi_y) = 0, \qquad B = Q(\psi_x, \psi_y) = 0, \qquad (3.44)
$$

for the hyperbolic case.

It is clear that the characteristic equation associated with (3.40) is

$$
AZ^2 + 2BZ + C = 0, \qquad \text{where} \quad Z = \frac{d\beta}{d\alpha}. \qquad (3.45)
$$

Thus we see that *the characteristics are preserved* for a transformation into characteristic coordinates. This means that the two families of characteristic lines in the (x, y) plane given by (3.38) are mapped by (3.39) into the two families of characteristics in the characteristic plane (α, β) which are the solutions of the transformed characteristic equation (3.45). Using (3.44), the PDE (3.40) becomes

$$
u_{\alpha\beta} + \frac{D}{2B} = 0, \qquad B \neq 0. \qquad (3.46)
$$

Equation (3.46) is the canonical form of the second-order hyperbolic PDE. The principal part of (3.46) is $u_{\alpha\beta}$. For the homogeneous wave equation $D/2B = 0$ yielding $u_{\alpha\beta} = 0$ as the canonical form.

It is of interest to consider the other two types of second-order PDEs. For the *elliptic case*, we have $b^2 - ac < 0$ so that the roots are complex conjugates. We may define the characteristic coordinates (α, β) by the following relations:

$$y - z_+ = \alpha + i\beta, \qquad y - z_- = \alpha - i\beta.$$

This is a real transformation yielding

$$y - \text{Re}(z)x = \alpha, \qquad -\text{Im}(z)x = \beta. \tag{3.47}$$

It is easily seen that the canonical form of the PDE for the elliptic case is

$$u_{\alpha\alpha} + u_{\beta\beta} + (\text{terms including first-order derivatives}) = 0. \tag{3.48}$$

For the *parabolic case* the roots z are equal so that $b^2 - ac = 0$. This leads to the following canonical form of the parabolic PDE:

$$u_{\alpha\alpha} + (\text{terms including first-order derivatives}) = 0. \tag{3.49}$$

The canonical forms of the second-order linear PDE with *variable coefficients* follow the same treatment as above, except that the two families of characteristics are not straight lines as given by (3.38) or (3.39). This is clear, since the characteristic equations

$$\frac{dy}{dx} = z_+(x, y), \qquad \frac{dy}{dx} = z_-(x, y),$$

which may be nonlinear, yield as solutions the two-parameter families of curves given by

$$\phi(x, y) = \alpha, \qquad \psi(x, y) = \beta, \tag{3.39'}$$

where $\phi(x, y)$ and $\psi(x, y)$ are not linear functions of x and y. This means that the PDE with the variable coefficients (a, b, c) yield a two-parameter set of curvilinear curves, which remain curvilinear when transformed to the (α, β) plane by the transformation into characteristic coordinates given by (3.39′). A, B, and C have the same quadratic forms in $\phi_x, \ldots, \psi_y$ as (3.41), but the extreme right-hand sides of the expressions given by (3.41) are clearly not valid. Equation (3.44) is valid at each field point where the PDE is hyperbolic so that the canonical form is still given by (3.46). The same canonical forms hold for the elliptic and parabolic cases.

For the quasilinear case the situation is complicated by the fact that ϕ and ψ are, in addition, functions of u, u_x, and u_y, so that the transformation to characteristic coordinates is difficult and, in general, can only be done by numerical methods, since the characteristic net in the (x, y) plane, as well as the transformed characteristic net, depends on the solution.

3.7. Conservation Laws, Weak Solutions

In Chapter 2 mention was made of the three conservation laws of physics: mass or continuity, momentum, and energy. The two representations, those of Lagrange and Euler, were described, and a rough physical interpretation of these laws was given. In this section, we get more quantitative and describe these laws in the setting of the so-called *weak solutions* of PDEs. This is explained as follows: In solving quasilinear PDEs of the hyperbolic type (the Cauchy IV Problem) by the method of characteristics, it was shown how to construct solutions at field points in the neighborhood of the IV curve, and, by a stepwise process, continue developing the solution field in the interior of the domain. This procedure allows us to construct solutions "locally", since they depend on solutions at neighboring field points. Such a method of constructing *local solutions* has the disadvantage that errors in the computations may lead to singularities or discontinuities in the solution as time increases. In addition, shock waves may occur which yield certain discontinuities. These discontinuities terminate the region of the solution field, and we eventually reach a region where the solution is not valid. The underlying principle behind constructing local solutions is that we are treating the Cauchy IV Problem in a local or differential manner by solving the PDE point by point from data supplied on the IV curve. The method of weak solutions gets around this difficulty of not being able to extend the solution field beyond a discontinuity, by using a different principle, namely, an integration process directly from the conservation laws. This is actually a more natural procedure. It arose, in large part, from the necessity of developing a rational approach to the treatment of shock waves; it allows us to obtain *global solutions* which accounts for discontinuities across shock waves. The conservation laws will be formulated in integral form, rather than the differential form which leads to the PDE. This integral representation of the conservation laws leads to a mathematical structure from which we can obtain a more general class of solutions, called *weak solutions* which accounts for discontinuities. These weak solutions, of course, reduce to the solutions we have already discussed from characteristic theory which depend on the differential treatment of the Cauchy IV Problem.

As an example of how discontinuities arise in hyperbolic systems, we consider the following Cauchy IV Problem applied to a first-order quasilinear PDE for the scalar $u(x, t)$. (We replace y by t):

$$\begin{aligned} &u_t + au_x = 0, \qquad -\infty < x < \infty, \quad t > 0, \\ &u(x, 0) = f(x) \qquad \text{on IV curve } \Gamma\text{: } t = 0, \end{aligned} \tag{3.50}$$

where a is a given function of u, $a = a(u)$, and the initial data $f(x)$ is given on the x axis. The characteristic ODE gives the slope of a characteristic curve in the (x, t) plane

$$\frac{dt}{dx} = \frac{1}{a(u)}. \tag{3.51}$$

We, of course, see that the one-parameter family of characteristics depends on the solution, since the PDE is quasilinear. However, in this case, there is a simple interpretation of this family of characteristics. From (3.50) we have $du/dx=0$ along each characteristic curve, whose slope is given by (3.51). This means u is constant along a characteristic, the constant depending on the characteristic. Since $dt/dx=z(u)$ the conclusion is: the characteristics consist of a single family of straight lines.

We now prescribe an $f(x)$ that leads to a *singular point* at a field point. This means the solution $u(x, t)$ is not unique at that field point. To do this, consider the points x_1, x_2 on $t=0$ such that $0<x_1<x_2$. Let $f(x_1)=f_1$, $f(x_2)=f_2$. Suppose f_1 and f_2 are such that $0<a_2<a_1$. If we now construct characteristics issuing from x_1 and x_2 into the region $t>0$ we see that, as time increases, these characteristics will intersect. Let P_I: (x_I, t_I) be the field point of intersection of these characteristics. Then we get $x_I=(a_1x_2-a_2x_1)/(a_1-a_2)$, $t_I=(x_2-x_1)/(a_1-a_2)$. Now, the two characteristics from x_1 and x_2 carry different values of u, namely f_1 and f_2. Therefore the P_I is a *singular point* in the sense that the solution is not unique there (since $f_1 \neq f_2$) Moreover, we cannot obtain u as a continuous solution for $t>t_I$. This example illustrates the fact that continuous solutions of quasilinear PDEs do not in general exist "in the large" (globally). The operative word here is "continuous"; for if we relax the condition of continuity by allowing for discontinuous solutions, then we admit a wider class of solutions (weak solutions) to the quasilinear system. In the study of shock wave phenomena it will be seen that the shock waves are curves on which discontinuities occur, the so-called "jump conditions".

Weak solutions are obtained by representing the Cauchy Problem by a set of "conservation laws". To get a feel for what we mean by conservation laws in this setting, we interpret u as representing mass, momentum, energy, or any other physical quantity with which we associate a flux or flow of that quantity across a surface in a given region. It is clear from vector analysis that a flux of u involves the divergence of u. This is described mathematically as

$$\frac{d}{dt}\int_{x_1}^{x_2} u\,dx=f_2-f_1, \tag{3.52}$$

where $f_i=f(u(x_i), x_i, t)$, $i=1, 2$. Equation (3.52) expresses the fact that the total quantity represented by u contained in the closed interval $[x_1, x_2]$ changes at a rate equal to the flux f of u through the endpoints of the interval. Therefore (3.52) is a *general conservation law*, and thus can be applied to the conservation laws of mass, momentum, and energy, where u may be associated with mass, momentum, or energy.

We first consider the case where $u(x, t)$ is a differentiable solution of the general conservation law (3.52), meaning that there are no interior points of discontinuity in u in the closed interval $[x_1, x_2]$. Then, in the limit as $x_2 \to x_1$ for $x_1<x_2$, the conservation law (3.52) becomes the quasilinear PDE given by

$$u_t=\left(\frac{\partial}{\partial x}\right)f(u, x, t)=f_u u_x+f_x. \tag{3.53}$$

Equation (3.53) is the differential form of the conservation law. It does not allow for discontinuous solutions. It is the type of equation we have been working with all along when studying characteristic theory for a first-order system.

Next we consider the case where $u(x, t)$ is a solution of (3.52) that is *piecewise differentiable* in the interval. By this we mean the following: Suppose $\bar{x}$ is an interior point in the closed interval $[x_1, x_2]$ at which there is a discontinuity in u. This means $\bar{u}_+ - \bar{u}_- = [\bar{u}] \neq 0$ where $\bar{u}_+$ is the value of u as it approaches $\bar{x}$ from the right and $\bar{u}_-$ is that value when u approaches x from the left. We assume a finite jump discontinuity $[u]$. We now turn to (3.52) and apply the operator (d/dt) to the integration over the separate intervals $[x_1, \bar{x} - \varepsilon]$ and $[\bar{x} + \varepsilon, x_2]$, where ε is a small positive number which we later set equal to zero. We assume the endpoints are differentiable functions of t. Performing the integrations over these intervals, using the jump $[u]$ on the interval $[\bar{x} - \varepsilon, \bar{x} + \varepsilon]$, letting $\varepsilon \to 0$ and $x_1 \to x_2$, we get the *jump condition*

$$[f] = -[u]U, \tag{3.54}$$

where $U = d\bar{x}/dt$ is the speed of the propagation of the discontinuity at $\bar{x}$.

The chapter on fluid dynamics will go into these jump conditions more thoroughly from a shock wave point of view.

3.8. Divergence Theorem, Adjoint Operator, Green's Identity, Riemann's Method

We shall investigate in the subsequent section a very important method of solving PDEs of the hyperbolic type by G.F.B. Riemann, the great German mathematician. Riemann devised this method in his classic studies of wave propagation in gases. In order to get an appreciation of his approach to hyperbolic systems, we first present some mathematical concepts.

Divergence Theorem

We first review the divergence theorem. For simplicity, it is only necessary to consider the two-dimensional case. Roughly speaking, the main idea of this important theorem is that it transforms the divergence of a vector in a region to the normal component of the vector integrated over the surface bounding the region. Specifically, let $\mathbf{V}(x, y)$ be a vector, let $\mathbf{i}, \mathbf{j}$ be unit vectors in the x, y directions, and let P, Q be the x, y components of $\mathbf{V}$. Then

$$\mathbf{V} = \mathbf{i}P + \mathbf{j}Q.$$

Let $\mathcal{R}$ be a simply connected region in the (x, y) plane bounded by a closed curve C, and let $\mathbf{n}$ be the unit outward normal along C. Then the divergence theorem is

$$\iint_{\mathcal{R}} \nabla \cdot \mathbf{V}\, dA = \oint_C \mathbf{V} \cdot \mathbf{n}\, ds, \qquad \nabla = \mathbf{i}\frac{\partial}{\partial x} + \mathbf{j}\frac{\partial}{\partial y}, \qquad \mathbf{n} = \left(\frac{dy}{ds}\right)\mathbf{i} - \left(\frac{dx}{ds}\right)\mathbf{j}, \tag{3.55}$$

where dA is the element of area, ds is the element of arc length of C, and dy/ds, $-dx/ds$ are the direction cosines of $\mathbf{n}$. Then (3.55) becomes

$$\iint_{\mathcal{R}} (P_x + Q_y)\, dx/dy = \oint_C (P\, dy - Q\, dx). \tag{3.56}$$

This is one form of the divergence theorem. Now consider the scalar functions $u(x, y)$, $v(x, y)$ with continuous second derivatives. $\nabla = \mathbf{i}\, \partial/\partial x + \mathbf{j}\, \partial/\partial y$ is the gradient operator. Suppose we choose a $\mathbf{V}$ such that $\mathbf{V} = u\nabla v$. Using the vector identity

$$\nabla \cdot u\nabla v = u\nabla^2 v + (\nabla u) \cdot (\nabla v),$$

in (3.55) we obtain

$$\iint_{\mathcal{R}} [u\nabla^2 v + (\nabla u) \cdot (\nabla v)]\, dA = \oint_C \mathbf{n} \cdot u\nabla\, ds. \tag{3.57}$$

Equation (3.57) is sometimes called the *first form of Green's identity.* To obtain a more symmetric form of Green's identity we interchange u and v in (3.57) and subtract the result from (3.57). We express the result as

$$\iint_{\mathcal{R}} v\nabla^2 u\, dA = \iint_{\mathcal{R}} \nabla^2 v\, dA + \oint_C \left[v\frac{\partial u}{\partial n} - u\frac{\partial v}{\partial n}\right] ds. \tag{3.58}$$

Equation (3.58) is sometimes called the *second form of Green's identity.* The equation represents a sort of integration by parts in which the Laplacian operator in two dimensions ∇^2 got shifted from u to v. Let us rewrite (3.58) as

$$\iint_{\mathcal{R}} [v\nabla^2 u - u\nabla^2 v]\, dA = \oint_C \left[v\frac{\partial u}{\partial n} - u\frac{\partial v}{\partial n}\right] ds, \tag{3.58$'$}$$

where the left-hand side involves integration over the region $\mathcal{R}$ and the right-hand side involves integration over the closed curve C bounding $\mathcal{R}$. Comparing (3.58′) with the divergence theorem (3.55), we want to find a vector $\mathbf{V}$ such that the integrand of the left-hand side of (3.58′) equals div($\mathbf{V}$) or $\nabla \cdot \mathbf{V}$. This is easily obtained by writing out this integrand in extended form, obtaining the following identity:

$$v(u_{xx} + u_{yy}) - u(v_{xx} + v_{yy}) = (vu_x - uv_x)_x + (vu_y - uv_y)_y = P_x + Q_y. \tag{3.59}$$

We have thus found a vector $\mathbf{V} = \mathbf{I}P + \mathbf{j}Q$, where

$$P = vu_x - uv_x, \qquad Q = vu_y - uv_y. \tag{3.60}$$

This gives

$$v\nabla^2 u - u\nabla^2 v = \nabla \cdot \mathbf{V} = P_x + Q_y, \tag{3.61}$$

and (3.58′) can be then identified with the divergence theorem given in the form (3.56). The Laplacian ∇^2 in (3.61) is an example of a second-order linear operator L, so that (3.58′) can be written as

$$\iint_{\mathscr{R}} (vL[u] - uL[v])\, dA = \oint_C \left[v\frac{\partial u}{\partial n} - u\frac{\partial v}{\partial n} \right] ds, \tag{3.62}$$

where

$$L[u] = u_{xx} + u_{yy}, \qquad L[v] = v_{xx} + v_{yy}.$$

Adjoint Operator L^*

We saw that $L[u] = L[v]$ for $L = \nabla^2$. However, the Laplacian is not the only second-order linear differential operator. For a general second-order linear differential operator it is not necessarily true that $L[v] = L[u]$ in order for the integrand $vL[u] - uL[v]$ to be of the form $\nabla \cdot \mathbf{V}$. Then $L[v]$ is replaced by a more general linear differential form $L^*[v]$ such that the integrand is of a divergence form, or (3.62) is satisfied. $L^*[v]$ is called the *adjoint* operator operating on v corresponding to $L[u]$. If $L[u]$ is such that $L^*[v] = L[v]$, then L is called a *self-adjoint operator*. If $L[u]$ is such that this equality does not hold, then L is *not* self-adjoint. As mentioned above, L^* operates on v such that the following equation is satisfied:

$$\iint_{\mathscr{R}} (vL[u] - uL^*[v])\, dA = \oint_C \left[v\frac{\partial u}{\partial n} - u\frac{\partial v}{\partial n} \right] ds. \tag{3.63}$$

This means that we want the integrand in the left-hand side of (3.63) to be the divergence of a vector so that (3.63) is actually the divergence integral given by (3.55) or (3.56).

The important point is: If we can construct $L^*[v]$ that depends on $L[u]$, then we can make use of the divergence integral (3.63) to integrate along the boundary C surrounding the region $\mathscr{R}$. This means that there exists a function $v(x, y)$ such that the adjoint operator L^* operating on v enjoys the above-mentioned property. $v(x, y)$ is called *Riemann's function.* Its purpose is that it allows us to determine the solution $u(x, y)$ to the PDE. Later on we shall show how to construct this function.

An example of a non-self-adjoint operator is the one-dimensional unsteady Fourier heat transfer operator L operating on u given by

$$L[u] = u_{xx} - u_y, \tag{3.64}$$

where x represents space and y represents time. Let us assume $L^*[v]$ takes the form

$$L^*[v] = v_{xx} + (\alpha v)_x + (\beta v)_y, \tag{3.65}$$

where α and β are functions of x and y to be determined. Then we want to find the components $P(x, y)$ and $Q(x, y)$ of $\mathbf{V}$ such that $vL[u] - uL^*[v] = P_v + Q_y$, or

$$v(u_{xx} - u_y) - u[v_{xx} + (\alpha v)_x + (\beta v)_y] = P_x + Q_y,$$

which gives

$$P = vu_x - uv_x, \qquad Q = -uv, \qquad \alpha = 0, \quad \beta = 1,$$

yielding

$$L^*[v] = v_{xx} + v_y. \tag{3.66}$$

Physically, the adjoint expression $L^*[v]$ for the heat transfer operator has the following rather curious physical interpretation: If $L[u] = 0$ is the one-dimensional unsteady heat conduction equation for the temperature u, then the *adjoint problem* is the solution of the PDE $L^*[v] = 0$, where the "temperature" $v(x, y)$ is the solution of the heat conduction equation with time y "running backwards". The author has investigated and successfully used this method in the more general study of unsteady thermal stresses, where $v(x, y)$ plays the role of a Green's function.

For a general linear second-order hyperbolic PDE we write the canonical form (3.46) as

$$L[u] = u_{xy} + au_x + bu_y + cu = f, \tag{3.67}$$

where a, b, c, f are known functions of x, y. For the special case $a = b = c = f = 0$, (3.67) reduces to the one-dimensional homogeneous wave equation in the characteristic coordinates (x, y) for constant wave speed. For this case it is easily seen that the operator L is self-adjoint.

Returning to the general canonical form (3.67) we now construct L^* by assuming the following expression for $L^*[v]$:

$$L^*[v] = v_{xy} + (\alpha v)_x + (\beta v)_y + cv. \tag{3.68}$$

Recall that $L^*[v]$ must have the property that $vL[u] - uL^*[v] = P_x + Q_y$. Our task is now to determine P and Q which depend on u and v. We have

$$\begin{aligned} vL[u] - uL^*[v] &= v(u_{xy} + au_x + bu_y + cu) - u[v_{xy} + (\alpha v)_x + (\beta v)_y + cv] \\ &= (-uv_y + auv)_x + (vu_x + buv)_y = P_x + Q_y. \end{aligned} \tag{3.69}$$

This gives

$$P = -uv_y + auv, \qquad Q = vu_x + buv, \qquad \alpha = -a, \quad \beta = -b, \tag{3.70}$$

and $L^*[v]$ becomes

$$L^*[v] = v_{xy} - (av)_x - (bv)_y + cv. \tag{3.71}$$

$L[u]$ need not be second order. As an example, consider $L[u]$ as the first-order form

$$L[u] = au_x + bu_y,$$

where a, b are known functions of x, y. It is easily seen that

$$L^*[v] = -(av)_x - (bv)_y.$$

This is verified as follows:

$$\begin{aligned} vL[u] - uL^*[v] &= v[au_x + bu_y] + u[(av)_x + (bv)_y] \\ &= (auv)_x + (buv)_y = P_x + Q_y. \end{aligned}$$

Now the following question arises: What are the conditions on $a(x, y)$, $b(x, y)$ such that $L[u]$ is self-adjoint? We must have $au_x + bu_y = -(av)_x - (bv)_y = -av_x - bv_y - v(a_x + b_y)$, so that $a_x + b_y = 0$, where use was made of the PDE $L[u] = 0$. $a = x^2 - y^2$, $b = -2xy$ is a possible set of values of (a, b) that satisfies this condition.

Riemann's Method of Integration

We now have all the mathematical apparatus at our disposal to describe Riemann's method of solving second-order PDEs. We restrict the discussion to a linear system. Specifically, we investigate Riemann's method of solving the Cauchy IV Problem, which is given by the canonical form of the hyperbolic PDE (3.67). Clearly (x, y) are the characteristic coordinates. Note that the principal part of $L[u]$ is u_{xy}. The formulation of the Cauchy Problem is: Determine the solution $u(x, y)$ which satisfies (3.67) and the initial data u, p, q which are prescribed along an IV curve Γ which has nowhere the characteristic direction. Figure 3.5 shows a portion of the Γ curve in the (x, y) or characteristic plane on which are prescribed u, p, q. We consider points A and B on Γ near each other. Let the domain of the solution be the region above Γ. From A draw the characteristic line segment $y = \text{const.}$ into the solution domain, and from B draw the characteristic line segment $x = \text{const.}$ These characteristics will intersect at the field point P whose coordinates are (ξ, η). The *domain of dependence* of the field point P is the triangular region D bounded by the characteristics AP and BP and the portion of the IV curve Γ intercepted by these characteristics stemming from P. The *range of influence* is this portion AB of Γ.

The Cauchy IV Problem as described in Fig. 3.5 is to obtain the solution u_P at the field point given the data on the range of influence AB. This was obtained by Riemann and will now be described. The essence of the Riemann method is to calculate u_P by multiplying the PDE (3.67) by the Riemann function $v(x, y)$, which we now call a *test function*, integrate over the domain of dependence D, transform the integral by Green's identity so that u appears as a factor of the integrand, and try to determine $v(x, y)$ to satisfy the conditions of the problem. The resulting algorithm involves integrating over D making use of the test function v.

To carry out this procedure we make use of the divergence theorem in the form (3.56). Inserting (3.69) and (3.70) into (3.56) yields

$$\int\int_D (vL[u]-uL^*[v]\,dx\,dy=\oint_C [(-uv_y+auv)\,dy-(vu_x+buv)\,dx]. \tag{3.72}$$

Referring to Fig. 3.5, the closed curve C is the path $AB+BP+PA$ bounding the region D. The objective is to use (3.72) to determine u_P. At this stage the test function v is undefined. We shall define it in such a way that the line integral in (3.72) will only involve the path AB along Γ. This means that we shall construct v such that this line integral depends only on the initial data. To simplify matters further we shall choose v as the solution to the following PDE:

$$L^*[v]=0. \tag{3.73}$$

Now $dx=0$ on BP, $dy=0$ on PA, and $L[u]=f$, so that (3.72) becomes

$$\int\int_D vf\,dx\,dy=\int_{AB}[-uv_y+auv)\,dy-(vu_x-buv)\,dx]$$
$$+\int_{BP}(-uv_y+auv)\,dy+\int_{PA}(vu_x+buv)\,dx. \tag{3.74}$$

We now integrate the line integral along the path PA by parts in order to separate v from u in the integrand and isolate u_P. We obtain

$$\int_{PA} v(u_x+bu)\,dx=v_Au_A-v_Pu_P-\int_{PA} u(v_x-bv)\,dx.$$

Inserting this expression into (3.74) and isolating u_P gives

$$v_Pu_P=v_Au_A+\int_{AB}[v(u_x+bu)\,dx+u(v_y-av)\,dy]$$
$$+\int_{BP}u(v_y-av)\,dy-\int_{PA}u(v_x-bv)\,dx+\int\int_D vf\,dx\,dy. \tag{3.75}$$

The plan is to get u_P in terms of the integral of the area D and the initial data on Γ. This means we would like the line integrals along the characteristics BP and PA to vanish. Since v is not fully defined we can do this by defining v such that

$$\begin{aligned} &v_y-av=0 \text{ on } BP \quad \text{where} \quad x=x_B=\xi,\\ &v_x-bv=0 \text{ on } PA \quad \text{where} \quad y=y_A=\eta, \quad \text{also set} \quad v_P=1. \end{aligned} \tag{3.76}$$

Inserting (3.76) into (3.75) gives

$$u_P=v_Au_A+\int_{AP}[v(u_x+bu)\,dx+u(v_y-av)\,dy]+\int\int_D vf\,dx\,dy. \tag{3.77}$$

This is not the symmetric form we want since it involves $u = u_A$ at only one of the endpoints of the range of influence. We want a symmetric formula in the sense that it would also involve u_B. To this end, we manipulate the line integral in (3.77), recognizing that $vu_x\,dx + uv_y\,dy = d(uv) - uv_x\,dx - vu_y\,dy$. Calling this line integral I we obtain

$$I = u_B v_B - u_A v_A - \int_{AB} [u(v_x - bv)\,dx + v(u_y + au)\,dy].$$

Inserting this expression into (3.77) yields

$$u_P = v_B u_B - \int_{AB} [u(v_x - bv)\,dx + v(u_y + au)\,dy] + \iint_D vf\,dx\,dy. \quad (3.78)$$

Finally, adding (3.77) and (3.78) gives the desired symmetric expression for u_P.

$$u_P = u(\xi, \eta) = \tfrac{1}{2}(v_A u_A + v_B u_B) + \tfrac{1}{2}\int_{AB} [(vu_x - uv_x + 2buv)\,dx + (uv_y - vu_y - 2auv)\,dy] + \iint_D vf\,dx\,dy. \quad (3.79)$$

Equation (3.79) represents the solution u_P in terms of the Riemann function v on Γ and in D. If the forcing function $f = 0$, then u_P can be obtained by integrating the data and v along the range of influence AB. Otherwise, we have the additional integral of vf in D.

The defining equations for the Riemann function (3.76) show that v is a continuous differentiable function of two sets of variables: (x, y), the coordinates of the IV curve; and (ξ, η), the coordinates of the field point P. We may therefore rewrite the defining equations of v as

$$\begin{aligned} L_{(x,y)}[v(x, y; \xi, \eta)] &= 0, \\ v_y(\xi, y; \xi, \eta) - a(\xi, y)v(\xi, y; \xi, \eta) &= 0 \quad \text{on characteristic } BP, \\ v_x(x, \eta; \xi, \eta) - b(x, \eta)v(x, \eta; \xi, \eta) &= 0 \quad \text{on characteristic } PA, \\ v(\xi, \eta; \xi, \eta) &= 1 \quad \text{at field point } P, \end{aligned} \quad (3.80)$$

where we have used the obvious notation $L_{(x,y)}$ to mean differentiation with respect to x, y. The second and third equations of system (3.80) are ODEs along the respective characteristics. Their solutions are

$$\begin{aligned} v(\xi, y; \xi, \eta) &= \exp\left(\int_n^y a(\lambda, \xi)\,d\lambda\right) \quad \text{along } BP, \\ v(x, \eta; \xi, \eta) &= \exp\left(\int_\xi^x b(\lambda, \eta)\,d\lambda\right) \quad \text{along } PA. \end{aligned} \quad (3.81)$$

In summary, the system (3.79), (3.80), and (3.81) allows us to solve the Cauchy IV Problem in characteristic coordinates x, y in terms of the Riemann function $v(x, y; \xi, \eta)$.

As our first example we take the one-dimensional homogeneous wave equation in characteristic coordinates in the canonical form $L[u] = u_{xy} = 0$ (where $a = b = c = f = 0$). It is immediately obvious that L is self-adjoint so that $L^*[v] = v_{xy} = 0$. Then v has the form

$$v(x, y, \xi, \eta) = F(x; \xi, \eta) + G(y; \xi, \eta).$$

It is easily seen from the defining equations for v that the Riemann function for the one-dimensional wave equation is

$$v(x, y; \xi, \eta) = 1.$$

Then the solution for u in characteristic coordinates, as given by (3.79), becomes

$$u(\xi, \eta) = \tfrac{1}{2}[u_A + u_B] + \tfrac{1}{2} \int_\Gamma (u_x\, dx - v_y\, dy).$$

As another example we consider the equation

$$L[u] = u_{xy} + cu = 0,$$

where c is a given constant. It is easily seen that L is self-adjoint. Therefore $v_{xy} + cv = 0$. Since this PDE has constant coefficients v depends on the relative positions of P with respect to A and B. In other words, v depends on the coordinates $\bar{x} = x - \xi$, $\bar{y} = y - \eta$. Moreover, the characteristic equations for v, given by (3.81), tell us that v is constant along the characteristics since $a = b = 0$. Also $v(0, 0) = 1$. This means that v must be symmetric in x and y. The simplest assumption is that v has the form

$$v = w(z) \qquad \text{where} \quad z = \bar{x}\bar{y},$$

which finally yields

$$\frac{z\, d^2 w}{dz^2} + \frac{dw}{dz} + cw = 0.$$

This equation can be reduced to Bessel's equation of the first kind of order zero by setting $s = (4cz)^{1/2}$.

Then the solution for w that has no singularities is

$$w = J_0((4cxy)^{1/2}).$$

CHAPTER 4

Transverse Vibrations of Strings

Introduction

To introduce this chapter we can do no better than to quote from [32, Vol. I, Chap. VI]. "Among vibrating bodies there are none that occupy a more prominent position than Stretched Strings. From the earliest times they have been employed for musical purposes.... To the mathematician they must always possess a peculiar interest as a battle-field on which were fought out the controversies of D'Alembert, Euler, Bernoulli, and Lagrange relating to the nature of the solutions of partial differential equations. To the student of Acoustics they are doubly important." Davis (1986) emphasized finite difference methods of solving the wave equation, and the vibrating string was discretized by a finite number of degrees of freedom approximation. The reader is referred to this work for details. In this chapter we shall be concerned with the representation of a string as a perfectly flexible filament of a solid *continuum*. In Chapter 2 we derived the wave equation for the vibrating string. In this chapter we shall investigate methods of solving this one-dimensional wave equation, including the method of characteristics, and describe some properties of the lateral vibrations of strings.

4.1. Solution of the Wave Equation, Characteristic Coordinates

Let $u(x, t)$ be the lateral displacement of a taut string from its equilibrium position. Then the wave equation for the vibrating string is

$$c^2 u_{xx} - u_{tt} = 0, \qquad c^2 = \frac{T}{\rho}, \tag{4.1}$$

where c is the speed of the propagating wave, T is the tension of the string, and ρ is the linear density (mass per unit length of string). The *wave operator* $L = c^2\, \partial^2/\partial x^2 - \partial^2/\partial t^2$ is a linear second-order operator with constant coefficients. It can be factored into the two operators $L_1 = c\,\partial/\partial x - \partial/\partial t$ and $L_2 = c\,\partial/\partial x + \partial/\partial t$ so that

$$Lu = c^2 u_{xx} - u_{tt} = L_1 L_2 u = (cu_x - u_t)(cu_x + u_t) = 0.$$

This is the product of two *directional derivatives* in the (x, t) plane, the first in the direction with direction numbers $(c, -1)$ and the second in the direction with direction numbers $(c, 1)$. This suggests that we introduce these two directions as new coordinate directions, which are actually the characteristic coordinates. Therefore, we set

$$\xi = x - ct, \qquad \eta = x + ct. \tag{4.2}$$

ξ, η are the *characteristic coordinates* for the one-dimensional wave equation, and (4.2) represents the transformation from rectangular to characteristic coordinates. We now use (4.2) to transform the wave equation given by (4.1) to characteristic coordinates. By the chain rule, using (4.2) we have

$$u_x = u_\xi + u_\eta, \qquad u_{xx} = \left(\frac{\partial}{\partial \xi} + \frac{\partial}{\partial \eta}\right)^2 u = u_{\xi\xi} + 2u_{\xi\eta} + u_{\eta\eta}.$$

Similarly, we find

$$u_{tt} = c^2(u_{\xi\xi} - 2u_{\xi\eta} + u_{\eta\eta}).$$

Substituting these expressions for u_{xx} and u_{tt} into (4.1) gives

$$u_{\xi\eta} = 0. \tag{4.3}$$

Equation (4.3) is the expression for the wave equation in characteristic coordinates. It is actually the canonical form of the hyperbolic partial differential equation (PDE) and was given by (3.46) in Chapter 3.

We now determine the general solution of the canonical equation (4.3). Writing (4.3) as $(\partial/\partial\eta)u_\xi = 0$ we see that u_ξ is independent of η. However, it depends on ξ in an arbitrary way, meaning

$$u_\xi = f(\xi),$$

where f is an arbitrary function of ξ. Integrating this expression and letting $\int f(\xi)\, d\xi = F(\xi)$ we obtain

$$u(\xi, \eta) = F(\xi) + G(\eta). \tag{4.4}$$

Equation (4.4) is the general solution of the canonical equation in characteristic coordinates. Using (4.2), (4.4) becomes

$$u(x, t) = F(x - ct) + G(x + ct), \tag{4.5}$$

which is the general solution of (4.1). The function F represents progressing waves for $x > 0$ and G represents regressing waves. All solutions of (4.1) must be of the form (4.5) and any u that is not of this form is not a solution.

An important point about (4.4) is the following: Suppose $G = 0$ for all x, t. If we set $\xi = x - ct = \text{const.}$, then $u = F(\xi) = \text{const.}$ This means that the

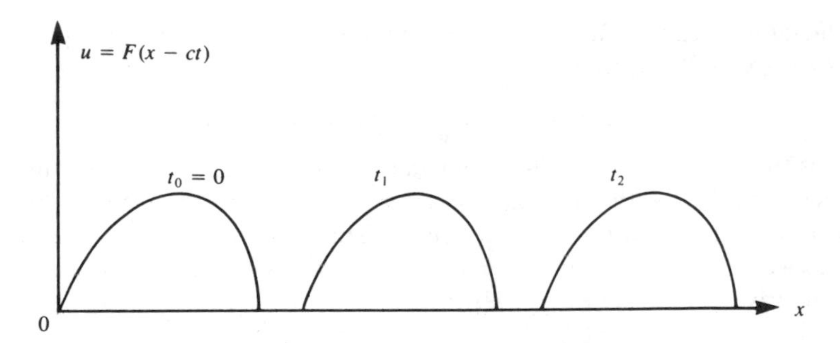

Fig. 4.1. Profiles of a progressing wave.

solution is constant on the straight line $x - ct = \text{const.}$ Indeed, the family of straight lines $x - ct = \text{const.}$ is a family of C_+ characteristics in the (t, ψ) plane. Similarly, if we set $F = 0$ for all x, t then $u = G(\eta)$ is constant on each characteristic of the C_- family of characteristics $x + ct = \text{const.}$ Figure 4.1 shows profiles of a progressing wave $u = F(\xi)$ at $t_0 = 0$, t_1, and t_2. At $t = 0$, $u = F(x)$ which is represented by a pulse. At $t = t_1$, the pulse travels a distance equal to ct_1 to the right, and is unchanged in form, since each point on the pulse travels the same distance. At t_2 the pulse travels a distance ct_2. A similar geometric construction can be performed for the regressive wave represented by $u = G(\eta)$. *Note*: since there is no damping or dispersion, the wave form does not change its shape.

4.2. D'Alembert's Solution

We now investigate D'Alembert's method of solving the one-dimensional homogeneous wave equation for the vibrating string. The physical model consists of a taut string infinite in length. Since the boundaries are at infinity, all we demand from the solution is that it be bounded at infinity, so that the boundary conditions are not considered explicitly. We thus have a pure Cauchy IV Problem. The formulation is: Find $u(x, t)$ which satisfies the following Cauchy Problem:

$$\begin{gathered} c^2 u_{xx} - u_{tt} = 0, \qquad -\infty < x < \infty, \quad t > 0, \qquad c^2 = \frac{T}{\rho}, \\ u(x, 0) = f(x), \qquad u_t(x, 0) = g(x), \qquad t = 0. \end{gathered} \tag{4.6}$$

The initial lateral displacement of the string is the prescribed $f(x)$ and the initial particle velocity (lateral velocity of string) is the prescribed $g(x)$. This is a well-defined initial value (IV) problem and has a unique solution.

The uniqueness of the solution is proved as follows: Suppose u and v are two different solutions that satisfy (4.6). Then, by the linearity of the

PDE $w = v - u$ is also a solution of the wave equation. But w satisfies the initial conditions (ICs) $w(x, 0) = w_t(x, 0) = 0$. We know that all solutions are of the form $w = F(x - ct) + G(x + ct)$. To satisfy the ICs we have $F(x) + G(x) = 0$ and $-F'(x) + G('(x) = 0$. Integrating the second equation gives $-F(x) + G(x) = 0$, which shows that $F = G = 0$ for all values of the arguments so that $w(x, t) = 0$, which is a contradiction. This completes the proof.

We continue with D'Alembert's proof. The general solutions for the displacement and particle velocity particle are

$$u(x, t) = F(x - ct) + G(x + ct),$$

$$u_t(x, t) = -c\acute{F}(x - ct) + c\acute{G}(x + ct), \qquad \acute{F} = \frac{dF}{d\xi}, \qquad \acute{G} = \frac{dG}{d\eta}.$$

Using the ICs in (4.6) we get

$$F(x) + G(x) = f(x),$$

$$-\acute{F}(x) + \acute{G}(x) = \frac{g(x)}{c}.$$

Integrating the second equation for 0 to x gives

$$-F(x) + G(x) = \frac{1}{c}\int_0^x g(z)\,dz - F(0) + G(0).$$

We thus have two algebraic equations for $F(x)$ and $G(x)$. Solving them yields

$$F(x) = \frac{f(x)}{2} - \frac{1}{2c}\int_0^x g(z)\,dz + F(0) - G(0),$$

$$G(x) = \frac{f(x)}{2} + \frac{1}{2c}\int_0^x g(z) - F(0) + G(0).$$

In these equations x is a generic variable. This means that the argument in F can be replaced by $x - ct$ and the argument in G can be replaced by $x + ct$. This allows us to get off the x axis. We get

$$F(x - ct) = \frac{f(x - ct)}{2} - \frac{1}{2c}\int_0^{x-ct} g(z)\,dz + F(0) - G(0),$$

$$G(x + ct) = \frac{f(x + ct)}{2} + \frac{1}{2c}\int_0^{x+ct} g(z)\,dz - F(0) + G(0).$$

Inserting these expressions into (4.5) yields

$$u(x, t) = \tfrac{1}{2}[f(x - ct) + f(x + ct)] + \frac{1}{2c}\int_{x-ct}^{x+ct} g(z)\,dz. \tag{4.7}$$

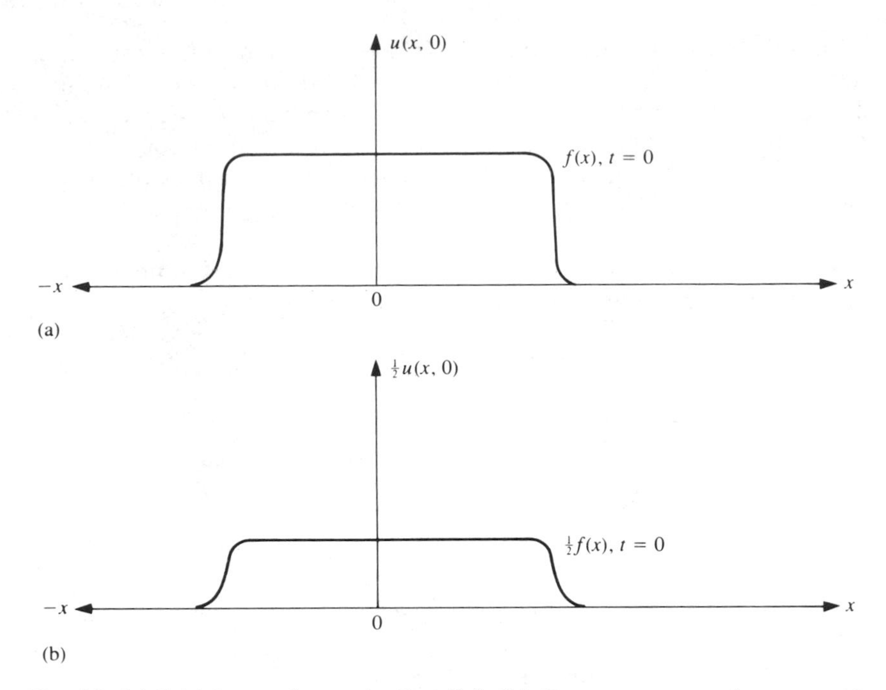

Fig. 4.2. (a) Initial wave form $u(x, 0) = f(x)$. (b) Component wave form at $t = 0$.

Equation (4.7) is called *D'Alembert's solution* to the one-dimensional wave equation.

Let us visualize the meaning of the solution given by (4.7) by giving a geometric interpretation of the traveling waves. We first consider the simplest case where the IC $g(x) = 0$ for all x. This means the initial particle velocity $u(x, 0)_t = 0$. Then the solution is reduced to

$$u(x, t) = \tfrac{1}{2}[f(x - ct) + f(x + ct)].$$

This means that the solution splits into two parts as shown below. Figure 4.2(a) shows a wave form $f(x)$ in the form of a pulse which is the sum of two equal pulses: $f(x) = \frac{1}{2}f(x) + \frac{1}{2}f(x)$. Figure 4.2(b) shows a component wave form $\frac{1}{2}f(x)$ at $t = 0$. As t increases these component wave forms split into a progressing and a regressing wave which travel in opposite directions with equal wave speeds c. Figure 4.3(a) shows the two component waves in their displaced positions at a small time t_1. The waves are distinguished by different dashed lines. In Fig. 4.3(b) these waves have been added to produce the resulting wave $u(x, t_1)$. t_1 is small enough so that the waves are not completely separated. Figure 4.3(c) shows the progressing and regressing waves for t_2 sufficiently large so that the two waves have separated completely. It is instructive to show an isometric drawing of the surface $u(x, t) = \frac{1}{2}[f(x - ct) + f(x + ct)]$. This is given in Fig. 4.4 where the different surfaces are clearly shown according to the x, t region.

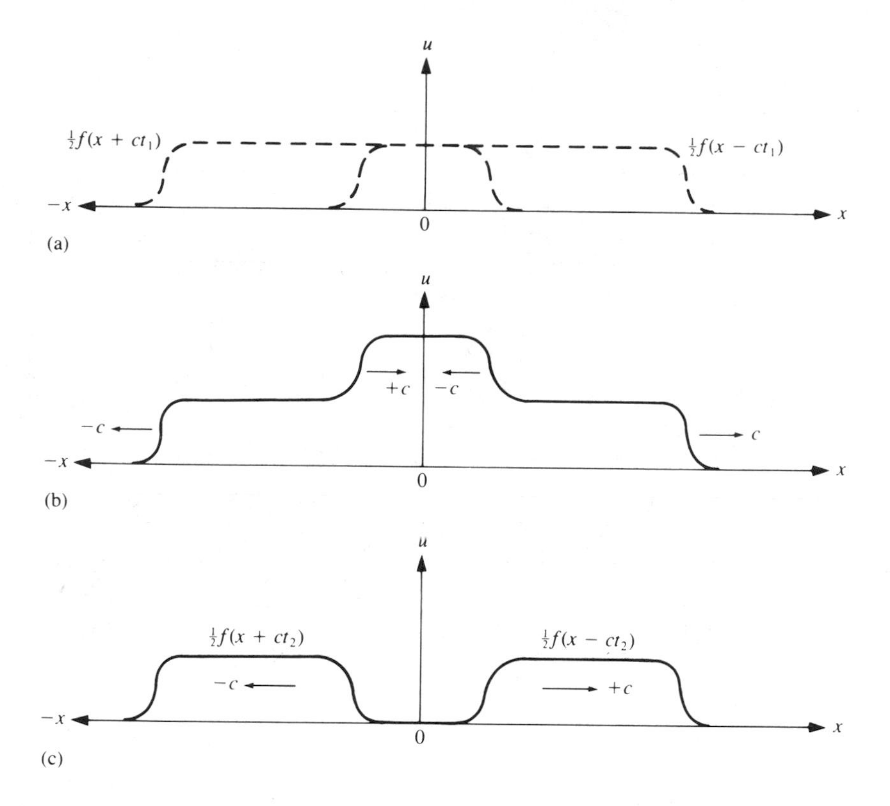

Fig. 4.3. (a) Two component waves in displaced positions at t_1. (b) Resultant wave form at t_1. (c) Resultant wave form at t_2.

We now consider the case where the string is initially at equilibrium with a given initial velocity distribution, so that the ICs are $u(x, 0) = f(x) = 0$, $u_t(x, 0) = g(x)$. Then the solution is

$$u(x, t) = \frac{1}{2c} \int_{x-ct}^{x+ct} g(z)\, dz.$$

Let us take $g(x)$ to be a pulse of unit intensity and pulse width $2a$ centered about the origin. Thus

$$g(x) = \begin{cases} 0, & |x| > |a|, \\ 1, & -a < x < a. \end{cases}$$

It is useful to describe the solution for this set of ICs by interpreting $u(x, t)$ geometrically as shown in Fig. 4.5. In the half-plane $t > 0$, draw the lines through $\pm a$ of slopes $\pm 1/c$; draw the line AB parallel to the x axis. For field points P between D and F the value of $u(x, t)$ is proportional to RS. This value will increase as t increases to a maximum when P is at the point K, when $u(x, t) = (1/2c) \int_{x-ct}^{x+ct} g(z)\, dz$. For points Q in AC or GB,

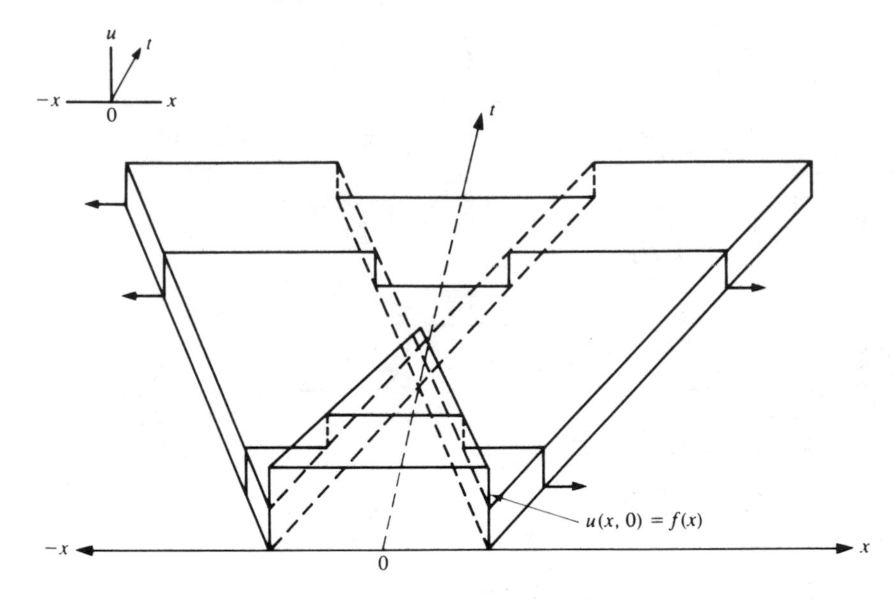

Fig. 4.4. Isometric drawing of surface $u(x, t) - \frac{1}{2}[f(x-ct)+f(x+ct)]$.

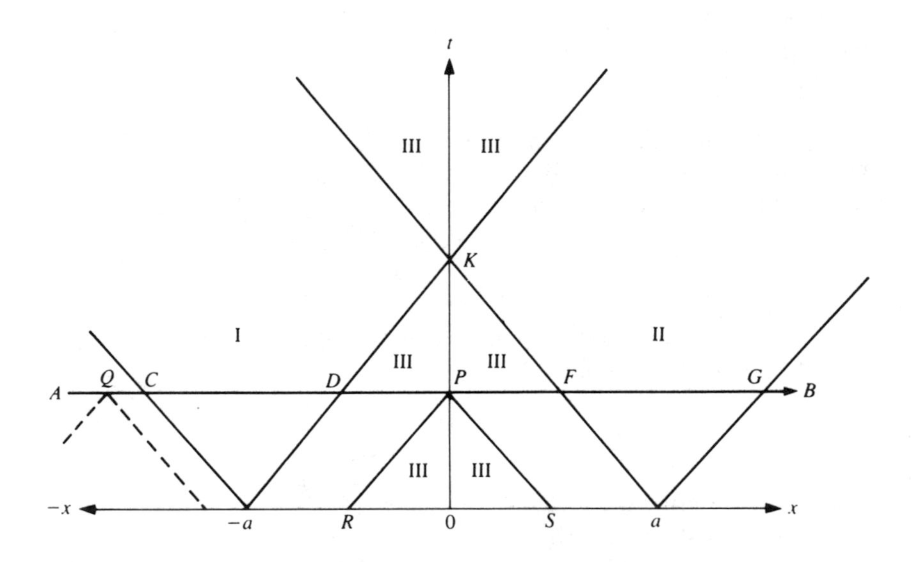

Fig. 4.5. Regions in (x, t) plane for initial condition $u_t = \begin{cases} 0, & |x|>|a| \\ 1, & -a<x<a \end{cases}$

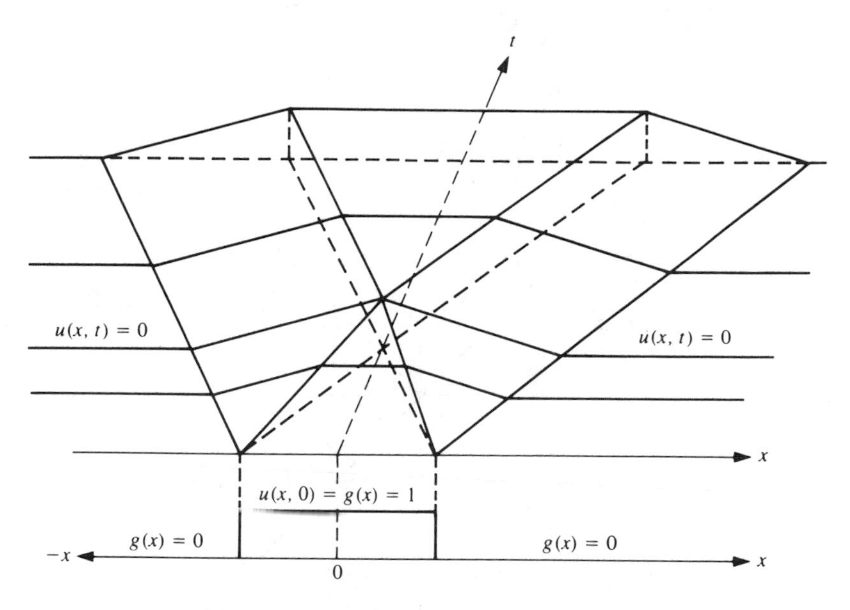

Fig. 4.6. Isometric drawing for the surface

$$u(x, t) = \frac{1}{2c} \int_{x-ct}^{x+ct} g(z)\, dz.$$

$u(x, t) = 0$. In regions I and II, $u(x, t)$ can be determined from the fact that u varies linearly on CD and FG, while $u(x, t) = u_K$ for all points in region III. We now use Fig. 4.5 to construct an isometric drawing for the surface $u = u(x, t)$ in (x, t, u) space, which is shown in Fig. 4.6.

The solution $u(x, t)$ for the Cauchy IV Problem (4.6) for the vibrating string for arbitrary ICs can easily be constructed from the two cases considered above. This is clearly seen by an application of the *superposition principle* which is valid for the linear wave equation. Suppose $v(x, t)$ is the solution to problem I defined by the ICs $v(x, 0) = f(x)$, $v_t(x, t) = 0$, and $w(x, t)$ is the solution to problem II defined by the ICs $w(x, 0) = 0$, $w_t(x, t) = g(x)$. The superposition principle tells us that, if we set $u(x, t) = v(x, t) + w(x, t)$, then $u(x, t)$ is the solution of (4.6). This is easily verified.

Cauchy Problem Revisited

It is instructive to set up and solve the Cauchy IV Problem for the vibrating string in the (ξ, η) or characteristic plane. To this end, we apply the transformation into characteristic coordinates given by (4.2) to (4.6). The formulation of the Cauchy Problem in the characteristic plane thereby

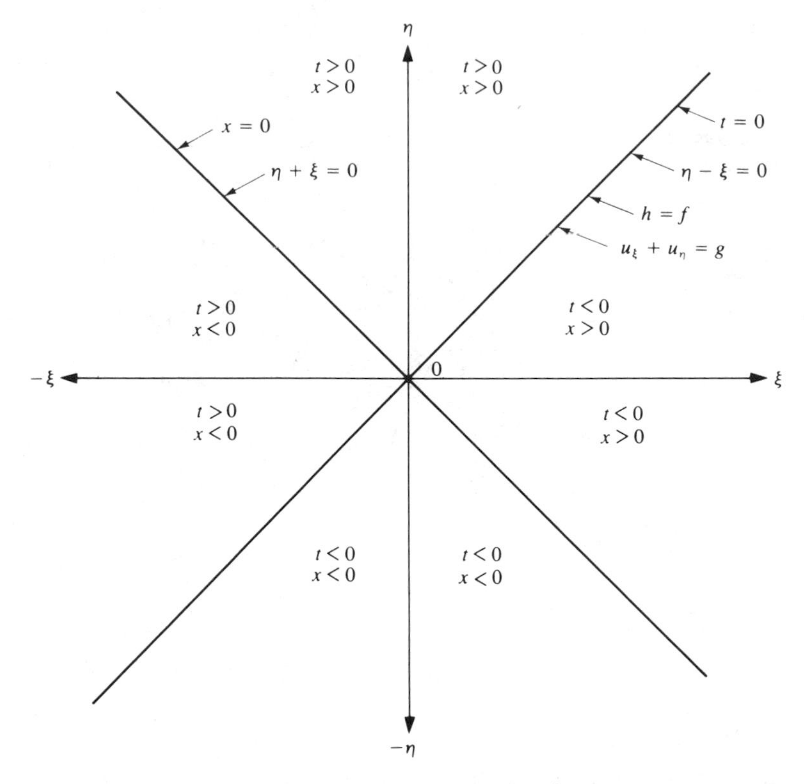

Fig. 4.7. Various regions in the (ξ, η) plane.

becomes

$$u_{\xi\eta} = 0 \qquad \text{for} \quad \eta > \xi, \tag{4.8}$$

$$u = f(\tfrac{1}{2}(\xi + \eta)), \qquad -u_\xi + u_\eta = \left(\frac{1}{c}\right) g(\tfrac{1}{2}(\xi + \eta)), \qquad \text{for} \quad \eta - \xi = 0,$$

since $u_t = c(-u_\xi + u_\eta)$. The line $t = 0$ maps into the line $\eta - \xi = 0$. The region $t > 0$ maps into the region $\eta > \xi$. Figure 4.7 shows the various regions in the characteristic plane. The general solution of (4.8) is given by (4.4),

$$u(\xi, \eta) = F(\xi) + G(\eta). \tag{4.4}$$

To satisfy the IC we must have

$$F(\xi) + G(\eta) = f(\xi) \qquad \text{for} \quad \eta - \xi = 0, \tag{4.9}$$

$$-F'(\xi) + G'(\eta) = \frac{g(\xi)}{c} \qquad \text{for} \quad \eta - \xi = 0. \tag{4.10}$$

Integrating (4.10) gives

$$-F(\xi)+G(\eta)=\frac{1}{c}\int_0^{\xi} g(z)\,dz+F(0)+G(0) \qquad \text{for} \quad \eta-\xi=0. \tag{4.11}$$

Solving (4.9) and (4.11) for $F(\xi)$ and $G(\eta)$ and using (4.4) gives the solution in the characteristic plane,

$$u(\xi,\eta)=\tfrac{1}{2}[f(\xi)+g(\eta)]+\frac{1}{2c}\int_{\xi}^{\eta} g(z)\,dz. \tag{4.12}$$

Equation (4.12) is D'Alembert's solution in characteristic coordinates.

4.3. Nonhomogeneous Wave Equation

We now consider the Cauchy IV Problem for the infinite vibrating string under the action of an applied external force. Let $h(x, t)$ be a given external force per unit mass. Then Newton's equation of motion, along with homogeneous ICs, becomes

$$\begin{gathered} c^2u_{xx}-u_{tt}=h(x,t), \\ u(x,0)=0, \qquad u_t(x,t)=0. \end{gathered} \tag{4.13}$$

Equation (4.13) is the one-dimensional *nonhomogeneous wave equation* in the (x, t) plane. For simplicity, we took the ICs to be homogeneous. It is clear that there is no loss in generality in solving this IV problem, because the general solution of the nonhomogeneous wave equation is equal to the general solution of the homogeneous wave equation (4.6), plus a particular solution of the nonhomogeneous equation which we can choose as the solution of (4.13). To solve this nonhomogeneous problem we recast (4.13) into characteristic coordinates by using the transformation (4.2). This yields the following formulation of the nonhomogeneous Cauchy IV Problem in characteristic coordinates:

$$\begin{gathered} u_{\xi\eta}=-h(\xi,\eta), \\ u=u_{\xi}=u_{\eta}=0 \qquad \text{on} \quad \eta-\xi=0, \end{gathered} \tag{4.14}$$

where

$$h(\xi,\eta)=\left(\frac{1}{4c^2}\right)h\left(\tfrac{1}{2}(\xi+\eta),\left(\frac{1}{2c}\right)(\eta-\xi)\right). \tag{4.15}$$

The transformation given by (4.2) maps the upper half of the (x, t) plane into the region above the line $\eta-\xi=0$. The homogeneous ICs map into $u=u_{\xi}=u_{\eta}=0$ on $\eta-\xi=0$. These three equations are actually equivalent to two conditions. This is seen as follows: From the IC $u(x,0)=0$ we obtain $u_x=u_{\xi}+u_{\eta}=0$ on $\eta-\xi=0$, while $u_t(x,0)=0$ is equivalent to $c(u_{\xi}-u_{\eta})=0$ on $\eta-\xi=0$, from which we obtain $u=u_{\xi}+u_{\eta}=0$ on $\eta-\xi=0$.

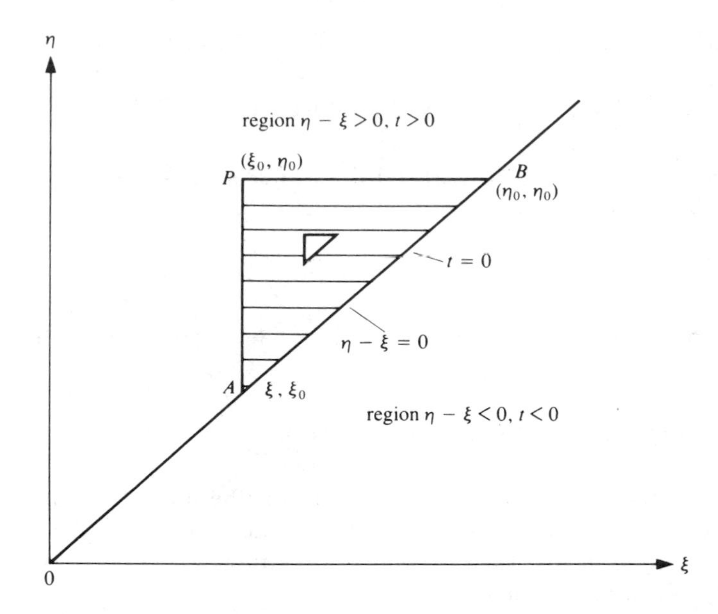

Fig. 4.8. Region in the characteristic plane for the nonhomogeneous wave equation.

Figure 4.8 shows the region in the (ξ, η) plane for the nonhomogeneous wave equation. The field point P: ξ_0, η_0 must lie in the region $\eta-\xi>0$ corresponding to the region $t>0$ in the (x, t) plane, since the transformation (4.2) maps the region of solution which is the upper half-plane of the (x, t) plane to the region $\eta>\xi$ which is above the line $\eta-\xi=0$. Consider Fig. 4.8. From the field point P draw a characteristic $\xi=\text{const.}=\xi_0$ down to where it intersects the line $\eta-\xi=0$ at the point A: ξ_0, η_0. Clearly the characteristic $\xi=\xi_0$ is the image in the (ξ, η) plane of corresponding C_+ characteristic in the (x, t) plane. From P (in the figure) draw the characteristic $\eta=\text{const.}=\eta_0$ (the image of a corresponding C_- characteristic) to where it intersects the line $\eta-\xi=0$ at the point B: η_0, η_0. Consider the wave equation in characteristic coordinates given by (4.14). Clearly $u(\xi_0, \eta_0)=u_P$ can be obtained by a double integration with respect to ξ and η in the triangle PAB bounded by the two characteristics from P and the intercepted portion of the line $\eta-\xi=0$. The solution of the nonhomogeneous Cauchy IV Problem in characteristic coordinates is therefore

$$u(\xi_0, \eta_0)=u_P=\int\int_{\nabla} h(\xi, \eta)\, d\xi\, d\eta=\int d\xi \int h\, d\eta. \tag{4.16}$$

To obtain this solution in the (x, t) plane refer to Fig. 4.9 which shows a mapping to the (x, t) plane of the region given by Fig. 4.8. In this figure the x axis is the image of the line $\eta-\xi=0$, P': x, t is the image of P: ξ_0, η_0, A': $x-ct$ is the image of A: ξ_0, η_0 and B': $x+ct$ is the image of B: η_0, η_0. The triangular region of solution $A'P'B'$ in the (x, t) plane is the image of

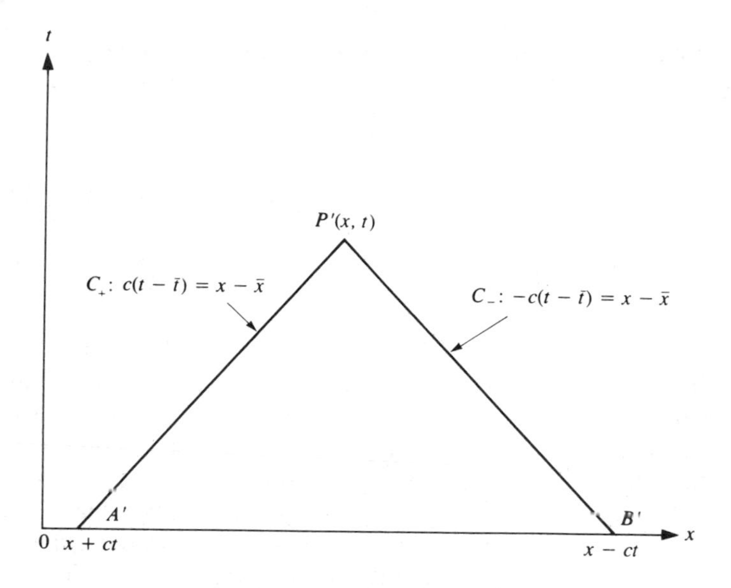

Fig. 4.9. Image in (x, t) plane of region given by Fig. 4.8.

the triangular region of solution APB in the (ξ, η) plane. The C_+ characteristic AP has the equation $c(t-\bar{t})=x-\bar{x}$, where $\bar{x}$, $\bar{t}$ are the running coordinates. It is the image of the $\xi=\xi_0$ or AP characteristic in the (ξ, η) plane. Similarly, the C_- characteristic BP has the equation $-c(t-\bar{t})=x-\bar{x}$, and is the image of PB in the (ξ, η) plane. A reminder: these results stem from the use of the transformation (4.2) which is assumed to have a unique inverse so that the Jacobian of the mapping $(x, y) \rightleftarrows (\xi, \eta)$ is nonsingular. Using this transformation we obtain the solution for the nonhomogeneous wave equation in the (x, t) plane,

$$u(x, t)=\frac{1}{2c}\int_0^t \int_{x-c(t-\bar{t})}^{x+c(t-\bar{t})} h(\bar{x}, \bar{t})\, d\bar{x}\, d\bar{t}. \tag{4.17}$$

The region of integration Δ is the triangle $A'P'B'$ bounded by the C_+ characteristic $A'P'$, the C_- characteristic $B'P'$, and the intercepted portion of the x axis, $A'B'$.

The following *example* illustrating the solution (4.17) was given in [12, p. 94]. It is repeated here for convenience. The externally applied force was given as

$$h(x, t)=\sin(\alpha x-\omega t). \tag{4.18}$$

This is a sinusoidally applied force of unit amplitude, where the wave number $\alpha=2\pi/\lambda$ (λ is the wave length), and ω is the angular frequency. It follows that $\omega/\alpha=v$, the wave speed of the external force. It is clear that (4.18) represents a progressing wave traveling in the direction of increasing

x. Inserting (4.18) into (4.17) gives the solution $u(x, t)$ for the nonhomogeneous wave equation. The integration yields two cases:

Case I. $v \neq c$.

$$u(x, t) = \left(\frac{1}{2\alpha^2 c}\right)\left[\left(\frac{1}{v-c}\right) \sin \alpha(x-ct) - \left(\frac{1}{v+c}\right) \sin \alpha(x+ct)\right]$$
$$+\frac{\sin \alpha(x-vt)}{\alpha^2(c^2-v^2)}. \tag{4.19}$$

The first term in the brackets of (4.19) represents a progressing wave with an amplitude as shown traveling with the wave speed c. The amplitude becomes infinite as $v \to c$. In practice, the term $v-c$ should be large enough so that the amplitude is of order unity because u is a solution of a linear PDE. The second term in brackets is a regressing wave. Thus these terms are of the form $F(x-ct)+G(x+ct)$. This expresses the fact that the external force contributes progressing and regressing waves of speed c to the solution (independent of the wave speed of the external force). The remaining term is a progressing wave traveling with the speed v of the external force, and does not depend on c.

Case II. $v = c$. As implied above this case represents a singularity in the solution for Case I and must therefore be treated separately. We may arrive at the solution for this case by a limiting process (letting $v \to c$). The result is

$$u(x, t) = \left(\frac{1}{4c^2\alpha^2}\right)[\sin \alpha(x-ct) - \sin \alpha(x+ct)]$$
$$+\left(\frac{t}{2c\alpha}\right) \cos \alpha(x-ct). \tag{4.20}$$

Again the terms in brackets represent progressing and regressing waves with wave speed c. However, the last term is an amplitude modulated progressing cosine wave whose amplitude is proportional to t, thus becoming nonlinear and finally unstable as t increases. For this reason this is called the *resonance term.*

4.4. Mixed Initial Value and Boundary Value Problem, Finite String

Up to now we have discussed a pure Cauchy IV Problem, since we considered an infinite string so that boundary conditions (BCs) were not considered, except for demanding the mild condition of boundedness at infinity (which comes out automatically from the solution to a linear equation). Now the more interesting problems that arise in the acoustics of the piano and stringed instruments involve vibrating strings of finite length so that BCs as well as ICs must be considered. That is why we shall discuss the so-called "mixed problem", involving both ICs and BCs. It is seen that wave propagation in a finite string is more complicated than in an infinite string since we must consider the BCs at each end of the string.

The Clamped String

The simplest type of problem of a finite string is one where the ends are kept clamped or fixed and the ICs are nonhomogeneous so that an initial displacement and particle velocity distribution of the string is prescribed. For a clamped string of length L we have essentially a vibration problem. Nevertheless, the solution may be interpreted by a superposition of multiply reflecting waves. Examples abound in acoustics. For example, the length of a violin string is the distance from the point on the string where the finger of the left hand stops the string to the bridge. If the string is plucked at $t=0$ (*pizzicato* technique) then for $t>0$ the string is essentially in free vibration, and our model applies where the initial velocity distribution is zero and the initial displacement is essentially a triangle due to the initial plucking. Another example is a piano string of finite length initially struck by a felt hammer which supplies an impulsive force.

This boundary value (BV) problem is formulated as follows: Find the lateral displacement $u(x, t)$ of the string that satisfies the one-dimensional wave equation with homogeneous BCs and nonhomogeneous ICs.

$$\begin{aligned} &c^2 u_{xx} - u_{tt} = 0 \qquad \text{for} \quad 0<x<L, \quad t>0, \\ &\text{BCs} \quad u(0, t) = 0, \qquad u(L, t) = 0 \qquad \text{for} \quad t>0, \\ &\text{ICs} \quad u(x, t) = f(x), \qquad u_t(x, 0) = g(x) \qquad \text{for} \quad 0<x<L, \end{aligned} \tag{4.21}$$

where f is the initial displacement distribution and g the initial velocity. The general solution is given by (4.5). To find the functions F and G we must make use of the condition that the ends of the string are fixed. This is done by extending the solution outside the region $(0, L)$ to the regions $(-L, 0)$, $(L, 2L)$, etc., and fixing up the general solution to account for the BCs. To do this we must first find $F(z)$ and $G(z)$ for $0<z<L$ from the ICs as before and then find these functions successively in adjacent intervals of length L from the two BCs. The result can best be expressed by reflecting (with change of sign to make the ends fixed) the IVs f, g successively at the end points of the intervals of length L. This leads to

$$\begin{aligned} f(-z) &= -f(z), \qquad f(2L-z) = -f(z), \quad \text{etc.}, \\ g(-z) &= -g(z), \qquad g(2L-z) = -g(z), \quad \text{etc.} \end{aligned}$$

Note that this extension is done in such a way that the extended functions f and g, together with their first derivatives f_x and g_x, are continuous at the ends of the string, provided f and g vanish at the endpoints and have continuous spatial derivatives there.

With these extended values of f and g we may imagine the medium extended successively to both sides for $t>0$ and then apply D'Alembert's formula (4.7) for the infinite string. Applying (4.7) in this manner satisfies the BCs since we extended the ICs f and g in a particular way to account for these BCs. Because of the periodicity of f and g with respect to $2L$ the solution $u(x, t)$ is also periodic in x with period $2L$. We can therefore expand the solution in a Fourier series:

$$u(x, t) = \sum_{n=1}^{\infty} c_n(t) \sin\left(\frac{n\pi x}{L}\right). \tag{4.22}$$

No terms involving $\cos(n\pi x/L)$ occur since the functions $u(x, 0)$ and $u_t(x, 0)$ are odd and periodic with period $2L$. The values of the coefficients $c_n(t)$ and their first derivatives at $t=0$ can be determined as the Fourier coefficients of the Fourier series expansions of $u(x, 0)=f(x)$ and $u_t(x, 0)=g(x)$. Inserting (4.22) into the wave equation $c^2u_{xx}-u_{tt}=0$ leads to the relation

$$\sum_{n=1}^{\infty} [\ddot{c}_n(t)+c^2n^2\pi^2L^{-2}c_n(t)] \sin\left(\frac{n\pi x}{L}\right)=0, \qquad \ddot{c}_n=\frac{d^2c_n}{dt^2}.$$

Since each term in the series must vanish we obtain the following infinite set of ODEs for the $c_n(t)$'s:

$$\ddot{c}_n+c^2n^2\pi^2L^{-2}c_n=0. \tag{4.23}$$

Clearly only a finite number of the c_n's need be considered because the series representation (4.22) is convergent. Note that the form of the solution given by (4.22) and (4.23) can be obtained by the elementary method of separation of variables applied to the wave equation. The solution of (4.23) is

$$c_n(t)=a_n \cos\left(\frac{n\pi ct}{L}\right)+(n\pi c)^{-1}Lb_n \sin\left(\frac{n\pi ct}{L}\right), \tag{4.24}$$

where $c_n(0)=a_n$ and $dc_n(0)/dt=b_n$. We now insert (4.24) into the series (4.22) and express the products of sine and cosine functions as sums of such functions of $n\pi(x\pm ct)/L$. We therefore obtain the solution in the form of a series of progressing and regressing waves.

$$\begin{aligned} u(x, t) = \tfrac{1}{2} \sum_{n=0}^{\infty} &\left[a_n \sin\left(\frac{n\pi(x-ct)}{L}\right)+(n\pi c)^{-1}Lb_n \cos\left(\frac{n\pi(x-ct)}{L}\right)\right] \\ &+\tfrac{1}{2}\sum_n \left[a_n \sin\left(\frac{n\pi(x+ct)}{L}\right)-(n\pi c)^{-1}Lb_n \cos\left(\frac{n\pi(x+ct)}{L}\right)\right]. \end{aligned} \tag{4.25}$$

Actually, the series solution as represented by (4.25) is not really infinite because the number of terms in each of the series depends on the number of reflections we must consider at the endpoints, which depends on how far out in time we take the solution.

As an example: Consider the case where the string is originally at rest and a localized displacement at $t=0$ (in the shape of half a sine wave) is confined to the neighborhood of the center of the string, namely in the interval (a, b) centered about $x=L/2$. Then the range of influence and domain of dependence is shown in Fig. 4.10. The range of influence is the interval (a, b). The domain of dependence consists of the two strips emanating from this interval and intersecting the $x=0$ and $x=L$ vertical lines at the same time. These strips are then multiply reflected from these boundaries in such a way as to satisfy the BCs. The strip from the interval (a, b) going toward $x=0$ carries in it regressing waves, and the strip going toward $x=L$ carries progressing waves. All the strips carrying incident waves to $x=0$ contain regressing waves, and all the reflected strips from $x=0$ consist of

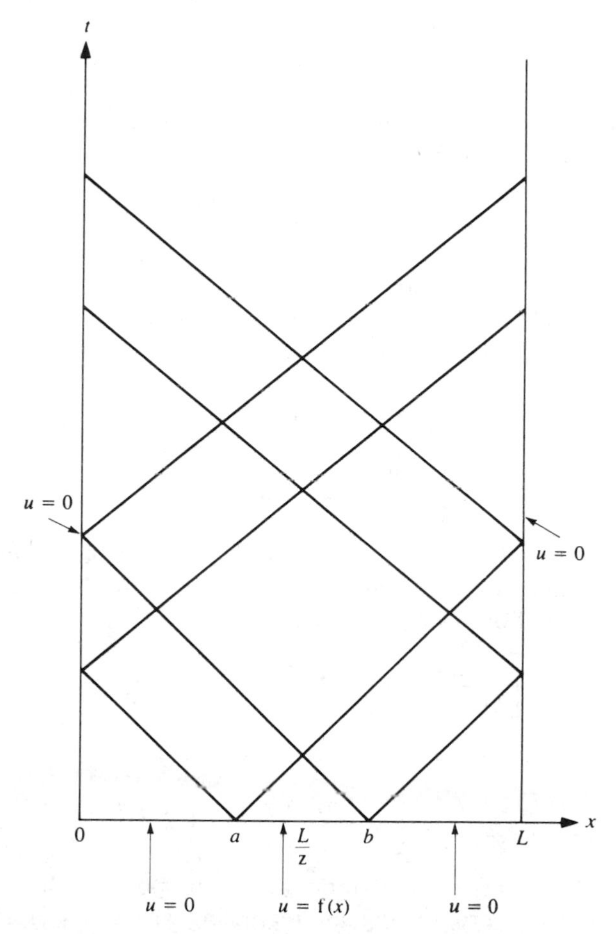

Fig. 4.10. Regions in (x, t) plane for finite string initially displaced in interval $[a, b]$.

progressing waves. The opposite statements apply to the boundary at $x = L$. The strips carrying progressing waves contain C_+ characteristics and the strips carrying regressing waves carry C_- characteristics.

The Plucked String

We mentioned above the plucking of a violin string, the so-called pizzicato effect. We shall now use the solution given by (4.25) to determine the free vibrations of a violin string resulting from plucking the string at $t = 0$. The ICs obtained from the series given by (4.25) are obtained by calculating from the series the expressions $u(x, 0) = f(x)$ and $u_t(x, 0) = g(x)$. We obtain

$$f(x) = \sum_n a_n \sin\left(\frac{n\pi x}{L}\right), \qquad g(x) = \sum_n b_n \sin\left(\frac{n\pi x}{L}\right). \tag{4.26}$$

This clearly means that the a_n's are the Fourier sine coefficients of $f(x)$ and the b_n's are the Fourier sine coefficients of $g(x)$.

We now suppose that the violin string of length L is initially at rest. This clearly means that all the g_n's vanish. From the orthogonality condition on the sine functions over the interval $(-L, L)$ we can easily deduce the following expression for the a_n's:

$$a_n = \frac{2}{L}\int_0^L f(x) \sin\left(\frac{n\pi x}{L}\right) dx. \tag{4.27}$$

We now assume that the initial displacement f is the triangle function as shown in Fig. 4.11. The string is plucked at some interior point b, and the amplitude of $f(x)$ is u_0 at $x = b$. The analytic expression for f is given by

$$f(x) = \begin{cases} \left(\dfrac{u_0}{b}\right)x, & 0 < x < b, \\ \left(\dfrac{u_0}{L-b}\right)(L-x), & b < x < L. \end{cases} \tag{4.28}$$

Inserting the expression given by (4.28) for the a_n's into (4.27) and performing the integration over the appropriate intervals yields the following expression for the Fourier coefficients:

$$a_n = \left[\frac{2u_0}{b(L-b)n^2}\right] \sin\left(\frac{n\pi b}{L}\right), \qquad n = 1, 2, \ldots. \tag{4.29}$$

Inserting (4.29) into (4.25) and setting $b_n = 0$ yields

$$u(x, t) = \frac{u_0}{b(L-b)} \sum_{n=1}^{\infty} \left(\frac{1}{n^2}\right)\left[\sin\left(\frac{n\pi(x-ct)}{L}\right) + \sin\left(\frac{n\pi(x+ct)}{L}\right)\right]. \tag{4.30}$$

Equation (4.30) shows that the vibrations of any point of the violin string can be interpreted as progressing and regressing waves that satisfy the BCs and the ICs of zero initial velocity and a triangle initial displacement. Note

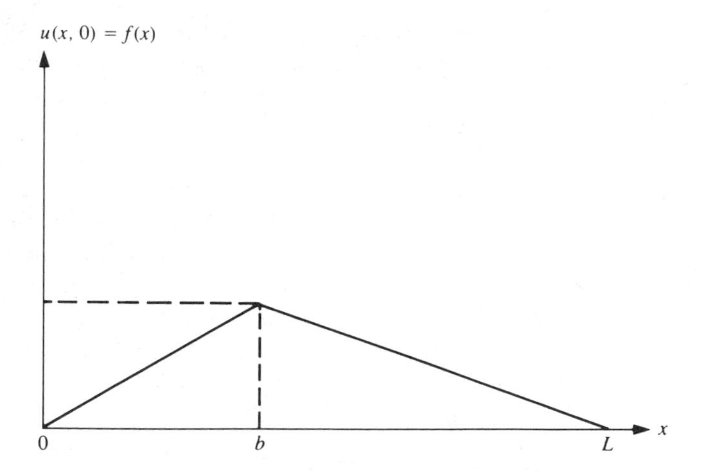

Fig. 4.11. Initial displacement of a plucked string.

the fact that the solution has a singularity if $b=0$ or $b=L$. This clearly means that the violinist cannot pluck the string at the point where his finger stops the string or at the bridge.

4.5. Finite or Lagrange Model for Vibrating String

The approach used above made use of the wave equation which was derived *ab initio* from Newton's second law. There is another approach, developed by Lagrange, which approximates the string (a continuous distribution of matter) by a finite number of linearly coupled masses and springs. This is the so-called "finite model" which involves a finite number of degrees of freedom. Each mass has associated with it one degree of freedom which is its lateral displacement from equilibrium—since we invoke a linear theory involving small amplitude oscillations. Clearly the sum of the masses equals the total mass of the original string, and the more masses per unit length the closer the approximation to the continuous mass distribution of the original string. In [12, Chap. 3, pp. 111 ff.] we discussed this approach under the heading, "Finite Number of Degrees of Freedom Approximation". In that work the point of view was in keeping with the purpose of the book, namely, to treat finite difference methods in connection with properties of PDEs. It was shown that the resulting system of ODEs, by a limiting process, was converted to a single PDE, the wave equation.

The reason for using this approach here is to show the power of the method in approximating the string as a limiting problem of vibrations of a coupled system of particles. Moreover, as shown in the above-mentioned work, we transform this system to an uncoupled system by introducing normal coordinates. This relates the method to the discussion of normal coordinates in Chapter 1, in connection with transforming a coupled system of oscillators to an uncoupled system by use of a similarity transformation to transform the original coordinates to normal coordinates. In the ensuing treatment we shall also make use of Lagrange's equations of motion from which we can derive the ODEs for the uncoupled system. This is another tie-in with Chapter 1 which was devoted to oscillatory phenomena.

We use the finite model and notation of [12]. To develop the model we start by representing the string of length L as a continuous filament of matter along a finite interval of the x axis such that $0 \leq x \leq L$. We divide this interval into N subintervals by letting an index $i = 0, 1, \ldots, N$ ($x = 0$ is $i = 0$, $x = L$ is $i = N$). The string is now approximated by $N-1$ evenly spaced interior mass points (particles)—the ith particle having mass m_i, $i = 1, 2, \ldots, N-1$, $m_i = m_j$. Each particle is connected to its nearest neighbors by weightless springs, each of spring constant K. According to the restrictions of the linear theory, each particle is constrained to move vertically with small amplitude so that axial motion is ignored. Also, the mass of each spring is ignored. K is related to the tension T of the string by

$$T = K\Delta x, \qquad \Delta x = \frac{L}{N},$$

where $\Delta x = x_{i+1} - x_i$. Let m be the total mass of the string. We then have $m_i = m/(N-1)$. This gives us the geometry of our finite degree of freedom system.

Since the motion of each particle is purely vertical it is only necessary to calculate this component of forces exerted by the two springs which connect the particle with its neighbors. Obviously the end particles (for $i = 0, L$) must be treated separately, as each is connected to only one spring, and appropriate boundary conditions must be imposed on these particles according to the nature of the problem. Newton's second law of motion yields the following system of coupled ODEs:

$$m_1 \ddot{u}_i = -K(u_i - u_{i-1}) - K(u_i - u_{i+1}), \qquad i = 1, 2, \ldots, N-1,$$
$$\ddot{u}_i = \frac{d^2 u_i}{dt^2}, \tag{4.31}$$

where $\ddot{u}_i$ is the acceleration of the ith particle. The system (4.31) is a set of $N-1$ coupled ODEs. Clearly, this process is a discretization only on the spatial coordinate, time remaining continuous. The right-hand side of (4.31) represents the net spring force acting on the ith particle due to the $(i-1)$st and $(i+1)$st particles. Note that this resultant force is a restoring force, hence the minus signs. By substituting $K = T/\Delta x$, $m_i = m/(N-1) = m/L = \rho$ (the density of the string) into (4.31) the system becomes

$$\ddot{u}_i = ((N-1)/N)c^2(u_{i-1} - 2u_i + u_{i+1})/(\Delta x)^2, \qquad i = 1, 2, \ldots, N-1,$$
$$\text{BC:} \quad x = 0, \quad u_0 = 0; \qquad \text{BC:} \quad x = L, \quad u_N = 0, \tag{4.32}$$

for the case of nonhomogeneous BCs (ends fixed). c is the wave speed of the string. Note that the expression $((N-1)/N)c^2 \to c^2$ as $N \to \infty$. For the case of the smallest N ($N=2$), which corresponds to one interior particle, the expression becomes $c^2/2$. This means that the "waves", generated by a single oscillating particle acted upon by two springs, travel with a wave speed $c/\sqrt{2}$.

We now uncouple the system (4.32) by transforming it to normal coordinates. To do this we first represent (4.32) by an equivalent matrix equation. We let $\mathbf{u}$ be the displacement vector whose ith component is u_i so that $\mathbf{u} = (u_1, u_2, \ldots, u_{N-1})$, and similarly for the particle acceleration $\ddot{\mathbf{u}}$. The system of ODEs is represented by the matrix equation

$$\ddot{\mathbf{u}}(t) = \mathbf{A}\mathbf{u}(t), \tag{4.33}$$

where

$$A = \left(\frac{c_N}{\Delta x}\right)^2 \begin{pmatrix} -2 & 1 & 0 & & & \cdots & 0 \\ 1 & -2 & 1 & 0 & & \cdots & 0 \\ 0 & 1 & -2 & 1 & 0 & \cdots & 0 \\ \cdots & \cdots & \cdots & \cdots & \cdots & \cdots & \cdots \\ 0 & \cdots & \cdots & \cdots & 0 & 1 & -2 \end{pmatrix}, \tag{4.34}$$

$$c_N^2 = \left(\frac{N-1}{N}\right)\left(\frac{T}{\rho}\right) = \left(\frac{N-1}{N}\right) c^2. \tag{4.35}$$

Since $\mathbf{A}$ is a symmetric matrix we can find an orthonormal matrix $\mathbf{P}$ such that

$$\mathbf{P}^{-1}\mathbf{A}\mathbf{P} = \mathbf{D}, \tag{4.36}$$

where $\mathbf{D}$ is a diagonal matrix whose (i, i)th element is d_i. The transformation given by (4.36) from $\mathbf{A}$ to $\mathbf{D}$ is a *similarity transformation* and was discussed in Chapter 1. Recall that $\mathbf{P}^{-1} = \mathbf{P}^*$ because of the orthogonality property of $\mathbf{P}$. The role of $\mathbf{P}$ is to uncouple the system. This can be seen more clearly by introducing the vector $\mathbf{v}(t)$ (of $N-1$ components) obtained from $\mathbf{u}(t)$ by the following transformation:

$$\mathbf{v} = \mathbf{P}^{-1}\mathbf{u} \qquad \text{or} \qquad \mathbf{u} = \mathbf{P}\mathbf{v}. \tag{4.37}$$

Substituting (4.37) back into (4.33), premultiplying by $\mathbf{P}^{-1}$, and using the definition of $\mathbf{D}$ from (4.36) gives

$$\ddot{\mathbf{v}}(t) = \mathbf{D}\mathbf{v}(t). \tag{4.38}$$

Equation (4.38) is a matrix ODE for $\mathbf{v}(t)$. It is an uncoupled system in the sense that each v_i does not depend on any of the other v_j's. Physically, this means that the similarity transformation given by (4.36) uncouples the particle motion; the solution of (4.38) for each v_i is a normal mode of oscillation with a given frequency independent of the other v_j's. Note that this process of diagonalization (yielding the diagonal matrix $\mathbf{D}$) can be performed only if **A** is symmetric. The ith equation of (4.38) is

$$\ddot{v}_i = d_i v_i, \qquad i = 1, 2, \ldots, N-1,$$

whose general solution is

$$\mathbf{v} = \exp(i\omega t)\mathbf{v}, \tag{4.39}$$

where ω is the frequency or eigenvalue corresponding to the ith mode of oscillation given by thc ith component of the eigenvector **v**, and **V** is a constant vector whose ith component is V_i. Inserting (4.39) into (4.38) gives

$$(\mathbf{D} + \omega^2\mathbf{I})\mathbf{V} = 0. \tag{4.40}$$

By the usual argument, in order to obtain a nontrivial solution for **V** we must have

$$\det(\mathbf{D} + \omega^2\mathbf{I}) = 0. \tag{4.41}$$

The solution of (4.41) clearly gives the elements of **D** as

$$d_i = \omega^2, \qquad i = 1, 2, \ldots, N-1. \tag{4.42}$$

Note that (4.42) does not give ω in terms of the physical constants contained in the original system of ODEs (4.32) or (4.33). To obtain these eigenvalues we could transform **u** into **v** by (4.37) and the similarity transformation (4.36). This involves constructing the orthonormal matrix **P** which is a little tedious. We shall bypass having to construct **P** by working directly with **u**. We attempt to obtain the general solution for $\mathbf{u}(t)$ by finding a family of special solutions such that for each solution of the family all the particles oscillate with the same frequency. The family will consist of elements, each of which will have different frequencies. In other words, each element of the family is a normal mode of oscillation. Then a linear combination of these normal modes is taken which will be consistent with the ICs of a particular problem. We therefore look for normal modes for **u** in the form

$$\mathbf{u}(t) = \exp(\pm i\lambda t)\mathbf{U}, \tag{4.43}$$

where **U** is a constant vector whose ith component is U_i, and λ is an eigenvalue (to be determined below) corresponding to the eigenvector **u**. Inserting (4.43) into (4.33) yields

$$(\mathbf{A} + \lambda^2\mathbf{I})\mathbf{U} = 0. \tag{4.44}$$

In order to have nontrivial solutions for **U** we must have

$$\det(\mathbf{A} + \lambda^2\mathbf{I}) = 0. \tag{4.45}$$

Equation (4.45) is the secular equation for the eigenvalues λ. When expanded, it yields a polynomial of $(N-1)$st degree in λ^2. Since the matrix $\mathbf{A} + \lambda^2\mathbf{I}$ is symmetric all the eigenvalues are real. We may, of course, solve for the roots of this polynomial which can be written as

$$(\lambda^2 - \lambda_1^2)(\lambda^2 - \lambda_2^2) \ldots (\lambda^2 - \lambda_{N-1}^2) = 0.$$

However, it is easier to solve for λ^2 directly from the system (4.32). To do this we insert (4.43) into (4.32) and obtain the following set of equations for U_j:

$$U_{j+1} + aU_j + U_{j-1} = 0, \qquad j = 1, 2, \ldots, N-1, \tag{4.46}$$

where

$$a = \lambda^2 \left(\frac{\Delta x}{c_N} \right)^2 - 2, \tag{4.47}$$

and the BCs are $U_0 = U_N = 0$. The system (4.46) is called a set of *finite difference equations.* Such finite difference systems are treated more thoroughly in [12] with respect to their relationship to the PDEs occurring in continuum dynamics.

To determine the U_j's we assume a solution of the form

$$U_j = r^j. \tag{4.48}$$

To find r we substitute (4.48) into (4.46) and obtain the following quadratic equation:

$$r^2 + ar + 1 = 0. \tag{4.49}$$

We recall that each U_j is the spatially dependent part of the corresponding u_j, since U_j depends on $x_j = j\Delta x$. Therefore if U_j is given by (4.48), then $r \leq 1$ in order to insure stability, i.e., to insure that U_j does not grow with the index j. This means that the roots of (4.49) must not be real, but must be complex conjugates. It follows that the following inequality must be valid: $a^2 - 4 \leq 0$. Using (4.47) this inequality yields the following restriction on λ^2 for stability:

$$\lambda^2 \leq 4 \left(\frac{c_N}{\Delta x} \right)^2. \tag{4.50}$$

Since the roots r are complex conjugates they can be written in the form

$$r = \exp(\pm i\theta), \tag{4.51}$$

where θ is to be determined. We shall see that there are $N-1$ values of θ. Using (4.51) in (4.48) we may write the solution for each eigenvector $\mathbf{U}_j$ in the two equivalent forms

$$\begin{aligned} \mathbf{U}_j &= \mathbf{C}_1 \exp(i\theta_j) + \mathbf{C}_2 \exp(-i\theta_j) \\ &= \mathbf{A} \sin(\theta_j) + \mathbf{B} \cos(\theta_j), \qquad j = 1, 2, \ldots, N-1. \end{aligned} \tag{4.52}$$

The BC $\mathbf{U}_0 = 0$ gives $\mathbf{B} = 0$; the BC $\mathbf{U}_N = 0$ gives

$$\theta = \theta_j = \frac{j\pi}{N}, \qquad j = 1, 2, \ldots, N-1. \tag{4.53}$$

Equation (4.53) tells us that there are $N-1$ values of θ_j corresponding to each of the interior points x_j.

Inserting (4.51) into (4.49) yields

$$2\cos\theta_j + a = 0.$$

Using the definition of a from (4.47) we solve this equation for λ^2, obtaining

$$\lambda_j^2 = 2\left(\frac{c_N}{\Delta x}\right)^2\left(1-\cos\left(\frac{j\pi}{N}\right)\right). \tag{4.54}$$

This gives the $N-1$ eigenvalues λ_j^2 which are the roots of the secular equation (4.45). We then have

$$\mathbf{U}_j = \mathbf{C}\sin\left(\frac{j\pi}{N}\right). \tag{4.55}$$

Inserting (4.55) into (4.43) we see that the general solution of (4.32) or (4.33) can be represented as a linear combination of the solutions

$$\mathbf{u}_j(t) = [\mathbf{A}_j \exp(i\lambda_j t) + \mathbf{B}_j \exp(-i\lambda_j t)]\sin\left(\frac{j\pi}{N}\right). \tag{4.56}$$

The general solution consists of a linear combination of the $N-1$ modes. This is obtained by summing (4.56) over j as follows:

$$\mathbf{u}(t) = \sum_{j=1}^{N-1} \mathbf{u}_j = \sum_j [\mathbf{A}_j \exp(i\lambda_j t) + \mathbf{B}_j \exp(-i\lambda_j t)]\mathbf{U}_j. \tag{4.57}$$

The $\mathbf{A}_i$'s and $\mathbf{B}_i$'s can be determined from the ICs. Instead of determining these coefficients by using finite Fourier series expansions for specific ICs, we take a different tack by passing in the limit to the wave equation.

Continuous String as Limiting Case

We recall that the system of finite difference equations in the u_i's given by (4.32) or (4.33) is a finite degree of freedom system in space, but continuous in time. We focus on the spatially dependent part of the u_i's. To do this we now study the convergence to a continuum of the finite degree of freedom system represented by a finite number of mass-spring couplings. Recall that we have $N-1$ interior masses making up the string of length L. Fixing the length we let $N \to \infty$. This means that the number of interior x_i's also becomes infinite. Let $h = \Delta x = L/N \to 0$ so that $h \to dx$. Considering (4.32) we manipulate the term $(u_{i-1} - 2u_i + u_{i+1})/h^2$ and pass to the limit as $h \to 0$.

$$\begin{aligned}\lim_{h\to 0} \frac{u_{i+1} - 2u_i + u_{i-1}}{h^2} &= \lim_{h\to 0} \frac{1}{h}\left[\frac{u_{i+1} - u_i}{h} - \frac{u_i - u_{i-1}}{h}\right] \\ &= u_{xx}.\end{aligned}$$

We used the forward difference approximation to u_x at x_i given by $(u_{i+1} - u_i)/h$ and the same approximation to u_x at x_{i-1}. We then took $\lim_{h \to 0} [u_x(x_i, t) - u_x(x_{i-1}, t)]/h$ to get u_{xx}. See [12] for details. Using these results (4.32) converges to the wave equation

$$u_{tt} = c^2 u_{xx}, \qquad c^2 = \frac{T}{\rho},$$

where $\ddot{u}_i \to u_{tt}$ and $u = u(x, t)$.

CHAPTER 5

Water Waves

Introduction

In the Introduction of Chapter 2 it was pointed out that, "since a musical instrument generates sound waves, it appears that the next class of wave propagation to discuss would be this problem of sound wave propagation". We have already discussed the fundamentals of wave propagation in a vibrating string which supplies a background for the acoustics of musical instruments. To continue with our systematic treatment of wave propagation, in a logical progression from the simple to the more complex phenomena, it would appear that the next stage in our investigation is to study the mathematical physics of wave propagation in a one-dimensional unsteady compressible gas; for this is the type of phenomenon involved in the study of sound waves. However, for this topic we need a fundamental knowledge of the hydrodynamics of inviscid, incompressible, irrotational fluids as a basis for the more complicated treatment of compressible flow phenomena. It appears natural, then, to present, at this stage the chapter on water waves, which is essentially the hydrodynamics of incompressible flow. Indeed, this might very well be the title of this chapter, except that we take advantage of our study of this branch of hydrodynamics to study water wave phenomena, which has its own intrinsic interest. We therefore treat the subject of sound waves in air in the next chapter.

By an inviscid, incompressible, irrotational fluid we mean a continuous fluid medium (water or air) whose viscosity can be neglected, where there is no volume change—the dilatation is zero and the *curl* of the particle velocity vanishes so that a velocity potential exists. In Chapter 2 the conservation laws of mass, momentum, and energy were discussed qualitatively. The Lagrange and Euler representation of describing the coordinates of each particle were also discussed. In fluid mechanics we shall use the Euler representation. As pointed out in Chapter 2, in this representation the observer stations himself at laboratory fixed points in space (the Euler coordinates) and observes the flow of fluid particles. The position of each particle is defined by its Lagrange coordinates. Each particle carries with it the dynamical variables such as velocity, pressure, and density, and thermodynamic variables such as temperature and entropy. These dynamic and thermodynamic variables are functions of the Euler coordinates and time. Mathematically, the solution of a problem in fluid flow is given by a mapping from the Lagrangian coordinate to the Euler coordinate system, time mapping into itself. Physically, the solution of a problem in fluid flow consists of a one-parameter family of trajectories of Euler coordinates versus time, where the parameter is the set of Lagrange coordinates.

Wave motion in water and other fluids involves a free surface (no external forces acting on the surface) on which surface waves act. The water particles are subject to the action of gravity which is an external force, so that this external force cannot be neglected in the equations of motion. Surface wave problems have interested many mathematical physicists: Euler, the Bernoullis, Lagrange, Cauchy, Poisson, Poincaré, the British school: Airy, Stokes, Kelvin, Rayleigh, Lamb, and the American school of Friedrichs, F. John, J. Keller, Stoker, etc. The literature is vast and the more modern literature concentrates on nonlinear problems. In this chapter we can only briefly explore the basic foundations of this extensive theory.

5.1. Conservation Laws

We now describe the three conservation laws in detail in the Euler representation.

Conservation of Mass

We consider some volume V_0 of our fluid in a three-dimensional space bounded by a closed surface S_0. Assuming that there are no sources or sinks of fluid in the volume then the mass is conserved (does not change) during the time history of the fluid flow. The mass of fluid in this volume is $\int_{V_0} \rho \, dV$, where ρ is the fluid density, and the integration is taken over the volume element V_0. Consider an element of surface area dS on the surface S_0. dS is a vector having the direction of the outward drawn normal to the surface. The mass of fluid flowing per unit time across this element of surface is $\rho\mathbf{v} \cdot d\mathbf{S}$, where $\mathbf{v}$ is the velocity of a particle of fluid and the scalar or dot product is taken to yield flow normal to the surface element. Since the outward drawn normal to the surface is positive $\rho\mathbf{v} \cdot d\mathbf{S}$ is positive if the fluid is flowing out of the volume, and, of course, negative if the flow is into the volume. The total mass flow of fluid out of the volume V_0 per unit time is therefore

$$\int_{S_0} \rho\mathbf{v} \cdot d\mathbf{S},$$

where the integration is taken over the whole surface S_0 bounding the volume V_0.

Next, if the fluid is flowing out across S_0 then there is a decrease per unit time in the mass of fluid flowing out of V_0, and conversely, if the fluid is flowing into V_0 across S. The decrease per unit time in mass flow across V_0 can be written as

$$-\frac{\partial}{\partial t}\rho\, dV.$$

Equating the two expressions yields

$$\frac{\partial}{\partial t}\int_{V_0}\rho\, dV = -\int_{S_0}\rho\mathbf{v}\cdot d\mathbf{S}. \tag{5.1}$$

The divergence theorem of Gauss transforms the surface integral to a volume integral:

$$\int_{S_0}\rho\mathbf{v}\cdot d\mathbf{S} = \int_V \operatorname{div}(\rho\mathbf{v})\, dV.$$

The term ρv is called the *mass flux density*. It represents the mass of fluid flowing across a unit surface area per unit time. Its dimensions are M/L^2T (in obvious notation). The above representation of the divergence theorem tells us that the surface integral of the mass flux normal to S_0 is equal to the divergence of the mass flux out of V_0. Then (5.1) becomes

$$\int_{V_0}\left[\frac{\partial\rho}{\partial t}+\operatorname{div}(\rho\mathbf{v})\right] dV = 0.$$

Since this equation must hold for any volume however small, the integrand must vanish yielding

$$\rho_t + \operatorname{div}(\rho\mathbf{v}) = 0. \tag{5.2}$$

Equation (5.2) is the *continuity equation*. It expresses the law of conservation of mass. It is clear from the nature of the divergence operator that $\operatorname{div}(\rho\mathbf{v}) = (\rho u)_x + (\rho v)_y + (\rho w)_z$ where $\mathbf{v} = (u, v, w)$. Expanding $\operatorname{div}(\rho\mathbf{v})$, we may also write (5.2) as

$$\rho_t + \operatorname{div}\rho\mathbf{v} + \mathbf{v}\cdot\mathbf{grad}\,\rho = 0, \tag{5.2$'$}$$

where $\mathbf{grad}\,\rho = \mathbf{i}\rho_x + \mathbf{j}\rho_y + \mathbf{k}\rho_z$, $(\mathbf{i}, \mathbf{j}, \mathbf{k})$ are unit vectors in the (x, y, z) directions.

Conservation of Momentum, Euler's Equation

By momentum here we mean linear momentum (mass time particle velocity). The conservation of angular momentum is automatically satisfied since we assume an irrotational fluid. We know that the law of conservation of momentum as expressed by Newton's second law states that the time rate of change of the linear momentum of a body is equal to the resultant external force acting on the body. This is the vector equation of motion of a rigid body. The extension of this law to an inviscid fluid was derived by L. Euler in 1755, and is therefore called *Euler's equation of motion.* We now derive this vector equation of motion. Consider again our volume V_0 containing an inviscid fluid and bounded by the surface S_0. We first assume that V_0 is acted on by a pressure force which is a compressive stress p (the pressure) acting inward on S_0 in the normal direction. Therefore the force due to p acting on V_0 is

$$-\int_{S_0} p\, d\mathbf{S},$$

the integral being taken around the closed surface S_0. Transforming this integral to a volume integral we obtain

$$-\int_{S_0} p\, d\mathbf{S} = -\int_{V_0} (\mathbf{grad}\; p)\, dV.$$

This equation tells us that the fluid surrounding any volume element dV exerts a force equal to $-(\mathbf{grad}\; p)\, dV$ on that element. This means that a force $-\mathbf{grad}\; p$ acts on a unit volume of fluid. We now write down the equation of motion of a volume element by equating this expression for the force per volume element to the density ρ (mass per volume element) times the particle acceleration $d\mathbf{v}/dt$.

$$\rho \frac{d\mathbf{v}}{dt} = -\mathbf{grad}\; p. \tag{5.3}$$

Equation (5.3) is Euler's equation of motion. The assumption is that the only force acting on the volume element of inviscid fluid is due to the pressure.

We can extend (5.3) to account for external force other than that due to p. Let the resultant of these forces per unit mass be $\mathbf{F} = \mathbf{i}X + \mathbf{j}Y + \mathbf{k}Z$, where (X, Y, Z) are the (x, y, z) components of $\mathbf{F}$. Then the generalization of (5.3) is

$$\frac{d\mathbf{v}}{dt} = -\left(\frac{1}{\rho}\right) \mathbf{grad}\; p + \mathbf{F}. \tag{5.4}$$

Written out in extended form (5.4) becomes

$$\begin{aligned} \frac{du}{dt} &= -\frac{p_x}{\rho} + X, \\ \frac{dv}{dt} &= -\frac{p_y}{\rho} + Y, \\ \frac{dw}{dt} &= -\frac{p_z}{\rho} + Z. \end{aligned} \tag{5.4'}$$

Kinematics

We now discuss the physical meaning of the particle acceleration $d\mathbf{v}/dt$ independent of the system of forces that produce this acceleration. It is clear that two time integrations of $d\mathbf{v}/dt$ yields the trajectory of each fluid particle as a function of the Euler coordinates $\mathbf{x}$. The branch of mechanics that treats these matters, independent of the system of forces, is called kinematics. Let us take a closer look at the meaning of $d\mathbf{v}/dt$ from the kinematic point of view. We assume that $\mathbf{v}$ or (u, v, w) are differentiable functions of $\mathbf{x}=(x, y, z)$ and t and that the first partial derivatives $u_t, \ldots, w_t, u_x, \ldots, w_z$ are continuous. Exceptional cases such as vortex motion and shock wave propagation will be treated separately. In $\mathbf{x}$ space consider any point P: (x, y, z). Imagine a spherical surface of infinitesimal radius surrounding P. As P moves with the fluid the volume enveloped by this surface contains the same fluid particles. Associated with this infinitesimal volume are all the dynamical and thermodynamic variables which are assumed to be constant at a particular time. Let $f(x, y, z, t)$ stand for any of these variables. Then f_P depends only on t. Consider a neighboring point P': $(x+u\,dt, y+v\,dt, z+w\,dt)$. We now calculate the rate at which f varies for a moving particle. We assume that f_P goes to $f_{P'}$ in a time dt. Therefore

$$f_{P'}=f(x+u\,dt, y+v\,dt, z+w\,dt)=f_P+u\,dt\,f_x+v\,dt\,f_y+w\,dt\,f_z.$$

The differential operator d/dt as used here is sometimes called the *Stokesian derivative* after Sir G.G. Stokes. When operating on f it denotes the total derivative of f *following the motion of the fluid.* The above expression then becomes (in the limit as P' approaches P):

$$\frac{f_{P'}-f_P}{dt}=\frac{df}{dt}=f_t+uf_x+vf_y+wf_z=f_t+\mathbf{v}\cdot\mathbf{grad}\,f. \tag{5.5}$$

The term f_t represents the time rate of change of f where f is fixed at P. The term $\mathbf{v}\cdot\mathbf{grad}\,f$ represents the spatial derivative of f (as f goes from P to P') at a fixed time.

One of the dynamical variables associated with the spherical element of fluid surrounding P is the particle velocity. Now f can stand for a scalar or vector quantity. Therefore, if we set $f=\mathbf{v}$ then (5.5) becomes

$$\frac{d\mathbf{v}}{dt}=\mathbf{v}_t+(\mathbf{v}\cdot\mathbf{grad})\mathbf{v},$$

so that Euler's equation with an external force per unit mass $\mathbf{F}$ becomes

$$\frac{d\mathbf{v}}{dt}=\mathbf{v}_t+(\mathbf{v}\cdot\mathbf{grad})\mathbf{v}=-\left(\frac{1}{\rho}\right)\mathbf{grad}\,p+\mathbf{F}. \tag{5.6}$$

The second term of the left-hand side of (5.6) is called the *convective* term since it represents that part of the acceleration that is spatially dependent.

This term involves **grad(v)** which is a tensor of rank two (meaning that each element of the expression can be defined by a double index in tensor notation) defining the velocity component and the direction of the differentiation. The scalar product of this tensor with **v**, i.e., (**v** · **grad**)**v**, yields a vector. This is easily visualized by casting (5.6) into the matrix equation

$$\mathbf{u}_t + \mathbf{M}\mathbf{u} = -\left(\frac{1}{\rho}\right) p_x \mathbf{I} + \mathbf{F}, \tag{5.7}$$

where

$$\mathbf{u} = \begin{pmatrix} u_t \\ v_t \\ w_t \end{pmatrix}, \quad \mathbf{M} = \begin{pmatrix} u_x & u_y & u_z \\ v_x & v_y & v_z \\ w_x & w_y & w_z \end{pmatrix}, \quad \mathbf{F} = \begin{pmatrix} X \\ Y \\ Z \end{pmatrix}, \quad \mathbf{I} = \begin{pmatrix} 1 & 0 & 0 \\ 0 & 1 & 0 \\ 0 & 0 & 1 \end{pmatrix}. \tag{5.8}$$

Writing out (5.7) in extended form gives

$$\begin{aligned} u_t + uu_x + vu_y + wu_z &= -\left(\frac{1}{\rho}\right) p_x + X, \\ v_t + uv_x + vv_y + wv_z &= -\left(\frac{1}{\rho}\right) p_y + Y, \\ w_t + uw_x + vw_y + ww_z &= -\left(\frac{1}{\rho}\right) p_z + Z. \end{aligned} \tag{5.9}$$

We now give another derivation of Euler's equations given by (5.9), which is instructive because of the geometric nature of the proof. This is essentially the derivation given by Sir H. Lamb [25, pp. 3 ff.]. Let the point P: (x, y, z) be the center of an elementary parallelepiped whose edges are dx, dy, dz. The time rate of increase of the x component of momentum of the fluid particle in the parallelepiped is $\rho\, dx\, dy\, dz(du/dt)$. This expression must equal the x component of the resultant force acting on the parallelepiped. Let the "pressure force" p act on this area in the x direction. Now consider two elementary areas I and II parallel to the (y, z) plane at $x - dx/2$ and $x + dx/2$, respectively. Expanding p in a Taylor series about P, we have for the pressure on II ultimately: $-p + p_x(dx/2)$, and the pressure on I becomes $-p - p_x(dx/2)$, where we neglected second-order terms. The net x component of the resultant force due to p is then equated to the x component of the particle acceleration yielding

$$dx\, dy\, dz\left(\frac{du}{dt}\right) = -\left(\frac{1}{\rho}\right) p_x\, dx\, dy\, dz + X\, dx\, dy\, dz,$$

where we have added the x component of the external force $X\, dx\, dy\, dz$ per unit density. Performing the same analysis on the other two pairs of parallel faces, we thereby obtain (5.9).

For the case of a body of water acted upon by gravity, we take a coordinate system where z is vertically upward so that the components of $\mathbf{F}$ are $(0, 0, -g)$. These values for X, Y, Z must be inserted in (5.9) when treating water waves, since gravity is assumed to be the only external force.

Since water is assumed to be an incompressible fluid the density ρ can be taken as a known constant ρ_0 so that the continuity equation (5.2) reduces to

$$\operatorname{div} \mathbf{v} = u_x + v_y + w_z = 0. \tag{5.10}$$

Since $\rho = \rho_0$ is assumed to be known there are four unknown variables, namely, u, v, w, p, and the three equations of motion (5.9) and the continuity equation (5.10). These equations are sufficient to solve for $\mathbf{v}$ and p as functions of $\mathbf{x}$ and t once appropriate ICs and BCs are imposed. The equations of motion (5.9) form a set of nonlinear PDEs. Note that in this case the energy equation is not used explicitly.

If we want to consider the slight compressibility of water in connection with wave propagation on the surface we need to consider an adiabatic equation of state of the form $p/p_0 = (\rho/\rho_0)^\gamma$ where $\gamma = C_P/C_V = 7$ (instead of 1.4 for air). C_P and C_V are the heat capacities at constant pressure and volume, respectively. It was shown experimentally that for wave propagation in fluids the temperature changes are rapid enough to give an adiabatic condition (no heat exchange with the environment) during the thermodynamic process.

Conservation of Circulation, Kelvin's Theorem

The notion of circulation in a fluid is defined as follows: Let us consider a closed curve C which moves with the fluid. This means C always contains the same fluid particles. The circulation Γ around C is defined by the line integral

$$\Gamma = \oint \mathbf{v} \cdot d\mathbf{s} = \oint [u\,dx + v\,dy + w\,dz], \tag{5.11}$$

where $d\mathbf{s}$ is the element of arc length tangent to c and the integral is taken around C.

We shall now investigate how Γ behaves with time, i.e., we study $(d/dt) \oint \mathbf{v} \cdot d\mathbf{s}$. To avoid confusion here, we shall temporarily denote differentiation with respect to the coordinates by the symbol δ, retaining d for differentiation with respect to time. Next, we observe that the element of arc length $d\mathbf{s}$ of C can be written as the difference $\delta\mathbf{r}$ between the radius vectors $\mathbf{r}$ at the ends of the element. This means that we can write the circulation as $\oint \mathbf{v} \cdot \delta\mathbf{r}$. Since the contour C is moving with time, we note that in differentiating the integral with respect to time, we must take into account this changing contour. This means that, on taking the time derivative under the integral sign we must not only differentiate $\mathbf{v}$ but also $\delta\mathbf{r}$. We get

$$\frac{d}{dt}\oint \mathbf{v}\cdot\delta\mathbf{r}=\oint\frac{d\mathbf{v}}{dt}\cdot\delta\mathbf{r}+\oint\mathbf{v}\cdot d\frac{\delta\mathbf{r}}{dt}.$$

Since $\mathbf{v}=d\mathbf{r}/dt$, we have

$$\mathbf{v}\cdot d\frac{\delta\mathbf{r}}{dt}=\mathbf{v}\cdot\delta\frac{d\mathbf{r}}{dt}=\mathbf{v}\cdot\delta\mathbf{v}=\delta(\tfrac{1}{2}v^2).$$

Since the integral of a total differential along the closed contour c vanishes, we obtain

$$\frac{d}{dt}\oint\mathbf{v}\cdot\delta\mathbf{r}=\oint\frac{d\mathbf{v}}{dt}\cdot\delta\mathbf{r}.$$

We wish to show that the integral of the right-hand side vanishes for the case where the force on the system is due to p and g. Then the equation of motion (5.4) for an incompressible fluid can be written as

$$\frac{d\mathbf{v}}{dt}=-\left(\frac{1}{\rho_0}\right)\mathbf{grad}\,p+(0,0,-g).$$

This gives

$$\oint\frac{d\mathbf{v}}{dt}\cdot\delta\mathbf{r}=\int\int_{S_0}\mathbf{curl}\left(\frac{d\mathbf{v}}{dt}\right)\cdot\delta\mathbf{S}=\int\int_{S_0}\mathbf{curl}\left(-\left(\frac{1}{\rho_0}\right)\mathbf{grad}\,p-\mathbf{F}\right)=0,$$

upon using Stokes's theorem. The surface integral of the right-hand side is taken around the closed surface S_0 having C as its boundary, and we have used the fact that $\mathbf{curl}(\mathbf{grad}\,p)=0$. It is clear that $\mathbf{curl}(\mathbf{F})=0$ since $\mathbf{F}=(0,0,-g)$. Gathering our results, we have shown that $(d/dt)\oint\mathbf{v}\cdot d\mathbf{s}=(d/dt)\Gamma=0$ which means

$$\oint\mathbf{v}\cdot d\mathbf{s}=\Gamma=\text{const.},$$

for the case where the force on our fluid system is proportional to **grad** p. This statement about the constancy of the circulation around a closed curve C bounding the region of fluid where the force on the system is due to p is called *Kelvin's theorem* or the *law of conservation of circulation*. The key element in the proof is that the force on the system is due to p which was written as **grad** p. We may generalize and state that this conservation law holds for any external force $\mathbf{F}$ that is *derivable from a potential function*. This means that $\mathbf{F}=\mathbf{grad}\,\Phi$, where Φ is a potential function. For in this case $\mathbf{curl}(\mathbf{F})=\mathbf{curl}(\mathbf{grad}\,\Phi)=0$. It should be noted that this result was obtained for the pressure force being proportional to **grad** p, which means that we have assumed isentropic flow, i.e., a unique relationship between p and ρ where the entropy is constant. Also, this law of conservation of circulation does not hold where the external force $\mathbf{F}$ is not derivable from a potential function. If the flow is *irrotational*, by definition we have **curl** $\mathbf{v} = 0$ so that $\mathbf{v}$ is derivable from a potential (the velocity potential) which we shall explore in detail below. Therefore, for an irrotational flow field the circulation around any closed curve bounding the surface is constant. In other words, the conservation of circulation holds for an irrotational flow field. We shall now explore the important consequences of such a flow.

5.2. Potential Flow†

By potential flow we mean a flow field such that the particle velocity $\mathbf{v}$ is the gradient of the *velocity potential* ϕ,

$$\mathbf{v} = \mathbf{grad}\ \phi, \qquad \text{or} \qquad u = \phi_x, \qquad v = \phi_y, \qquad w = \phi_z. \tag{5.12}$$

If we define the *vorticity* $\boldsymbol{\omega}$ by $\boldsymbol{\omega} = \mathbf{curl}\ \mathbf{v}$, then potential flow means that $\boldsymbol{\omega} = 0$, since $\boldsymbol{\omega} = \mathbf{curl}(\mathbf{grad}\ \phi) = 0$. By using a well-known formula in vector analysis, namely

$$(\mathbf{v} \cdot \mathbf{grad})\mathbf{v} = \tfrac{1}{2}\,\mathbf{grad}\ v^2 - \mathbf{v} x \boldsymbol{\omega}, \tag{5.13}$$

(5.6) becomes

$$\mathbf{v}_t + \tfrac{1}{2}\,\mathbf{grad}\ v^2 = -\left(\frac{1}{\rho_0}\right)\mathbf{grad}\ p + \mathbf{F}, \tag{5.14}$$

for an irrotational flow field, since $\boldsymbol{\omega} = 0$.

Streamlines and Bernoulli's Equation

Euler's equation is simplified for the case of a *steady state flow field*, which is defined as one in which $\mathbf{v}$ and p are independent of t so that the velocity and pressure distribution in space are the same at all times. Then (5.14) becomes

$$\tfrac{1}{2}\,\mathbf{grad}\ v^2 = -\left(\frac{1}{\rho_0}\right)\mathbf{grad}\ p + \mathbf{F}, \tag{5.15}$$

which is Euler's equation of motion for a steady state, incompressible, irrotational flow field.

We now introduce the concept of *streamlines*. These are defined as a family of curves in space such that $\mathbf{v}$ is tangent to each curve. In other words, a streamline is a curve which has everywhere the direction of the velocity field $\mathbf{v}$. For steady state flow the streamlines are independent of time. In any case, the streamlines are determined by solving the following

† See also Chapter 7, Section 10 on potential flow.

system of ODEs:

$$\frac{dx}{u} = \frac{dy}{v} = \frac{dz}{w}. \tag{5.16}$$

It is clear that for steady state flow these equations are independent of time. If we have solved a particular steady state problem then we know $\mathbf{v} = \mathbf{v}(\mathbf{x})$, so that the solution of (5.16) yields the family of streamlines. In steady state flow the streamlines (since they do not vary with time) coincide with the trajectories of the field particles—using the definition of streamlines. However, for nonsteady flow this coincidence no longer occurs: the tangents to the streamlines give the directions of the motion of particles at various points in space at a given time, whereas the tangents to the particle trajectories give the directions of the velocities of the fluid particles at various times.

We continue with our assumption of steady state flow, and suppose that we have a family of *equipotential surfaces* meaning that there exists a family of surfaces in space such that the velocity potential $\phi = \phi_i$, where ϕ_i is a constant defining a particular surface. Then on each surface we have

$$d\phi = \phi_x\,dx + \phi_y\,dy + \phi_z\,dz = u\,dx + v\,dy + w\,dz = \mathbf{v} \cdot d\mathbf{t} = 0, \tag{5.17}$$

where $\mathbf{t}$ is a unit vector whose direction cosines are proportional to (dx, dy, dz), so that $\mathbf{t}$ lies in the tangent plane of the surface $\phi_i = \text{const.}$ Then (5.17) tells us that $\mathbf{v}$ is normal to this surface. This important result can be expressed as: *The streamlines are everywhere orthogonal to the family of equipotential surfaces.*

Let $\mathbf{s}$ be a unit vector tangent to a streamline at any point in space. We now consider (5.15) where $\mathbf{F} = (0, 0, -g)$. We now form the scalar product of $\mathbf{s}$ with (5.15). This means that we project each term of (5.15) in the direction of $\mathbf{s}$. We therefore obtain from (5.15) the following:

$$\frac{\partial}{\partial s}\left[\tfrac{1}{2}v^2 + \left(\frac{1}{\rho_0}\right)\mathbf{grad}\,p + gz\right] = 0, \tag{5.18}$$

where the direction of z is vertically upward, and we used the fact that the cosine of the angle between the directions of $-\mathbf{k}g$ and $\mathbf{s}$ is equal to $-dz/ds$, so that the projection of $-\mathbf{k}g$ on $\mathbf{s}$ is $-g\,dz/ds$. It follows that the term in the brackets of (5.18) is constant along a streamline, or

$$\tfrac{1}{2}v^2 + \left(\frac{1}{\rho_0}\right)\mathbf{grad}\,p + gz = \text{const.} \tag{5.19}$$

This is *Bernoulli's equation* for the case where the external force is due to gravity. It tells us that the expression on the left-hand side of (5.19) is constant on each streamline, but may be different along different streamlines. Bernoulli's equation is an energy equation. It expresses the *conservation of energy* for an inviscid fluid since there is no dissipation of energy for such a fluid. To see this, we multiply (5.19) by ρ_0. Then the term $\rho_0 v^2/2$ is the kinetic energy of a unit volume of fluid, **grad** p represents the potential energy of a unit volume of fluid or the work done by the pressure forces, and $\rho_0 gz$ the work due to gravity. Equation (5.19) tells us that the sum of the kinetic and potential energies is a constant along a streamline. Therefore, we can say that each streamline in a steady, irrotational, incompressible, inviscid flow "conserves energy".

EXAMPLE. At this stage it may be appropriate to give the following example of such a flow for the two-dimensional case: Let $\phi = \frac{1}{2}(x^2 - y^2) = \phi_i$ be the family of equipotential surfaces. Then $u = x$, $v = -y$, and $\nabla^2\phi = 0$ where ∇^2 is the two-dimensional Laplacian. The ODEs of the streamlines (5.16) yield $dy/dx = v/u = -y/x$, whose solution is the family of hyperbolas $xy = \text{const.}$ which are the streamlines. This family is orthogonal to the given family of equipotential surfaces.

We *summarize* the most important results which stem, in part, from vector analysis. For an inviscid, isentropic, incompressible flow p is a function of ρ, and the pressure force is proportional to **grad** p. It follows that, for steady state flow, p is a *harmonic function* meaning that p satisfies Laplace's equation $\nabla^2 p = 0$ where ∇^2 is the Laplacian in three dimensions. The conservation of circulation or Kelvin's theorem tells us that the circulation of velocity around *any* closed contour in the fluid is zero which means that, if the vorticity is zero at one time in the flow field, it is always zero so that the flow is always irrotational. From vector analysis we know that for an irrotational flow field the particle velocity is derivable from the velocity potential or $\mathbf{v} = \mathbf{grad}\ \phi$. In other words, if $\mathbf{curl\ v = 0}$ *then* $\mathbf{v} = \mathbf{grad}\ \phi$. This means that ϕ is also an harmonic function for steady state flow. Also, from the conservation of circulation we deduce that the velocity potential persists in time so that potential flow exists for all subsequent times. Bernoulli's equation, which expresses the conservation of energy holds along each streamline where the term gz is added to account for the gravitational field.

5.3. Two-Dimensional Flow, Complex Variables

We now consider a two-dimensional steady state, inviscid, incompressible, irrotational flow field. For this case we use the powerful methods in the theory of functions of a complex variable. The reader is assumed to have some knowledge of complex variable theory.†

† There is, of course, a vast body of literature on this subject; [7, Chaps 2 and 9] is adequate to refresh the reader's knowledge.

In the notation of complex variables the particle velocity $\mathbf{v}$ for two-dimensional potential flow can be written as

$$\mathbf{v} = u + iv = \phi_x + i\phi_y, \qquad i = \sqrt{-1}.$$

Letting $z = x + iy$ we define $f(z)$ as an *analytic function* of z, where

$$f(z) = \phi(x, y) + i\psi(x, y). \tag{5.20}$$

Thus we see that the velocity potential $\phi(x, y)$ is the real part of $f(z)$ and the *stream function* $\psi(x, y)$ is the imaginary part. $f(z)$ is called the *complex potential.* We shall see that there is an important relation between the velocity potential and the stream function, which arises from the definition of the analyticity of $f(z)$. When we say that $f(z)$ is an analytic function of z we mean that $df(z)/dz$ exists and is finite everywhere in the region of analyticity. We differentiate with respect to the complex variable z. In order for df/dz to be unique at a given point in the z plane, this derivative must be invariant to the direction of differentiation. This means that $\partial f/\partial x = -i\,\partial f/\partial y$, upon using $z = x + iy$. We get

$$\frac{df}{dz} = \left(\frac{\partial}{\partial x}\right)(\phi + i\psi) = \phi_x + i\psi_y = -i\left(\frac{\partial}{\partial y}\right)(\phi + i\psi) = -i\phi_y + \psi_y.$$

Equating real and imaginary parts we get

$$\phi_x = \psi_y, \qquad \phi_y = -\psi_x. \tag{5.21}$$

These are the *Cauchy–Riemann* equations. From them it is easily seen that ϕ and ψ are harmonic functions, or

$$\nabla^2\phi = \phi_{xx} + \phi_{yy} = 0, \qquad \nabla^2\psi = \psi_{xx} + \psi_{yy} = 0.$$

The Cauchy–Riemann equations also tells us that curves of constant $\phi(x, y)$ are orthogonal to curves of constant $\psi(x, y)$. This is easily seen as follows: Suppose we have the specific functions $\phi = \phi(x, y)$ and $\psi = \psi(x, y)$. Let $\phi(x, y) = \phi_i$ and $\psi(x, y) = \psi_j$, where ϕ_i and ψ_j are constants that depend on i, j, respectively. Then

$$d\phi = \phi_x\,dx + \phi_y\,dy = 0, \qquad d\psi = \psi_x\,dx + \psi_y\,dy = 0,$$

along the respective curves $\phi = \phi_i$ and $\psi = \psi_j$. This means that $dy/dx = -\phi_x/\phi_y$ is the slope of a curve of constant ϕ_i, and the slope of the curve of constant ψ_j is $dy/dx = -\psi_x/\psi_y$. Using the Cauchy–Riemann equations (5.21) we easily see that the curve of constant ϕ_i is orthogonal to the curve of constant ψ_j. Therefore the family of curves of constant velocity potentials is orthogonal to the family of curves of constant stream functions. We saw above that the streamlines ψ cut the curves of constant velocity potential ϕ orthogonally. Indeed, the purpose of defining the complex potential $f(z)$, such that the real part is the velocity potential and the imaginary part the stream function, is to take advantage of the Cauchy–Riemann equations which tell us that the family of curves of constant real part of $f(z)$ cut the family of curves of constant imaginary part orthogonally. In the language of function theory we say that the real and imaginary parts of $f(z)$ are *conjugate harmonic functions* since they enjoy this orthogonality property and both satisfy the two-dimensional Laplace equation, providing $f(z)$ is analytic.

EXAMPLE. Suppose $f(z) = z^2$. $f(z)$ is analytic for all finite values of z. We have

$$f(z) = x^2 - y^2 + 2ixy,$$

which means $\phi(x, y) = x^2 - y^2$, $\psi(x, y) = 2xy$, and we have the example given on page 118 where we showed that curves of constant ϕ and ψ are orthogonal hyperbolas.

Another property of $f(z)$ is seen from its definition (5.20) and (5.21). We have

$$\frac{df}{dz} = f'(z) = \phi_x + i\phi_x = \phi_x - i\psi_y = u - iv.$$

From this we get the result that

$$\bar{f}' = \mathbf{v}, \tag{5.22}$$

where $\bar{f}' = \overline{df/dz}$ is defined as the complex conjugate of $\bar{f}'$. From the example $f(z) = z^2$ we have $\bar{f}' = 2\bar{z} = 2(x - iy)$, yielding $u = 2x$, $v = -2y$. Since $\phi_x = u$, $\phi_y = v$, and $\phi = x^2 - y^2$ we get $\mathbf{v} = 2x - 2iy$ which verifies the result.

We now give a geometric interpretation of the stream function $\psi(x, y)$. Suppose we connect two points in the z plane P_0: (x_0, y_0) and P: (x, y) by any curve C of finite length. Since $d\psi = \psi_x\, dx + \psi_y\, dy$ we have (upon setting $\psi(x_0, y_0) = 0$) the line integral

$$\begin{aligned} \psi(x, y) &= \int_{P_0}^{P} [\psi_x\, dx + \psi_y\, dy] = \int_{P_0}^{P} [-\phi_y\, dx + \phi_x\, dy] \\ &= \int_{P_0}^{P} [-v\, dx + u\, dy], \end{aligned} \tag{5.23}$$

where the path of integration is any curve C connecting P_0 and P. The integrand $-v\, dx + u\, dy = \mathbf{v} \cdot \mathbf{n}\, ds$ where ds is the element of arc length of C and $\mathbf{n}$ is the unit normal to C. This is seen as follows: This integrand is the scalar product of $\mathbf{v} = u + iv$ and $-i\, dz = dy - i\, dx$ (which is an elementary vector of length ds obtained by rotating the tangent vector dz through $-90°$). The integrand is therefore the product of $\mathbf{v}$ and $-i\, dz$ and the cosine of the angle between these vectors, which yields $\mathbf{v} \cdot \mathbf{n}\, ds$. We can therefore write (5.23) as

$$\psi(x, y) = \int_{P_0}^{P} \mathbf{v} \cdot \mathbf{n}\, ds.$$

The physical interpretation of this expression is that ψ represents the time rate of flow of the fluid across any curve C connecting P_0 and P. We may further interpret this as the rate of flow per unit volume across a cylinder of unit height standing perpendicularly to the xy plane on the curve C.

Flow Around a Cylinder

Let a long circular cylinder of unit radius be placed in a body of water whose free-stream velocity is $\mathbf{v}_\infty = (U, 0)$. The free stream velocity is the constant velocity far enough away from the cylinder so as not to be affected by it. Let the axis of the cylinder be perpendicular to the complex or z plane, so that we consider a cross section in the z plane, and take the origin at the center. Then we are interested in flow in the z plane around the circular disk whose boundary is $x^2+y^2=1$. We want to determine the two-dimensional flow field in the neighborhood of the disk by finding the streamlines. The flow does not penetrate the disk and there is the boundary condition (BC) of slip on the boundary (unit circle) which is the streamline $\psi=0$. First we observe that the complex potential for the free stream velocity is $f(z) = U\ \mathrm{Re}[z]$ since $\phi_x = U$ and $\phi_y = 0$.

The flow field is symmetric about the y axis so that we need only consider flow in the upper half-plane ($y>0$). We look for a mapping that will yield approximately $f(z) = Uz$ for very large z and will have the unit circle C and the part of the x axis exterior to C to be the streamline $\psi=0$ so that $f(z)=\phi$ is real on these boundaries. We therefore consider the complex potential $f(z)$ defined by

$$f(z) = \phi + i\psi = U\left(z + \frac{1}{z}\right). \tag{5.24}$$

Letting $z = re^{i\theta}$ we see that for r large $f(z) = Uz$ approximately, and $f(z) = \phi$ for $z\bar{z} = 1$ and for $y=0$. The mapping function $w = z + 1/z = x(1+1/r^2) + iy(1-1/r^2)$ so that

$$\phi = Ux\left(1+\frac{1}{r^2}\right) = U\left(r+\frac{1}{r}\right)\cos\theta, \qquad \psi = Uy\left(1-\frac{1}{r^2}\right) = U\left(r-\frac{1}{r}\right)\sin\theta, \tag{5.25}$$

which also shows that $\psi=0$ for $r=1$ (on C) and for $y=0$. Obviously, for flow we must have $r>1$ so that $f(z)$ defined by (5.24) maps the region of flow in the z plane to the upper half-plane. The interior of the upper semicircle $r<1$ maps into the lower half-plane $\mathrm{Im}(w)<0$.

From (5.24) we have $f'(z) = U(1-1/z^2)$ so that $\mathbf{v} = U(1-1/z^2)$ which yields

$$\mathbf{v} = U\left(1 - \frac{z^2}{r^4}\right) = U[1 - r^{-4}(x^2 - y^2 + 2ixy)] = u + iv,$$

so that

$$u = U[1 - r^{-4}(x^2 - y^2)], \qquad v = -2Ur^{-4}xy. \tag{5.26}$$

The velocity components given by (5.26) can also be obtained by calculating ϕ_x and ϕ_y from (5.25). The streamlines can be plotted from (5.25) which shows us that they are symmetrical with respect to the y axis, for changing the sign of x does not alter the equation for ψ. Also the equation shows $\psi(x, y) = -\psi(x, -y)$, which means that the streamlines in the lower half-plane are the reflections of those in the upper half-plane. Thus we have "fore-and-aft" and "up-and-down" symmetry, so that the same streamline pattern is obtained if U is replaced by $-U$ (the free stream flow $\mathbf{v}$ can be directed either to the right or left). From (5.26) we can calculate the velocity on the boundary C; we obtain

$$u = U(1 - \cos 2\theta), \qquad v = -U \sin 2\theta, \qquad \sqrt{\mathbf{v} \cdot \mathbf{v}} = 2U \sin \theta, \qquad \text{for} \quad r = 1.$$

The *stagnation points* are those points on C for which $\mathbf{v} = 0$. Those are the values for which $\theta = 0$ and $\theta = \pi$. Also the speed is greatest at the points where $\theta = \pm\pi/2$; the speed being $2U$ at those points. The pressure distribution can be obtained from Bernoulli's equation in the form

$$p/\rho_0 + \tfrac{1}{2}\mathbf{v} \cdot \mathbf{v} = \text{const.},$$

where we neglect external forces.

It will be shown that there is no drag and lift on the cylinder in potential flow, so that the pressure forces exerted on the body by the fluid would balance out (a result known as "D'Alembert's paradox"). This will be described in the next section where we discuss the drag force in potential flow past a body in a more general setting.

5.4. The Drag Force Past a Body in Potential Flow

We now consider three-dimensional potential flow past a solid body. By supplying a negative free stream velocity, and thereby changing to a coordinate system in which the fluid is at rest at infinity, it is clear that the problem is equivalent to that of the motion of a fluid when the same body moves through it. In what follows we shall consider the case where the body is moving through the fluid with a velocity equal to the negative of the free stream velocity.

We now determine the velocity of the fluid in a potential flow field due to the moving body. We thereby attempt to find the velocity potential ϕ which satisfies the BC at infinity of zero free stream velocity and slip on the surface of the body. This means we want to find the solution of Laplace's equation $\nabla^2\phi = 0$ in three dimensions that satisfies the appropriate BCs. We take the origin of our coordinate system inside the body (at the center of gravity, for convenience) and shall determine the steady state flow field. As we know, in three dimensions, Laplace's equation has solutions of the form $1/\mathbf{r}$, where $\mathbf{r}$ is the radius vector from the origin to the field point (x, y, z) in the flow field. We also know that linear combinations of the gradient and higher space derivatives of $1/\mathbf{r}$ are also solutions which vanish at infinity (satisfy the BC there). Hence the general form of the required solution of Laplace's equation that satisfies the BC at infinity is

$$\phi = -\frac{a}{r} + \mathbf{A} \cdot \mathbf{grad}\,\frac{1}{r} + \left(\frac{\partial}{\partial \mathbf{r}}\right) \mathbf{A} \cdot \mathbf{grad}\,\frac{1}{r} + \ldots,$$

where a and $\mathbf{A}$ are constants independent of the coordinates. We shall now show that the constant a must be zero. Consider ϕ due to the first term of the right-hand side, $\phi = -a/r$. The velocity due to this term is $\mathbf{v} = -\mathbf{grad}(a/r) = a\mathbf{r}/r^3$. We now calculate the mass flux due to this velocity across some closed surface in the flow field, say a sphere of radius R. On the surface of this sphere the velocity is constant and has the magnitude $v = a/R^2$, so that the total mass flux across this surface is $4\pi R^2 \rho_0 (a/R^2) = 4\pi\rho_0 a$, which must be zero, since the continuity equation for an incompressible fluid is $\operatorname{div} \mathbf{v} = 0$. Hence we conclude that $a = 0$.

Therefore ϕ contains terms of order $1/r^2$; the higher-order terms may be neglected since we are trying to find the velocity at a large distance from the body. Thus we have the approximation

$$\phi = \mathbf{A} \cdot \mathbf{grad}\left(\frac{1}{r}\right) = -\frac{\mathbf{A} \cdot \mathbf{n}}{r^2}, \tag{5.27}$$

where $\mathbf{n}$ is the unit vector in the $\mathbf{r}$ direction. Since $\mathbf{v} = \mathbf{grad}\,\phi$ we have

$$\mathbf{v} = \mathbf{A} \cdot \mathbf{grad}\left[\mathbf{grad}\left(\frac{1}{r}\right)\right] = \frac{3(\mathbf{A} \cdot \mathbf{n})\mathbf{n} - \mathbf{A}}{r^3}. \tag{5.28}$$

This tells us that the velocity is of the order $1/r^3$ at large distances. The vector $\mathbf{A}$ depends on the shape of the body, and can only be determined by completely solving $\nabla^2 \phi = 0$ at all distances, taking into account the BC of slip on the surface of the body.

We now calculate the total kinetic energy T of the fluid. We have $2T = \rho_0 \int v^2\, dV$, where the integration is taken over the volume V_E external to the body. Let $\mathbf{u}$ be the velocity of the body. Then $\mathbf{u}$ is independent of the coordinates. We may write T as a volume integral involving $\mathbf{u} \cdot \mathbf{u}$ and $\mathbf{v} \cdot \mathbf{v}$ by the following identity:

$$2T = \rho_0 \int \mathbf{v} \cdot \mathbf{v}\, dV = \rho_0 \int \mathbf{u} \cdot \mathbf{u}\, dV + \rho_0 \int (\mathbf{v} + \mathbf{u}) \cdot (\mathbf{v} - \mathbf{u})\, dV,$$

where the integration is over V_E. Since $\mathbf{u}$ is independent of $\mathbf{r}$ the first integral of the right-hand side is simply $\rho_0 \mathbf{u} \cdot \mathbf{u} V_E$. Using the fact that $\operatorname{div} \mathbf{v} = 0$ (incompressible fluid) and $\operatorname{div} \mathbf{u} = 0$, the integrand of the second integral may be written as

$$(\mathbf{v}+\mathbf{u}) \cdot (\mathbf{v}-\mathbf{u}) = [\mathbf{grad}(\phi + \mathbf{u} \cdot \mathbf{r})] \cdot [\mathbf{v}-\mathbf{u}] = \operatorname{div}[(\phi + \mathbf{u} \cdot \mathbf{r})(\mathbf{v}-\mathbf{u})].$$

Using these results we get

$$2T = \rho_0 \mathbf{u} \cdot \mathbf{u} V_E + \rho_0 \int \operatorname{div}[(\phi + \mathbf{u} \cdot \mathbf{r})(\mathbf{v}-\mathbf{u})]\, dV.$$

Using the divergence theorem we transform the volume integral over V_E to an integral over the surface S_R of the sphere of radius R and the surface S_B of the body. We get

$$2T = \rho_0 \mathbf{u} \cdot \mathbf{u} V_E + \rho_0 \int (\phi + \mathbf{u} \cdot \mathbf{r})(\mathbf{v}-\mathbf{u}) \cdot d\mathbf{S},$$

where $d\mathbf{S}$ is the element of surface area directed toward the outward normal to the surfaces. On the surface S_B the normal components of $\mathbf{v}$ and $\mathbf{u}$ are equal since there is obviously no flow across S_B. Therefore the surface integral over S_B vanishes identically. On the remote surface S_R we use (5.2) and (5.28) for ϕ and $\mathbf{v}$, and neglect terms which vanish as $R \to \infty$. Writing the surface element on the sphere S_R as $d\mathbf{S} = \mathbf{n}R^2\, d\Omega$ where $d\Omega$ is an element of solid angle, we obtain

$$2T = \rho_0 \mathbf{u} \cdot \mathbf{u}[\tfrac{4}{3}\pi R^3 - V_B] + \rho_0 \int [3(\mathbf{A} \cdot \mathbf{n})(\mathbf{u} \cdot \mathbf{n}) - (\mathbf{u} \cdot \mathbf{n})^2 R^3]\, d\Omega.$$

Upon performing the integration we obtain

$$T = \tfrac{1}{2}\rho_0(4\pi \mathbf{A} \cdot \mathbf{u} - V_B \mathbf{u} \cdot \mathbf{u}), \tag{5.29}$$

where V_B is the volume of the body. Equation (5.29) essentially represents the total energy of the system since all we need do is add the internal energy to T which is a constant for an incompressible inviscid fluid. Even though the specific formula for $\mathbf{A}$ depends on a complete solution of Laplace's equation, as mentioned above, $\mathbf{A}$ depends on $\mathbf{u}$ which is determined from the BC on the surface S_B of slip for an inviscid fluid. Since this BC is linear in $\mathbf{u}$ it follows that $\mathbf{A}$ is a linear function of $\mathbf{u}$. Therefore (5.29) tells us that T is a quadratic function of the components of $\mathbf{u}$. Let $\mathbf{u} = (u_1, u_2, u_3)$. Then T can be written in the form

$$T = \tfrac{1}{2} m_{ij} u_i u_j, \qquad i, j = 1, 2, 3, \tag{5.30}$$

where the tensor summation notation is used (summed over i, j). m_{ij} is the ijth element of some constant symmetric tensor, which can be calculated from $\mathbf{A}$; it is called the *induced mass tensor.*

Knowing T allows us to obtain an expression for the total momentum of the fluid $\mathbf{P}$. To calculate $\mathbf{P}$ we use the differential relationship between $\mathbf{P}$ and T, namely

$$dT = \mathbf{u} \cdot d\mathbf{P}.$$

We note that the reason for using this form rather than the integral form $\mathbf{P} = \rho_0 \int \mathbf{v}\, dV$, is that, in this case, the integration is taken over a very large region; moreover, the integration depends on the volume of the specific body under consideration, and computations thereby get fairly messy. Since T is given by (5.30) it follows that the components of $\mathbf{P}$ must be of the form

$$P_i = m_{ij} u_j \quad (\text{sum over } j). \tag{5.31}$$

Finally, using (5.29) and (5.30) we get the following expression for $\mathbf{P}$ in terms of $\mathbf{A}$ from (5.31):

$$\mathbf{P} = 4\pi\rho_0 \mathbf{A} - \rho_0 V_B \mathbf{u}. \tag{5.32}$$

Equation (5.32) tells us that the total momentum of the fluid is a finite quantity (independent of the volume V_E—as it should be), which depends on the induced mass tensor and the volume and velocity of the body.

Let $\mathbf{F}$ be the resultant of the external force transmitted to the fluid by the body. Then $-\mathbf{F}$ is the reaction of the fluid, i.e., the force acting on the body. This is given by Newton's second law, so that

$$\mathbf{F} = -\frac{d\mathbf{P}}{dt}. \tag{5.33}$$

The component of $\mathbf{F}$ parallel to $\mathbf{u}$ is called the *drag force*, and the component normal to $\mathbf{u}$ directed upward is called the *lift force*. We are now in a position to show that $\mathbf{F} = 0$ for potential flow. For the body moving uniformly in an incompressible inviscid fluid, so that potential flow exists, we have $\mathbf{P} =$ const., which yields $\mathbf{F} = 0$. Clearly this means that, since there is no drag or lift, the pressure forces exerted by the fluid on the body would balance out. This result is called *D'Alembert's paradox*. The physical reason for this paradox is: Suppose we do have a drag force acting on the body in uniform motion. Then, in order to maintain the motion, work must be continuously done by some external mechanism. This work must be either dissipated in the fluid as heat or converted to kinetic energy of the fluid; in any case, the result is that there is a continual flow of energy to infinity in the fluid. However, for potential flow we have an inviscid fluid so that no dissipation of energy occurs; moreover, the velocity of the fluid caused by the moving body diminishes so rapidly with increasing distance that there can be no flow of energy to infinity. In this argument we consider only the case of a body moving in an infinite fluid. Consider the case where the fluid has a free surface (the net force or traction on the surface is zero). Then a body moving uniformly parallel to this surface will experience a drag, so that D'Alembert's paradox does not hold. This drag is called the *wave drag* and is due to the occurrence of a system of waves which propagate on the free surface. These surface waves continually remove energy to infinity.

Oscillatory Motion

Suppose we again consider a body moving in a potential flow field. However, suppose we have an external force **f** which produces oscillatory motion of the body. We calculate the motion of the body by again using Newton's second law which equates the rate of change of total momentum to the external force. However, in this case the velocity **u** of the body is not constant so that the total momentum is the sum of **P** (obtained previously) and the momentum of the body $M\mathbf{u}$ (where M is the mass of the body). The equation of motion is then given by

$$M\frac{d\mathbf{u}}{dt}+\frac{d\mathbf{P}}{dt}=\mathbf{f}.$$

Using (5.31), this equation becomes (in tensor form)

$$\frac{du_j}{dt}(M\delta_{ij}+m_{ij})=f_i, \qquad i,j=1,2,3. \tag{5.34}$$

EXAMPLE. To illustrate the theory we treat the classical case of a sphere of radius R moving with a velocity **u** in an incompressible, inviscid fluid of infinite extent. Therefore the fluid velocity must vanish at infinity. The solution of Laplace's equation in three dimensions which satisfy this BC is given by (5.27). **A** is determined from the BC on the spherical surface which says that $\mathbf{u}\cdot\mathbf{n}=\mathbf{v}\cdot\mathbf{n}$ for $r=R$. Taking the scalar product of (5.28) and equating the result to $\mathbf{u}\cdot\mathbf{n}$ for $r=R$ allows us to solve for **A**, which is $\mathbf{A}=\frac{1}{2}R^3\mathbf{u}$. Using this result gives

$$\phi=-\tfrac{1}{2}\frac{R^3}{r^2}\mathbf{u}\cdot\mathbf{n}, \qquad \mathbf{v}=\frac{1}{2}\left(\frac{R}{r}\right)^3[3\mathbf{n}(\mathbf{u}\cdot\mathbf{n})-\mathbf{u}].$$

The total momentum **P** transmitted to the fluid by the sphere is given by (5.32) and becomes

$$\mathbf{P}=\tfrac{2}{3}\pi\rho_0R^3\mathbf{u},$$

so that the induced mass tensor becomes

$$m_{ij}=\tfrac{2}{3}\pi\rho_0R^3\delta_{ij}.$$

The drag **F** on the moving sphere becomes

$$\mathbf{F}=-\tfrac{2}{3}\pi\rho_0R^3\frac{d\mathbf{u}}{dt}.$$

The equation of motion of the sphere in the oscillating fluid is

$$\tfrac{4}{3}\pi R^3(\tfrac{1}{2}\rho_0+\rho_B)\frac{d\mathbf{u}}{dt}=\mathbf{f},$$

where ρ_B is the density of the sphere. The coefficient of $d\mathbf{u}/dt$ is called the *virtual mass* of the sphere. It consists of the actual mass of the sphere plus the induced mass which is half the mass of the fluid displaced by the sphere.

If the external force $\mathbf{f}$ is oscillatory we may set $\mathbf{f}=\bar{\mathbf{f}}\exp(i\omega t)$ where $\bar{\mathbf{f}}$ is the amplitude and ω is the radial frequency. The equation of motion of the sphere then becomes

$$\mu\frac{d\mathbf{u}}{dt}=\bar{\mathbf{f}}\exp(i\omega t),$$

where μ is the virtual mass of the sphere. Setting $\mathbf{u}=\bar{\mathbf{u}}\exp(i\omega t)$ and inserting this expression into the equation of motion gives the solution for the velocity of the sphere:

$$\mathbf{u}=-\left(\frac{1}{\omega\mu}\right)i\mathbf{f}\exp(i\omega t)=\left(\frac{\bar{\mathbf{f}}}{\omega\mu}\right)\exp\left(i\left(\omega t-\frac{\pi}{2}\right)\right),$$

which shows that there is a 90° phase difference between $\mathbf{f}$ and $\mathbf{u}$.

5.5. Energy Flux

In discussing Bernoulli's equation for steady state potential flow in a gravitational field it was pointed out that this equation is an energy equation expressing the conservation of energy for an inviscid fluid. In our future discussion of gravity waves in water it is important to analyze in some detail the flow of energy in the fluid past a given surface. We start by choosing some volume element fixed in three-dimensional space and determine how the energy of the fluid contained in this volume element varies with time. The energy per unit volume of fluid is

$$\tfrac{1}{2}\rho\mathbf{v}\cdot\mathbf{v}+\rho e=T+\rho e,$$

where T is the kinetic energy and ρe is the internal energy per unit volume. We do not necessarily assume an incompressible fluid so that ρ is a function of space and time. To find out how this energy changes with time we first calculate T_t and obtain

$$T_t=\tfrac{1}{2}(\rho\mathbf{v}\cdot\mathbf{v})_t=\tfrac{1}{2}\mathbf{v}\cdot\mathbf{v}\rho_t+\rho\mathbf{v}\cdot\mathbf{v}_t.$$

Using the continuity equation (5.2) and the equation of motion (5.6), the above equation becomes

$$T_t=-\tfrac{1}{2}\mathbf{v}\cdot\mathbf{v}\,\mathrm{div}(\rho\mathbf{v})-\mathbf{v}\cdot\mathbf{grad}\,p-\rho\mathbf{v}\cdot(\mathbf{v}\cdot\mathbf{grad})\mathbf{v}+\rho\mathbf{v}\cdot\mathbf{F}.$$

In this equation we set

$$\mathbf{v}\cdot(\mathbf{v}\cdot\mathbf{grad})\mathbf{v}=\tfrac{1}{2}\mathbf{v}\cdot\mathbf{grad}(\mathbf{v}\cdot\mathbf{v}),$$

$$\mathbf{grad}\,p=\rho\,\mathbf{grad}\,w-\rho\theta\,\mathbf{grad}\,S,$$

where θ is the temperature, w is the enthalpy, and S the entropy per unit mass, and we have the thermodynamic relation $dw = dS + (1/\rho)\, dp$. Using these relations we get

$$T_t = -\tfrac{1}{2}\mathbf{v}\cdot\mathbf{v}\,\mathrm{div}(\rho\mathbf{v}) - \mathbf{v}\cdot\mathbf{grad}(\tfrac{1}{2}\mathbf{v}\cdot\mathbf{v} + w) + \rho\theta\mathbf{v}\cdot\mathbf{grad}\,S + \rho\mathbf{v}\cdot\mathbf{F}. \quad (5.35)$$

We next work on $(\rho e)_t$. We assume an adiabatic fluid so that $dS/dt = 0$. Expanding this expression we get

$$S_t + \mathbf{v}\cdot\mathbf{grad}\,S = 0.$$

Using this equation and the fact that $w = e + pV = e + p/\rho$, and $de = \theta\, dS - p\, dV$, we get

$$d(\rho e) = e\, d\rho + \rho\, de = w\, d\rho + \rho\theta\, dS,$$

which yields (using the continuity equation)

$$(\rho e)_t = -w\,\mathrm{div}(\rho\mathbf{v}) - \rho\theta\mathbf{v}\cdot\mathbf{grad}\,S. \quad (5.36)$$

Adding (5.35) and (5.36) we find the change in the total energy of the fluid to be

$$[\tfrac{1}{2}\rho v^2 + \rho e]_t = -(\tfrac{1}{2}v^2 + w)\,\mathrm{div}(\rho\mathbf{v}) - \rho\mathbf{v}\cdot\mathbf{grad}(\tfrac{1}{2}v^2 + w) + \rho\mathbf{v}\cdot\mathbf{F}.$$

Combining terms we finally obtain

$$[\tfrac{1}{2}\rho v^2 + \rho e]_t = -\mathrm{div}(\rho\mathbf{v}(\tfrac{1}{2}v^2 + w - gz)), \quad (5.37)$$

For a gravitational field $\mathbf{F}$ is derivable from a potential $-gz$; we have added this term to the scalar w in the divergence term.

In order to see the meaning of (5.37) we integrate it over some volume V_0 enclosed by a surface S_0. By using the divergence theorem we then convert this volume integral over V_0 to a surface integral over S_0. We obtain

$$\begin{aligned}\frac{\partial}{\partial t}\int_{V_0} \rho[\tfrac{1}{2}v^2 + e]\, dV &= -\int \mathrm{div}[\rho\mathbf{v}(\tfrac{1}{2}v^2 + w - gz)]\, dV \\ &= -\int_{S_0} \mathbf{v}(\tfrac{1}{2}v^2 + w - gz)\cdot d\mathbf{S}. \quad (5.38)\end{aligned}$$

Equation (5.38) tells us that the rate of change of the total energy in a volume V_0 is equal to the amount of energy flowing out of this volume across the surface S_0 per unit time. For this reason we call the expression

$$\mathbf{v}(\tfrac{1}{2}v^2 + w - gz), \quad (5.39)$$

the *energy flux density* vector. Its magnitude is the amount of energy passing across a unit surface area perpendicular to the fluid velocity $\mathbf{v}$, per unit time. Putting $w = e + p/\rho$ we may rewrite the right-hand side of (5.38) as

$$-\int_{S_0} \mathbf{v}(\tfrac{1}{2}v^2 + e)\cdot d\mathbf{S} - \int_{S_0} p\mathbf{v}\cdot d\mathbf{S} - \int_{S_0} gz\mathbf{v}\cdot d\mathrm{S}. \quad (5.40)$$

The first term is the energy (kinetic plus internal) transported across S_0 per unit time by the fluid; the second term is the work done by the pressure forces on the fluid within the surface; and the third term is the work done by the gravitational field acting on the system.

5.6. Small Amplitude Gravity Waves

We first consider a body of water in equilibrium in a gravitational field and having a free surface (no tractions acting on the surface). The free surface is then a plane. We take the xy plane parallel to this equilibrium surface and the z axis vertically upward.† Now suppose that some external small amplitude perturbation is imposed at some point on this surface (say moving that point laterally to the surface). Then the surface is no longer in equilibrium, meaning that: From that point motion will occur and propagate over the surface of the water so that waves will be generated. They are called *gravity waves* since they are due to the action of the gravitational field on the water. These gravity waves appear mainly on the surface but also affect the interior, their amplitude decaying exponentially with distance from the surface.

The hypothesis of small amplitude wave generation is really an hypothesis of linearity. The particle velocity is so small that we neglect the nonlinear term (the convective term) $(\mathbf{v} \cdot \mathbf{grad})\mathbf{v}$ in Euler's equation (5.6) so that the particle acceleration is given by the linear term $\mathbf{v}_t$. We learned earlier that linearizing the equation of motion for the case where the forces are derivable from a potential leads to a potential flow field. Let us take a closer look at the consequences of such a linearization by discussing its physical significance. In order to do this we make an order of magnitude analysis of the convective term (nonlinear term) and the linear part of the particle acceleration, and determine under what conditions the nonlinear term is very small in comparison with the linear term. Consider the oscillations of amplitude A of a surface wave. During a time period of oscillation of the order τ the particles making up the wave travel a distance of the order A. The particle velocity is of the order A/τ. Let λ be the wavelength. The velocity gradient which has dimensions $(L/T)(1/L)$ is of the order v/λ while the $\mathbf{v}_t$ part of the acceleration is of the order v/τ. Therefore the condition that $(\mathbf{v}.\mathbf{grad})\mathbf{v} \ll \mathbf{v}_t$ leads to the following inequality condition:

$$\left(\frac{1}{\lambda}\right)\left(\frac{A}{\tau}\right)^2 \ll \left(\frac{A}{\tau}\right)\left(\frac{1}{\tau}\right) \quad \text{or} \quad A \ll \lambda,$$

i.e., the amplitude of the oscillations in the wave must be very small compared to the wavelength.

† The geometry will be described in detail below.

Another way of obtaining this condition is to directly study a traveling wave. Let ξ be the vertical displacement from equilibrium of a water particle on the surface making up a wave. We may set $\xi = A\cos((2\pi/\lambda)(x-ct))$. This expression represents a progressing wave in the x direction traveling with a wave speed c. The particle velocity is $u = \xi_t = -(2\pi/\lambda)cA\sin((2\pi/\lambda)(x-ct))$. The magnitude of u_t is $u_t = A(2\pi/\lambda)^2c^2$ and $u_x = A(2\pi/\lambda)^2c$, so that

$$\frac{|uu_x|}{|u_t|} = \frac{2\pi A}{\lambda},$$

which again shows that $A \ll \lambda$ in order to neglect the nonlinear term. We shall use this approximation and thereby investigate unsteady potential flow fields, since we are interested in wave propagation phenomena.

We now formulate the unsteady linearized or potential flow of an incompressible inviscid fluid in a gravitational field in three-dimensional space. The continuity equation is given by (5.10). The equation of motion is the linearization of (5.6), which becomes

$$\mathbf{v}_t = -\left(\frac{1}{\rho_0}\right)\mathbf{grad}\,p + (0, 0, -g). \tag{5.41}$$

For potential flow we, of course, have the velocity potential ϕ such that

$$\mathbf{v} = \mathbf{grad}\,\phi.$$

However, ϕ is a function of time as well as space since the flow field is unsteady. Since $\mathbf{F} = (0, 0, -g)$ there exists a potential function equal to $-gz$ such that $\mathbf{F} = -\mathbf{grad}(gz)$. Substituting these potentials for $\mathbf{v}$ and $\mathbf{F}$ into (5.41) gives

$$\mathbf{grad}\left(\phi_t + \frac{p}{\rho_0} + gz\right) = 0. \tag{5.42}$$

Integrating (5.42) yields

$$\phi_t + \frac{p}{\rho_0} + gz = f(t), \tag{5.43}$$

where $f(t)$ is an arbitrary function of time. Equation (5.43) is the first integral of the equations of unsteady potential flow, and is the linearized form of Bernoulli's equation for an unsteady flow in a gravitational field. It is easily seen that we may neglect $f(t)$ without any loss of generality. For, if we replace ϕ_t by $\phi_t + f(t)$, $\mathbf{v}$ remains the same and $f(t)$ thereby cancels out. For steady state flow (5.43) reduces to the linearized version of (5.19) where we neglect the nonlinear term $\frac{1}{2}v^2$.

Waves in a Two-Dimensional Uniform Body of Water

The above mathematical formulation of unsteady potential flow in a gravitational field allows us to continue with our program of investigating the properties of surface waves for a body of water. For simplicity we take as our model a two-dimensional body of water in the form of a straight canal of infinite extent with a horizontal floor and parallel vertical sides. To describe quantitatively the propagation of such waves we need a geometric framework. We construct a Cartesian coordinate system measured from an arbitrary origin on the floor, taking this as the xy plane with the x axis directed parallel to the length of the canal, the z axis is directed vertically upward. Let h be the height of the water surface at equilibrium. We take h to be constant which means that the floor is horizontal, as previously mentioned. Let $\zeta = \zeta(x, t)$ be the vertical position of a water particle on the wave surface measured from the equilibrium surface. ζ measures the vertical position of a surface particle with respect to its equilibrium position. Clearly ζ is a function of x, t since we are concerned with a two-dimensional model, which means that at a given time all wave profiles in planes normal to the y axis are the same. It is also clear that when the surface is in equilibrium we have $\zeta(x, t) = 0$. Clearly the height of the free surface, corresponding to x at time t, is denoted by $z = h + \zeta(x, t)$.

In our discussion of gravity waves or gravitational oscillations produced on the surface we shall assume that the motion of the fluid is mainly horizontal (in the x direction) and is the same for all particles in a vertical line, which means that we neglect the vertical component of the particle acceleration (we set $w_t = 0$). As mentioned above, the equation of the water surface is $z = h + \zeta(xt)$. Since the surface is composed of water particles, *the surface moves with the fluid* so that the Stokesian derivative (d/dt) can be applied to z, and we obtain

$$\frac{d}{dt}[\zeta(x, t) - z(x, t)] = 0,$$

which gives

$$\zeta_t + \left(\frac{dx}{dt}\right)\zeta_x = \frac{dz}{dt}. \tag{5.44}$$

We are concerned with the linear theory so that we need only consider first-order terms and thereby neglect products of small terms. The velocity term $dx/dt = u$ and the term ζ_x, which measures the slope of the wave profile, are both assumed small so that the nonlinear product term can be neglected; then the linearized version of (5.44) becomes

$$\zeta_t = \frac{dz}{dt} = w. \tag{5.45}$$

Equation (5.45) is the linearized kinematic BC on the surface. Since the flow field is two dimensional, the particle velocity is given by $\mathbf{v}(x, t) = (u, 0, w)$, where w enjoys the BC on the free surface, given by (5.45). Again we mention that the free surface is one defined by having zero tractions on it.

To get an expression for the pressure we differentiate Bernoulli's equation (5.43) with respect to z and use the condition that $w_t = 0$; since, as mentioned above, we neglecct vertical particle acceleration. We obtain $(1/\rho_0)p_z = -g$. Integrating with respect to z and using the BC that $p = p_0$ on the free surface $z = h + \zeta$, we get

$$p - p_0 = \rho_0 g(h + \zeta(x, t) - z), \tag{5.46}$$

where we interpret p_0 as the constant atmospheric pressure on the free surface. Differentiating equation (5.46) with respect to x yields

$$p_x = \rho_0 g \zeta_x. \tag{5.47}$$

The horizontal component of the linearized equation of motion is

$$\rho_0 u_t = -p_x. \tag{5.48}$$

Relating (5.47) to (5.48) and recognizing that $\zeta = \zeta(x, t)$, we conclude that the horizontal component of the particle acceleration u_t is independent of z, so that it is the same for all particles in a plane perpendicular to x. Again, from (5.47) and (5.48), we obtain

$$u_t = -g\zeta_x. \tag{5.49}$$

We take advantage of the fact that we have potential flow so that $\mathbf{v} = \mathbf{grad}\ \phi$ where $\phi = \phi(x, t)$. In Bernoulli's equation, given by (5.43), we set $\phi_t + p/\rho_0 = \phi'$ which makes no essential difference since $\mathbf{v} = \mathbf{grad}\ \phi = \mathbf{grad}\ \phi'$. Suppressing the superscript, (5.43) becomes the following equation which holds on the free surface where $z = h + \zeta$:

$$\phi_t = -gz.$$

Differentiating this equation with respect to t, and using the BC on the free surface given by (5.45), we get

$$\phi_{tt} = -gw = -g\phi_z \qquad \text{for} \quad z = 0, \tag{5.50}$$

where we translated the origin ($z = 0$) to the free surface to more conveniently handle the case of infinite depth. In this coordinate system it is clear that the floor of the body of water is at $z = -h$. Furthermore, since the oscillations are small we replaced the actual boundary $z = \zeta$ by $z = 0$.

We first consider the case of infinite depth where $h = \infty$. Then the fluid occupies the region $z \leqq 0$. The problem can now be formulated as follows: Find the solution to Laplace's equation for $\phi = \phi(x, z, t)$:

$$\nabla^2 \phi = 0, \tag{5.51}$$

subject to the BC (5.50). We are interested in solutions for $\phi(x, z, t)$ that have wave properties. Suppose we want to find a progressing wave solution traveling along the surface in the positive x direction. Then we may separate variables in the following special way peculiar to this type of progressing wave. We therefore assume a solution for (x, z, t) of the form

$$\phi = f(z) \sin(kx - \omega t), \tag{5.52}$$

where k is the wave number such that $k = 2\pi/\lambda$ (λ is the wavelength and ω is the frequency). We note that $f(z) \cos(kx - \omega t)$ is also a solution; and the general solution is a linear combination of these sinusoidal functions of $(kx - \omega t)$ to give progressing waves which satisfy prescribed ICs as well as the above BC. If, in addition, we want to take into account regressing waves we must add linear combinations of sinusoidal functions of $(kx + \omega t)$ multiplied by $f(z)$. A Fourier series expansion can be obtained if the ICs are periodic functions of x, otherwise Fourier transforms may be used. Substituting the above expression for ϕ into Laplace's equation yields the following ODE for $f(z)$:

$$\frac{d^2 f}{dz^2} - k^2 f = 0, \tag{5.53}$$

which has the solutions e^{kz} and e^{-kz}. We must eliminate the latter solution since ϕ must not increase exponentially as we go into the interior of the water. The velocity potential therefore becomes

$$\phi = A e^{kz} \sin(kx - \omega t). \tag{5.54}$$

Inserting (5.54) into the BC (5.50) yields

$$\omega^2 = kg, \tag{5.55}$$

which is the relation between the frequency and the wave number for a gravity wave traveling on the surface of a body of water of infinite depth.

The particle velocity distribution $\mathbf{v}$ is

$$u = \phi_x = kAe^{kz} \cos(kx - \omega t), \qquad w = \phi_z = kAe^{kz} \sin(kx - \omega t). \tag{5.56}$$

It is clear that $\sqrt{\mathbf{v} \cdot \mathbf{v}} = |\mathbf{v}| = kAe^{kz}$ so that the velocity exponentially decays as we go into the fluid so that on the floor (infinite depth) the velocity is zero, which satisfies the BC on the floor.

We now determine the paths of the fluid particles by integrating (5.56) with respect to time and recalling that $u = dx/dt$ and $w = dz/dt$, where x, z are the coordinates of a moving fluid particle (and not of a point fixed in space). We obtain

$$\begin{aligned} x - x_0 &= \left(\frac{k}{\omega}\right) A e^{kz} \sin(kx - \omega t), \\ z - z_0 &= \left(\frac{k}{\omega}\right) A e^{kz} \cos(kx - \omega t). \end{aligned} \tag{5.57}$$

Equation (5.57) tells us that the paths of the fluid particles are circles of radius $(k/\omega)Ae^{kz}$ about the points (x_0, z_0). It is clear that the radius decreases exponentially with increasing depth.

One-Dimensional Wave Equation for ϕ

It is easily seen that $\phi(x, z, t)$ satisfies the one-dimensional wave equation in x, t. The solution of Laplace's equation, as given by (5.51), shows that the function of z is separated from the functions of x and t which are bound up in the argument $kx - \omega t$. This means that ϕ satisfies the wave equation $c^2\phi_{xx} - \phi_{tt} = 0$ where

$$c^2 = \frac{\omega^2}{k^2}.$$

Upon using (5.55) this expression becomes

$$c = \sqrt{\left(\frac{g}{k}\right)} = \sqrt{\left(\frac{g\lambda}{2\pi}\right)}. \tag{5.58}$$

Equation (5.58) tells us that the wave speed for a gravity wave traveling on the surface of a body of water of infinite depth is proportional to the square root of the wavelength. This will be shown below as an approximation for the case $h/\lambda \gg 1$ in the section on simple harmonic progressing waves.

5.7. Boundary Conditions

We now discuss in some detail the BCs that occur on the free surface and on the floor, for the three-dimensional case. The floor is a fixed surface. The fluid has a free surface S separating it from another medium (water–air interface). Any particle on S remains on S. This property is a consequence of the basic assumption in continuum mechanics that the motion can be described as a topological deformation continuous in t. Suppose $f(x, y, z, t) = 0$ is the equation of S. Then

$$\frac{df}{dt} = uf_x + vf_y + wf_z + f_t = 0 \tag{5.59}$$

holds on S. (f_x, f_y, f_z) is normal to S so that the velocity potential ϕ has the condition

$$\frac{\partial\phi}{\partial n} = v_n = \frac{f_t}{((f_x)^2 + (f_y)^2 + (f_z)^2)^{1/2}},$$

where v_n is the common velocity of fluid and boundary surface in the direction of the outward normal to S.

For the special case where S is fixed (independent of t), say on the ocean floor or walls of an experimental water tank, we have the condition

$$\frac{\partial\phi}{\partial n} = v_n = 0 \quad \text{on fixed surface } S, \tag{5.60}$$

where v_n is the normal component of the particle velocity.

Another important special case is that in which S is a *free surface* on which the pressure p is prescribed, but the shape of S is not prescribed a priori. We shall, in general, assume that such a free surface S is given by the equation

$$z = \zeta(x, y, t).$$

On such a surface $f = z - \zeta(x, y, t) = 0$ for any particle on S, and hence we obtain

$$\phi_x \zeta_x + \phi_y \zeta_y - \phi_z + \zeta_t = 0 \quad \text{on free surface} \quad S\colon z = \zeta(x, y, t). \tag{5.61}$$

Since p is given on S, as a consequence of Bernoulli's law, which may be put in the form

$$\phi_t + \tfrac{1}{2}(u^2 + v^2 + w^2) + \frac{p}{\rho_0} + gz = 0,$$

we get the following condition:

$$g\zeta + \phi_t + \tfrac{1}{2}(\phi_x^2 + \phi_y^2 + \phi_z^2) + \frac{p}{\rho_0} = 0 \quad \text{on } S. \tag{5.62}$$

Thus the potential function ϕ must satisfy the two nonlinear BCs (5.61) and (5.62) on the free surface S. This is in sharp contrast to the single linear BC $\phi_n = 0$ for a fixed BC. However, it is not strange that two conditions should be prescribed in the case of a free surface since an additional unknown function $\zeta(x, y, t)$, the vertical displacement of the free surface, is involved in the latter case.

Concerning problems involving rigid bodies floating in water such as ships, S will be the portion of the body in contact with the water. In such cases the function $\zeta(x, y, t)$ will be determined by the motion of the rigid body, which in turn will be fixed (through the laws of rigid body dynamics) by the pressure p between the body and the water in accordance with Bernoulli's law.

5.8. Formulation of a Typical Surface Wave Problem

We now present a mathematical description of a typical surface wave problem such as that of breakers rolling on a sloping beach. We merely formulate this type of problem without solving it. The purpose is to put into focus some of the difficulties involved in this type of problem. Keeping the origin on the free surface at equilibrium, the water (assumed to be initially at rest) fills the space defined by

$$-h(x, y) < z < 0, \qquad -\infty < y < \infty, \quad 0 < x < \infty.$$

As mentioned above we consider the case of gravity waves impinging on a sloping beach. The xy plane is the equilibrium plane of the free surface S. $h(x, y)$, instead of being constant, is zero at the water–beach interface so that $h(x_s, y)=0$, and gradually increases as we go out in the water (as x increases). x_s is the abscissa of the water–shore line so that $x_s(y, t)$ is not known a priori.

For the two-dimensional case we neglect the y coordinate so that all variables are functions of x, z, t.

The formulation of this problem for the general case is:

$$\nabla^2\phi = \phi_{xx} + \phi_{yy} + \phi_{zz} = 0 \qquad \text{for} \quad x_s(y, t) < x < \infty,$$
$$-h(x, y) < z < \zeta(x, y, t), \quad -\infty < y < \infty.$$

This problem is nonlinear since neither $x_s(y, t)$ nor $\zeta(x, y, t)$, the free surface elevation, are known in advance, but must be determined as an integral part of the solution. The BC at the ocean floor is

$$\frac{\partial\phi}{\partial n} = 0 \qquad \text{for} \quad z = -h(x, y).$$

The free surface conditions are the kinematic condition (5.61) and (5.62) where p/ρ_0 is given by $F(x, y, t)$ so that

$$g\zeta + \phi_t + \tfrac{1}{2}(\phi_x^2 + \phi_y^2 + \phi_z^2) = F(x, y, t) \qquad \text{for} \quad z = \zeta(x, y, t) \quad \text{on } S,$$

where $F(x, y, t) = 0$ everywhere except over a region D where the disturbance is created. Obviously, we look for solutions that are bounded at infinity. Next we have the ICs

$$\zeta(x, y, 0) = 0, \qquad \phi_x = \phi_y = \phi_z = 0 \qquad \text{for} \quad t = 0,$$

which is the condition of rest in the equilibrium configuration. Finally, to prescribe F in D all we need do is prescribe p in D. This problem is difficult because it is nonlinear; the free surface is not known a priori and hence the domain in which the velocity potential is to be determined is not known in advance, and also its boundary is t varying. Another difficulty is the instability of the wave as it approaches the beach, as everyone knows that the waves do not come in smoothly toward the shore, but rather, the leading edge steepens up, curls over, and breaks—an instability sets in. Therefore any mathematical formulation of a wave traveling toward shore would necessitate the existence of singularities—moreover, of unknown location both in space and time. Because of the difficulty of solving such general nonlinear problems very little progress has been made in solving concrete problems. Some progress has been made in a specific theory: proving the existence of two-dimensional periodic progressing waves in water of uniform depth—Stoker [37] gives an account of this theory developed by Levi-Civita for infinite depth. Another problem for which some progress has been made is the problem of the solitary wave by Friedrichs and Hyers [37].

Some Observations on Waves on a Sloping Beach

Stoker [37] gives the following observations: Some distance from the shore line a train of nearly uniform progressing waves exists which have wavelengths of from about 50 to 100 ft. They are essentially small amplitude sinusoidal waves. As this ensemble of waves moves towards shore the lines of the crests and troughs become more and more nearly parallel to the shore line (even though they might not have been in deep water); also the distance between successive wave crests (the half wavelength) shortens slightly. In addition, as these waves travel toward shore, the wave height increases somewhat and the wave shape deviates more and more from a sinusoidal wave—the water in the vicinity of the crests tends to steepen, as does the leading edge, while the trailing edge (the vicinity of the troughs) tends to flatten out, until the wave profile becomes nearly vertical and an instability sets in so that the water curls over and the wave breaks. Therefore, clearly the linear theory cannot account for these observations of large amplitude waves which moreover become unstable. It is true that the linear theory is applicable in deep water where there is no surf or breaking waves. However, since in nature breakers do occur near shore we cannot neglect using a nonlinear theory which takes in account the steepening of the leading edge and the subsequent breaking of the wave.

Example of Instability in Stress Waves

We now give an example of the onset of instability in nonlinear elasticity; the principles being the same as those for a gravity wave breaking on a beach. The nonlinear one-dimensional wave equation for the displacement $u(x, t)$ of a particle of a bar from its equilibrium position due to an axial stress running through the bar is $c^2 u_{xx} - u_{tt} = 0$, where the wave speed c is a known function of the strain ε through the stress–strain curve (σ versus ε). $c^2 = (1/\rho_0)\, d\sigma/d\varepsilon$. In the nonlinear range of the stress–strain curve the slope $(d\sigma/d\varepsilon)$ changes with strain. Suppose the slope increases with increasing strain. Then, if we plot the large amplitude strain profile, for example (ε versus x), since ε increases with increasing x, the leading edge steepens (since the higher amplitude portion of ε travels with a higher speed). Therefore we have exhibited the phenomenon of steepening of the leading edge of a wavefront which eventually leads to an instability. In the case of our stress wave propagation through the bar, a shock wave must occur since in this case we cannot have an instability which would lead to a rupture of the bar; an instability means that for a given strain there occur three values of the stress.

5.9. Simple Harmonic Oscillations in Water at Constant Depth

Let the xy plane coincide with the equilibrium state of the free surface, z positive upward. The velocity potential $\phi(x, y, z, t)$ satisfies

$$\nabla^2\phi = \phi_{xx} + \phi_{yy} + \phi_{zz} = 0 \tag{5.63}$$

in the region $-h<z<0$. The ocean floor is at $z=-h$. We consider the case where h is constant. The free surface condition under the assumption of zero pressure (for atmospheric pressure p_0 just add p_0) is

$$\phi_{tt}+g\phi_z=0 \qquad \text{for} \quad z=0. \tag{5.64}$$

The condition at fixed boundary surfaces is $\phi_n=0$ so that

$$\phi_z=0 \qquad \text{for} \quad z=-h. \tag{5.65}$$

Once the velocity potential has been determined the elevation $\zeta(x, y, t)$ of the free surface is given by

$$\zeta=-\left(\frac{1}{g}\right)\phi_t(x, y, 0, t). \tag{5.66}$$

Conditions at infinity as well as appropriate ICs must also be prescribed.

Standing Waves

We first consider standing waves solutions. The most general standing wave solutions are of the form $\phi(x, y, z, t)=f(t)\bar{\phi}(x, y, z)$. Clearly this form is compatible with the definition of a standing wave, which says that the wave shape is fixed in space within a multiplying factor that depends on time; this tells us that the points of maxima and minima, etc. are time-independent. We are interested in those special types of standing waves that are simple harmonic in time. We therefore separate the time-harmonic part from the spatial part of the velocity potential, thus setting

$$\phi(x, y, z, t)=\exp(i\omega t)\bar{\phi}(x, y, z), \tag{5.67}$$

where $\bar{\phi}(x, y, z)$ is the real or imaginary part of the right-hand side and depends only on the spatial coordinates. We are thus interested in investigating the classical type of problems belonging to the theory of small oscillations of dynamical systems in the neighborhood of an equilibrium configuration. Inserting (5.67) into (5.63), (5.64), and (5.65) yields the following conditions on (x, y, z):

$$\nabla^2\bar{\phi}=0 \qquad \text{for} \quad -h<z<0, \quad -\infty<x<\infty, \quad 0<y<\infty. \tag{5.68}$$

$$\bar{\phi}_z-\left(\frac{\omega^2}{g}\right)\bar{\phi}=0, \qquad z=0. \tag{5.69}$$

$$\bar{\phi}_z=0, \qquad z=-h. \tag{5.70}$$

At infinity we assume that $\bar{\phi}$ and $\bar{\phi}_z$ are uniformly bounded. Since we have assumed our system to be simple harmonic in time we cannot prescribe arbitrary ICs. However, by expanding the ICs in Fourier series we can thereby obtain solutions in terms of Fourier series for period ICs in space.

Using (5.66) and (5.67) we get the free surface elevation as

$$\zeta = -\left(\frac{i\omega}{g}\right) \exp(i\omega t) \cdot \bar{\phi}(x, y, 0), \tag{5.71}$$

which tells us that ζ is a sinusoidal function of time at the equilibrium free surface xy.

Two-Dimensional Infinite Depth

We now seek standing wave solutions that are two dimensional so that $\bar{\phi} = \bar{\phi}(x, z)$. We first investigate the case of infinite depth so that $h = \infty$. It is easily seen that the functions

$$\bar{\phi} = \exp(mz) \cdot \cos mx, \qquad \text{and} \qquad \bar{\phi} = \exp(mz) \cdot \sin mx, \tag{5.72}$$

are harmonic functions (i.e., satisfy Laplace's equation in two dimensions) and satisfy the free surface condition (5.69) provided the constant m satisfies the relation

$$m = \frac{\omega^2}{g}. \tag{5.73}$$

The conditions of boundedness at infinity are also satisfied. Clearly (5.72) tells us that the oscillations die out with depth since as we go below the water surface z becomes more and more negative. Equation (5.72) represents the case of infinite depth. The free surface elevation ζ becomes

$$\zeta = -\left(\frac{i\omega}{g}\right) \exp(i\omega t) \cdot \begin{cases} \cos mx, \\ \sin mx. \end{cases} \tag{5.74}$$

We observe from (5.72) that the free surface waves are simple harmonic functions of x (the direction of wave propagation) as well as t, so that (5.67) can therefore be put in the form

$$\phi(x, y, z, t) = \exp(mz) \cdot \begin{cases} \cos(mx \mp \omega t), \\ \sin(mx \mp \omega t), \end{cases} \tag{5.75}$$

which show that the velocity potential can be represented as progressing and regressing waves traveling in the positive and negative x directions, respectively. They are exponentially damped in the $-z$ direction. Since the wave number m is given by (5.73) we obtain the important fact that the wavelength λ, which is given by

$$\lambda = \frac{2\pi}{m} = \frac{2\pi g}{\omega^2},$$

varies inversely as the square of the frequency of oscillations.

Two-Dimensional, Finite Depth

We now consider the case of finite depth so that h is finite. In contrast to the solution given by (5.72) for infinite depth, the solutions for $\bar{\phi}$ for water of finite depth are the harmonic functions

$$\bar{\phi} = \cosh m(z+h) \cos mx \qquad \text{and} \qquad \bar{\phi} = \cosh m(z+h) \sin mx. \quad (5.76)$$

Equation (5.76) represents the form of the solutions for $\bar{\phi}$ which satisfy the BC on the ocean floor. Substituting either one of these solutions into the BC on the free surface (5.69) yields the following expression for the wave number:

$$m = \left(\frac{\omega^2}{g}\right) \coth mh, \quad (5.77)$$

instead of (5.73) which holds for infinite depth. It is clear that (5.77) reduces to the case of infinite depth as we let h become infinite, since $\coth mh \to 1$ as $h \to \infty$.

Particle Trajectories

We now calculate the motion of the individual water particles. Let δx and δz be the small displacements from the mean position (x, z) of a given particle. Using the expression for $\phi(x, z, t)$ from (5.67) and (5.76) we have

$$d\frac{\delta x}{dt} = u(x, z, t) = \phi_x = -mA \cos \omega t \cosh m(z+h) \sin mx,$$

$$d\frac{\delta z}{dt} = v(x, z, t) = \phi_z = mA \cos \omega t \sinh m(z+h) \cos mx,$$

within the accuracy of our linear approximation. A is a constant determining the amplitude of the wave. Upon integration we obtain

$$\delta x = -\left(\frac{mA}{\omega}\right) \sin \omega t \cosh m(z+h) \sin mx,$$

$$\delta z = \left(\frac{mA}{\omega}\right) \sin \omega t \sinh m(z+h) \cos mx.$$

Eliminating ωt we get

$$\delta z = -\tanh m(z+h) \cot mx \delta x.$$

For the particles near the surface we have approximately $z = 0$ so that the particle trajectories become, upon using (5.77),

$$\delta z = -\left(\frac{\omega^2}{gm}\right) \cot mx \delta x.$$

Three-Dimensional Standing Waves

In investigating three-dimensional standing waves that are time harmonic we assume that the spatial dependence is only of the distance r from the z axis, i.e., we are interested in time-harmonic standing waves exhibiting cylindrical symmetry, the z axis being the axis of symmetry. We look for solutions of Laplace's equation in cylindrical coordinates for the velocity potentials that satisfy (5.69) and (5.70) for the case of infinite depth. Equation (5.70) is then replaced by the condition that the solution be bounded as $z \to -\infty$. We shall see below that all such bounded standing waves with cylindrical symmetry die out at infinity like the inverse square root of the distance. Laplace's equation with cylindrical symmetry is

$$(r\bar{\phi}_r)_r + r\bar{\phi}_{zz} = 0, \qquad 0 > z > -\infty, \quad 0 < r < \infty. \tag{5.78}$$

The boundary condition at $z=0$ is given by (5.69) which is

$$\bar{\phi}_z - m\bar{\phi} = 0 \qquad \text{for} \quad z = 0, \tag{5.69}$$

where m is given by (5.73). Assuming a solution of (5.78), that satisfies the BC at $z=0$, to be of the form

$$\bar{\phi}(r, z) = \exp(mz) f(r), \tag{5.79}$$

we arrive at the following ODE for $f(r)$:

$$(rf')' + m^2 rf = 0, \qquad y' \equiv \frac{dy}{dr}. \tag{5.80}$$

This is Bessel's equation of order zero so that

$$\bar{\phi}(r, z) = C \exp(mz) J_0(mr), \tag{5.81}$$

where we omitted the part of the function $Y_0(mr)$ which has a singularity at $r=0$ (since we want bounded solutions at $r=0$). C is an arbitrary constant. Inserting (5.81) into (5.79) and using (5.67) we find that the velocity potential for infinitely deep water for the case of cylindrical symmetry is

$$\phi(r, z, t) = C \exp(i\omega t) \exp(mz) J_0(mr). \tag{5.82}$$

As is well known, the asymptotic expansion of the right-hand side of (5.82) for large r is

$$\phi(r, z, t) \sim C \exp(i\omega t) \exp(mz) \left(\frac{2}{\pi m r}\right)^{1/2} \cos\left(mr - \frac{\pi}{4}\right), \tag{5.83}$$

which demonstrates that the velocity potential dies out as $1/r^{1/2}$, as asserted above.

Simple Harmonic Progressing Waves

We started out with time-harmonic standing waves of the form given by (5.67). We discuss two-dimensional wave propagation problems. It was pointed out that the velocity potential can be represented by (5.75) for progressing or regressing waves. The form given by (5.75) represents appropriate linear combinations of two-dimensional standing waves for the case of infinite depth. For water of finite depth h (5.75) is generalized, as indicated above, in the sense that the exponential term $\exp(mz)$ becomes the cosh $m(z+h)$ term to account for the BC (8) at $z=-h$. Therefore we may write the velocity potential for a simple harmonic progressing wave (for $x>0$) as

$$\phi(x, z, t) = A \cosh m(z+h) \sin(mx - \omega t + \alpha), \tag{5.84}$$

where A is the amplitude, m is related to ω by (5.77), and α is a phase angle. The argument of the sine term in (5.84) (not counting the constant phase angle) may be written as $mx - \omega t = m(x - ct)$, where c is the speed of the propagating wave. We have the following relations amongst the wave number, wavelength, and wave speed:

$$c = \frac{\omega}{m} = 2\pi\omega\lambda = \sqrt{\left(\frac{g\lambda}{2\pi}\right)\tanh\left(\frac{2\pi h}{\lambda}\right)}, \tag{5.85}$$

upon using (5.77). Equation (5.85) tells us that c depends on the dimensionless ratio h/λ. It is natural to consider two asymptotic cases: $h/\lambda \ll 1$ and $h/\lambda \gg 1$. For the first case, $\tanh(2\pi h/\lambda) \to 2\pi h/\lambda$ so that

$$c \approx \sqrt{gh} \qquad \text{for} \quad h/\lambda \ll 1. \tag{5.86}$$

This gives us the important result that a simple harmonic progressing wave traveling in shallow water, or with a long wavelength compared to the depth, has a wave speed given by the approximation (5.86), which tells us that it is independent of the wavelength but varies as the square root of the depth.

For the second case, we have $\tanh(2\pi h/\lambda) \to 1$ so that we obtain

$$c \approx \sqrt{\frac{g\lambda}{2\pi}} \qquad \text{for} \quad \frac{h}{\lambda} \gg 1, \tag{5.87}$$

which tells us that the wave speed is proportional to the square root of the wavelength for a progressing wave with small wavelength compared to the depth. Equation (5.87) is the approximation that corresponds to infinite depth (c is independent of h). Actually, a simple calculation shows us that the error in c, as computed by (5.87), is less than $\frac{1}{2}\%$ if $h/\lambda > \frac{1}{2}$. We therefore conclude that variations in the ocean floor will have slight effect on the speed of the progressing wave, provided that these variations do not result in depths which are less than half of the wavelength, and observations appear to bear out this statement. Another way of putting it is that these surface waves would not "feel the bottom" until the depth becomes less than half a wavelength.

Particle Paths

We now determine the particle paths as a result of the traveling progressing wave for this two-dimensional case. We use the same notation as for the standing wave case where we let δx, δz be the components of the displacement of a particle from its equilibrium position. Using (5.84) we have

$$d\frac{\delta x}{dt} = \phi_x = mA \cosh m(z+h) \cos(mx - \omega t + \alpha),$$

$$d\frac{\delta z}{dt} = \phi_z = mA \sinh m(z+h) \sin(mx - \omega t + \alpha).$$

Integrating this system yields

$$\delta x = \left(\frac{mA}{\omega}\right) \cosh m(z+h) \sin(mx - \omega t + \alpha),$$

$$\delta z = -\left(\frac{mA}{\omega}\right) \sinh m(z+h) \cos(mx - \omega t + \alpha).$$

Eliminating ωt in these equations we obtain

$$\frac{\delta x^2}{a^2} + \frac{\delta z^2}{b^2} = 1, \tag{5.88}$$

where the semi-axes a, b are given by

$$a = \left(\frac{mA}{\omega}\right) \cosh m(z+h), \qquad b = \left(\frac{mA}{\omega}\right) \sinh m(z+h).$$

Equation (5.88) tells us that the particle trajectories are ellipses. Both axes of the ellipse shorten with increasing depth. On the ocean floor we have $z = -h$ so that the particle trajectories degenerate into horizontal straight lines, as we would expect. We can readily verify that the particle paths become circles at infinite depth. The displacement of the particles dies out exponentially with depth. This explains, for example, why a submarine need only submerge a small distance below the surface—about half a wavelength—in order to be practically unaffected by severe storms.

5.10. The Solitary Wave

There have been many observations of wave phenomena other than the type studied up to now. For example, Scott Russell reported [37] his observations on what has since been called the solitary wave. It is a wave having a symmetrical form with a single hump and which propagates at uniform velocity without change of form. The situation as presented by Stoker [37] is as follows: The only continuous waves furnished by the theory presented thus far, which progress unchanged in form, are the motions with uniform velocity and a horizontal free surface. These above-mentioned waves would, quoting Stoker, "seem to be in crass contradiction with our intention to discuss the solitary wave in terms of the shallow water theory, and it has been regarded by some authors as a paradox." Some investigators cast doubt on the validity of the shallow water theory in general since it supposedly does not give rise to the solitary wave. Stoker points out that there is no paradox when the situation is properly examined. What is involved is a matter of accuracy of a given approximation theory. A perturbation or iteration scheme of universal applicability does not exist, so that one must develop schemes appropriate for specific problems.

Solution of the Solitary Wave Problem

We now discuss the solution of the solitary wave problem using the method of Friedrichs and Hyers [37]. The problem was motivated by the desire to formulate the problem in terms of the velocity potential and stream function as independent variables, in order to work in the fixed domain between the two constant streamlines corresponding to the floor of the ocean and the free surface instead of working with the partially unknown domain in the physical plane. We therefore consider the general theory of irrotational waves in water having a free surface, and construct a coordinate system that moves with the same velocity as the wave, hence yielding a steady state flow in this coordinate system.

We start with a two-dimensional (x, y) coordinate system where y is directed vertically downward such that $y = 0$ is at the equilibrium surface and $y = -h$ is at the floor. (The reason for using y as the vertical coordinate is that we reserve z for the complex variable.) We recall that the complex velocity potential is $f(z) = \phi + i\psi$. Let w be the velocity vector, then

$$w = \frac{df^*(z)}{dz} = u + iv, \tag{5.89}$$

where $\phi_x = \psi_y$, $\phi_y = -\psi_x$. $f^*(z)$ is the complex conjugate of $f(z)$. The BC at infinity is $w = U$, the velocity of the wave. We introduce dimensionless variables and parameter γ defined by

$$\bar{z} = \frac{z}{h}, \qquad \bar{w} = \frac{w}{U}, \qquad \bar{f} = \frac{f}{hU}, \qquad \gamma = \frac{gh}{U^2}. \tag{5.90}$$

Let the stream function $\psi = 0$ at the floor. Then $\psi = 1$ at the free equilibrium surface, since the total flow through a curve from the floor to the free surface is hU. We now formulate the BCs as follows:

$$v = \text{Im}(w) = 0 \qquad \text{at} \quad \psi = 0, \tag{5.91}$$

$$\tfrac{1}{2}|w|^2 + y = \text{const.} \qquad \text{at} \quad \psi = 1. \tag{5.92}$$

The condition given by (5.92) stems from Bernoulli's equation on taking the pressure to be constant on the free surface in equilibrium and the density to be unity. The BC at infinity is

$$w \to 1 \qquad \text{as} \quad |x| \to \infty. \tag{5.93}$$

We now assume that the physical or (x, y) plane is mapped by means of the complex potential $f(z)$ onto the $f(z)$ or (ϕ, ψ) plane in such a way that the entire flow is mapped in a one-to-one manner on the strip bounded by $\psi = 0$ and $\psi = 1$ in the $f(z)$ plane. We then determine the velocity vector w, which is an analytic function $w(f(z))$, in the strip from the BCs given by (5.91), (5.92), (5.93). Having determined w we can then calculate $f(z)$ by integrating, and can thereby obtain the free surface of the curve as given by $\psi = 1$.

To determine w, which satisfies the BCs, we set

$$w = \exp(-i(\theta + i\lambda)) \qquad \text{or} \qquad |w| = e^{\lambda}, \qquad \theta = \arg w^*. \tag{5.94}$$

This means that $\lambda = \log|w|$ where $|w| = (u^2 + v^2)^{1/2}$. The BC on the fixed surface given by (5.91) clearly becomes $\theta = 0$ for $\psi = 0$. To handle the BC at $\psi = 1$ on the free surface we differentiate (5.92) with respect to ϕ obtaining

$$|w|(|w|)_\phi + \gamma y_\phi = 0 \quad \text{on} \quad \psi = 1. \tag{5.95}$$

Since x and y are conjugate harmonic functions of ϕ and ψ we have

$$\begin{aligned} x_\phi = y_\psi &= \frac{\phi_x}{(\phi_x^2 + \phi_y^2)} = \frac{u}{|w|^2}, \\ y_\phi = -x_\psi &= \frac{v}{|w|^2}, \end{aligned} \tag{5.96}$$

using the well-known rules for calculating the derivatives of functions determined implicitly. Using (5.96) we obtain the following from (5.95):

$$(|w|)_\phi = -\frac{\gamma v}{|w|^3},$$

which becomes

$$e^{\lambda}\lambda_\phi = -\gamma_e^{-2\lambda} \sin\theta,$$

since $\lambda_\phi = -\phi_\psi$, because λ and θ are harmonic conjugates we finally obtain

$$\theta_\psi = e^{-3\lambda} \sin\theta \qquad \text{for} \quad \psi = 1. \tag{5.97}$$

Stoker points out that in the case of a solitary wave the procedure of letting λ and θ be small, so that $\theta_\psi \approx \gamma\theta$ at $\psi = 1$, would not yield anything but a uniform flow. The procedure adopted by the Friedrichs–Hyer approach is to develop a perturbation theory near $\gamma = 1$, which is reminiscent of what he says about Keller's bifurcation theory [37] about $U^2 = hg$ $(\gamma = 1)$. A *stretching* of the horizontal axis is introduced, which is one of the characteristics of shallow water theory. It turns out that the approximating functions that are introduced are no longer harmonic in the new independent variables. Again, following Friedrichs–Hyer, we introduce a real parameter β defined by

$$\exp(-3\beta^2) = \gamma = \frac{gh}{U^2}, \tag{5.98}$$

which implies $\gamma < 1$ if β is not zero. We also introduce a function τ defined by

$$\tau = \lambda + \beta^2. \tag{5.99}$$

The BCs on θ and ψ are

$$\theta = 0 \quad \text{for} \quad \psi = 0, \tag{5.100}$$

$$\phi_\psi = e^{-3\tau} \sin\theta \quad \text{for} \quad \psi = 1, \tag{5.101}$$

$$\theta \to 0, \quad \tau \to \beta^2 \quad \text{for} \quad \phi \to \pm\infty. \tag{5.102}$$

We now introduce two new independent variables $\bar{\phi}$, $\bar{\psi}$ defined by

$$\bar{\phi} = \beta\phi, \qquad \bar{\psi} = \psi. \tag{5.103}$$

This involves a stretching of ϕ so that it grows large with respect to ψ for β small. The dependent variables θ and τ are now functions of $\bar{\phi}$ and $\bar{\psi}$. Friedrichs and Hyers showed that these functions can be expanded in the following power series:

$$\begin{aligned} \tau &= \beta^2\tau_1(\bar{\phi}, \bar{\psi}) + \beta^4\tau_2(\bar{\phi}, \bar{\psi}) + \ldots, \\ \theta &= \beta^3\tau_1(\bar{\phi}, \bar{\psi}) + \beta^5\theta_2(\bar{\phi}, \bar{\psi}) + \ldots. \end{aligned} \tag{5.104}$$

They then proved that the lowest-order terms in these series, as obtained formally by the use of the BCs, are the lowest-order terms in a convergent iteration scheme using β as a small parameter. Their convergence proof involves the explicit use of this stretching process; it is rather complicated and will not be given here. Inserting the series given by (5.104) into the appropriate equations which determine θ, τ and using the Cauchy–Riemann equations for θ and τ, we get relations for the coefficients in the right-hand side of the system (5.104). Thus the Cauchy–Riemann equations lead to

$$\theta_{\bar{\psi}} = -\beta\tau_{\bar{\phi}}, \qquad \tau_{\bar{\psi}} = \beta\theta_{\bar{\phi}}, \tag{5.105}$$

and the series (5.104) yield the following relationships among the first derivatives of the coefficients:

$$\tau_{1\bar{\psi}} = 0, \qquad \theta_{i\bar{\psi}} = -\tau_{i\bar{\phi}}, \qquad \tau_{2\bar{\psi}} = \theta_{1\bar{\phi}}, \qquad i = 1, 2, \ldots. \tag{5.106}$$

This means that τ_1 depends only on $\bar{\phi}$. Integration of the other equations of (5.106) yields

$$\begin{aligned} &\theta_1 = -\bar{\psi}\tau_1', \qquad \tau_2 = -\tfrac{1}{2}\bar{\psi}^2\tau_1'' + f(\bar{\phi}), \\ &\theta_2 = \tfrac{1}{6}\bar{\psi}^3\tau_1''' - \bar{\psi}f(\bar{\phi}), \end{aligned} \tag{5.107}$$

where f is an arbitrary function of $\bar{\phi}$ and the primes refer to differentiation with respect to $\bar{\phi}$. An additive arbitrary function of $\bar{\phi}$ was taken to be zero for the first equation in order to satisfy the BCs $\theta_1 = 0$ for $\psi = 0$. We now insert the series (5.104) into the BC on the free surface given by (5.101) and obtain the following series in odd powers of β:

$$\beta^3\theta_{1\bar{\psi}} + \beta^5\theta_{2}\bar{\psi} + \cdots = \beta^3\theta_1 - 3\beta^5\tau_1\theta_1 + \beta^5\theta_2 + \ldots.$$

Equating like powers of β yields

$$\begin{aligned} \theta_{1\bar{\psi}} &= \theta_1 \\ \theta_{2\bar{\psi}} &= \theta_2 - 3\tau_1\theta_1 \qquad \text{for} \quad \bar{\psi} = 1. \end{aligned} \tag{5.108}$$

Appealing to (5.107), the first equation is automatically satisfied while the second leads to the following ODE for τ_1:

$$\tau_1''' = 9\tau_1\tau_1'. \tag{5.109}$$

Once τ_1 is determined it is seen that θ_1 can then be calculated from the first of (5.107). We need the proper BCs for τ_1 in order to solve (5.109). They are

$$\begin{aligned} \tau_1'(0) &= 0, \\ \tau_1(\infty) &= 1, \\ \tau_1''(\infty) &= 0. \end{aligned} \tag{5.110}$$

Stoker points out that these BCs satisfy the physical needs of the solitary wave: The first arises from the assumption of a symmetric wave about its crest so that $\theta_1(0) = 0$; the second arises from the BC at infinity (5.102); and the third arises from the first of (5.107) and the BC $\theta_1(\infty) = 0$. Using the appropriate BC, the first integral of (5.109) is

$$\tau_1'' = \tfrac{9}{2}(\tau_1^2 - 1),$$

eventually yielding the solution for $\tau_1(\bar{\phi})$:

$$\tau_1(\bar{\phi}) = 1 - 3\,\text{sech}^2\left(\frac{3\bar{\phi}}{2}\right). \tag{5.111}$$

Knowing $\tau_1(\bar{\phi})$ from (5.111) we easily determine $\theta_1(\bar{\phi}, \bar{\psi})$ from the first equation of (5.107). The above results allow us to find the wave shape by finding the value of y at the free surface where $\psi = 1$, and also the horizontal component of the particle velocity u. The results are

$$y = 1 + 3\beta^2\,\text{sech}^2\left(\frac{3\beta x}{2}\right), \tag{5.112}$$

$$u = 1 - 3\beta^2\,\text{sech}^2\left(\frac{3\beta x}{2}\right). \tag{5.113}$$

As seen, these equations for y and u are good to the second-order terms (in β^2).

In summary: a symmetrical solitary wave has been found with an amplitude that increases with its speed U. Recall that, from (5.98), β^2 increases with increasing U.

5.11. Approximation Theories

Stoker [37, Chap. 2] gives two approximation theories: (1) theory of waves of small amplitude, and (2) shallow water theory. It is of interest to review Stoker's treatment of these important approximation theories. Some of the material has been covered above in different forms; however, it is best to examine here the treatment given by Stoker which takes advantage of various perturbation methods.

Theory of Waves of Small Amplitude

In this theory we assume $\mathbf{v}$ and the surface elevation $z=\zeta(x, y, t)$ are small quantities. We introduce the parameter ε which is a measure of the smallness of the velocity potential ϕ, from which we can derive the velocity field as well as ζ. We expand these functions in a power series in ε obtaining

$$\phi=\sum_j \varepsilon^j \phi_j, \tag{5.114}$$

$$\zeta=\sum_j \varepsilon^j \zeta_j, \tag{5.115}$$

where the unknown coefficients $\phi_j(x, y, z, t)$ and $\zeta_j(x, y, z, t)$ are called *expansion coefficients*, which are to be evaluated at $z=0$ (on the free surface in equilibrium).... We do not consider questions of convergence of these power series, but assume from the physics of the situation that these series will be sufficiently convergent so that only a few terms need be evaluated. From the fact that ϕ satisfies the three-dimensional Laplace equation it follows that

$$\nabla^2 \phi_j=0, \tag{5.116}$$

where ∇^2 is the three-dimensional Laplacian.

We next consider the BCs. Since $v_n=\partial\phi/\partial n=0$ on the fixed surface, we have

$$\frac{\partial \phi_j}{\partial n}=0, \tag{5.117}$$

where $\partial/\partial n$ means differentiation along the outward drawn normal to the surface. We next consider the free surface S: $z=\zeta(x, y, t)$. We have two BCs on this surface: (1) the pressure is the constant atmospheric pressure, $p=p_0$; (2) the water particles remain on the S.

(1) $p = p_0$ on free surface S and Bernoulli's law holds:

$$\phi_t + \tfrac{1}{2}(\phi_x^2 + \phi_y^2 + \phi_z^2) + g\zeta = 0 \qquad \text{on} \quad S\colon z = \zeta(x, y, t). \tag{5.118}$$

We insert the expansions (5.114) and (5.115) into (5.118), evaluating the partial derivatives for $z = \zeta(x, y, t)$. Using the expansion (5.115) for the elevation and equating coefficients of like powers of ε, we finally obtain the following expressions for the expansion coefficients:

$$\zeta_0 = 0, \tag{5.119}$$

$$g\zeta_1 + \phi_{1,t} = 0, \tag{5.120}$$

$$g\zeta_2 + \phi_{2,t} + \tfrac{1}{2}[(\phi_{1,x})^2 + (\phi_{1,y})^2 + (\phi_{1,z})^2] + \zeta_1\phi_{1,tz} = 0. \tag{5.121}$$

The obvious notation $\phi_{1,t} \equiv \partial\phi_1/\partial t$, etc. is used. Since in the definition of the series given by (5.114) and (5.115) the expansion coefficients ϕ_j and ζ_j are all evaluated at $z = \zeta_0$, and since (5.119) tells us that $\zeta_0 = 0$, we conclude from (5.119), (5.120) and the other equations that the infinite set of expansion coefficients are to be satisfied for the equilibrium configuration of the free surface where $z = 0$.

(2) The kinematic BC: the water particles remain on S. This is expressed by

$$\phi_x\zeta_x + \phi_y\zeta_y - \phi_z + \zeta_t = 0 \qquad \text{on} \quad S\colon z = \zeta(x, y, t). \tag{5.122}$$

Inserting the expansions (5.114) and (5.115) into (5.122), and equating coefficients of like powers of ε in the usual way, we get the following relationships:

$$\zeta_{1,t} = 0, \tag{5.123}$$

$$\phi_{1,x}\zeta_{0,x} + \phi_{1,y}\zeta_{0,y} - \phi_{1,z} + \zeta_{1,t} = 0, \tag{5.124}$$

$$\begin{aligned} \phi_{2,x}\zeta_{0,x} + \phi_{2,y}\zeta_{0,y} - \phi_{2,z} + \zeta_{2,t} &= -\phi_{1,x}\zeta_{1,x} - \phi_{1,y}\zeta_{1,y} \\ &\quad - \zeta_1(\phi_{1,xz}\zeta_{0,x} + \phi_{1,yz}\zeta_{0,y} - \phi_{1,zz}). \end{aligned} \tag{5.125}$$

This system is to be satisfied for $z = 0$.

From (5.119), (5.120), and (5.121) we write the nth equation for the expansion coefficients for the fixed surface as

$$g\zeta_n + \phi_{n,t} = F_{n-1}, \tag{5.126}$$

where F_{n-1} stands for a combination of the functions ϕ_k and ζ_k with $k \le n-1$, all calculations are done for $z = 0$.

Also, we set the derivatives of ζ_0 equal to zero in the other BC given by (5.123), (5.124), and (5.125). The nth equation is of the form

$$\zeta_{n,t} = \phi_{n,z} + G_{n-1}, \tag{5.127}$$

where G_{n-1} depends only on the functions ϕ_k and ζ_k with $k \le n-1$.

We now consider the first-order theory obtained by keeping only terms of order ε in the series expansions for ϕ and ζ. We therefore set $\phi = \varepsilon\phi_1$ and $\zeta = \varepsilon\zeta_1$. With this stipulation the free surface conditions become

$$g\zeta + \phi_t = 0 \tag{5.128}$$

$$\zeta_t - \phi_z = 0 \qquad \text{for} \quad z = 0. \tag{5.129}$$

By eliminating ζ from (5.128) and (5.129) we get

$$\phi_{tt} + g\phi_z = 0 \qquad \text{for} \quad z = 0. \tag{5.130}$$

It is clear that the same type of equation as (5.130) can be obtained for ζ at $z = 0$ by eliminating ϕ from (5.128) and (5.129). Coming back to (5.130). This is a BC at $z = 0$ which is used in solving ϕ from $\nabla^2\phi = 0$. After ϕ is obtained we can then calculate ζ from (5.128). If we take into account the atmospheric pressure p_0 then (5.129) is replaced by

$$g\zeta + \phi_t + \frac{p_0}{\rho_0} = 0. \tag{5.131}$$

It is seen that (5.128) and (5.131) are the linear approximations of the BCs on the free surface which are

$$\phi_x\zeta_x + \phi_y\zeta_y - \phi_z + \zeta_t = 0,$$

$$g\zeta + \phi_t + \tfrac{1}{2}(\phi_x^2 + \phi_y^2 + \phi_z^2) + \frac{p_0}{\rho_0} = 0.$$

There is a great simplification resulting from this first-order theory. The surface conditions are linearized, and also the domain of solution is fixed a priori, so that surface wave problems in this formulation belong to the classical problems of potential theory.

Shallow Water Theory

Instead of using the approximation of small amplitude as given above, a different kind of approximation to the full nonlinear theory of water waves is used when dealing with shallow water theory. This theory uses the approximation in which the depth h is small compared to another characteristic length which, for example, could be the radius of curvature of the water surface. However, it is customary to use the wavelength of the surface wave so that this is commonly called the theory of long waves in standard treatises on hydrodynamics such as Lamb's work. Stoker points out that the shallow water theory in its lowest approximation is the best theory used in hydraulics by engineers in dealing with flows in open channels or canals.

In this theory it is not necessary to assume that the displacement and slope of the water surface are small, so that the resulting theory is not a linear theory. Among such applications of this approximation theory are the theory of the tides in the oceans, the solitary wave, and the breaking of waves on shallow beaches. In addition, as mentioned above, there are many applications in hydraulics concerning flows in canals such as roll waves, flood waves in rivers, surges in channels due to a sudden influx of water, and other examples in nonlinear shallow water theory.

We now proceed to give a derivation of the theory of two-dimensional motion along the lines of Lamb's *Hydrodynamics* [25, p. 254]. The usual notation is used: x is the horizontal axis, z is the vertical axis directed upward so that $z=-h$ is the bottom, the particle velocity $\mathbf{v}(x, z, t)=(u, 0, w)$, and the surface displacement from the equilibrium position is $z=\zeta(x, t)$.

The equation of continuity in two dimensions for incompressible flow is

$$u_x + w_z = 0. \tag{5.132}$$

The kinematic condition that is to be satisfied on the free surface $z=\zeta(x, t)$ is

$$\frac{d\zeta}{dt} = u\zeta_x + \zeta_t = w \qquad \text{or} \qquad u\zeta_x + \zeta_t - w = 0 \quad \text{on } z=\zeta. \tag{5.133}$$

The dynamical condition to be satisfied on the free surface is

$$p = p_0 \qquad \text{on } z=\zeta. \tag{5.134}$$

The condition on the floor which is a fixed surface is

$$\frac{dh}{dt} = -w \qquad \text{or} \qquad uh_x + w = 0 \quad \text{on } z=-h. \tag{5.135}$$

We next integrate the continuity equation (5.132) with respect to z, use the BCs at ζ and $-h$ given by (5.133) and (5.134), and use the following condition

$$\frac{\partial}{\partial x}\int_{-h}^{\zeta} u\,dz = u\zeta_x + uh_x + \int_{-h}^{\zeta} u_x\,dz,$$

to obtain

$$\frac{\partial}{\partial x}\int_{-h}^{\zeta} u\,dz = -\zeta_t. \tag{5.136}$$

No approximations were used up to now. We now use an approximation in shallow water theory which says that the z component of particle acceleration has a negligible effect on the pressure. Therefore the z component of the equations of motion is $0=-p_z-\rho_0 g$, so that p is given by

$$p - p_0 = \rho_0 g(\zeta - z), \qquad -h \le z \le \zeta, \tag{5.137}$$

where p becomes the atmospheric pressure p_0 on the free surface where $z=\zeta$. From (5.137) we deduce that p_x is independent of z since $p_x=\rho_0 g\zeta_x$ and $\zeta(x, t)$ is independent of z. This means that the x component of the particle acceleration is independent of z and hence u is independent of z or $u = u(x, t)$. Therefore the equation of motion in the x direction can be written as

$$u_t + uu_x = -\rho_0 p_x = -g\zeta_x, \tag{5.138}$$

where we used the approximation $u_z = 0$. Another consequence of the approximation that u is independent of z is that $\int_{-h}^{\zeta} u\,dz = u(\zeta + h)$ so that (5.136) becomes

$$[u(\zeta + h(x))]_x = -\zeta_t. \tag{5.139}$$

Equations (5.138) and (5.139) are the equations of shallow water theory for u and ζ. Note that they arise from the assumption that we can neglect w_t which led to (5.137), and that they are nonlinear. Once the initial state of the water is prescribed (given $u(x, 0)$, $\zeta(x, 0)$) the equations (5.138) and (5.139) yield the subsequent motion of the fluid.

If, in addition to the assumption of shallow water theory used above, we now assume that u and ζ and their derivatives are small quantities such that all second-order terms can be neglected, the following system arises:

$$u_t = -g\zeta_x, \tag{5.140}$$

$$(uh)_x = -\zeta_t. \tag{5.141}$$

We can eliminate ζ from these equations and obtain the following linear PDE for u:

$$g(uh)_{xx} - u_{tt} = 0. \tag{5.142}$$

This is a linear wave equation for $u(x, t)$ with the variable coefficient $h(x)$. In addition, if h is constant (5.142) reduces to

$$c^2 u_{xx} - u_{tt} = 0, \qquad c^2 = gh. \tag{5.143}$$

c is the velocity of wave propagation for a shallow water wave for the above approximations. It is easily seen from (5.140) and (5.141) that ζ also satisfies (5.143) for constant h. The important result is that the wave speed is given by $c = \sqrt{gh}$ for this linearized shallow water theory. Stoker points out that in principle this linearized shallow water theory has always been used as the basis for the theory of tides. Clearly the tidal theory is not complete without the specification of the external forces acting on the water due to the gravitational attraction of the moon and, to a lesser degree, the sun, and also the Coriolis forces due to the rotation of the earth; but nevertheless from the mathematical point of view the theory of the tides rests on the approximations of the linearized shallow water theory. Although the ocean does not appear to be shallow, the ocean's depth is indeed very small compared to the curvature of the surface of the tidal wave, so that the approximation of shallow water theory of very small depth to curvature to wave length ratio is an excellent one. Also the small amplitudes of the tides compared to the wavelength of the tidal waves allows us to use the linear approximation with a great deal of confidence.

Shallow Water Theory Treated by a Perturbation Method

It is important to know under what circumstances the shallow water theory can be expected to furnish sufficiently accurate results. For the nonlinear shallow water theory means that the pressure can be given by (5.137) which stems from neglecting the z component of particle acceleration. Consequently, shallow water theory may be valid for large amplitude water waves provided the pressure law approximation given by (5.137) is not invalidated. This approximation does not account for the role played by depth on the surface pressure distribution. Lamb [25, p. 368] says that the solutions for steady progressing small amplitude waves (linearized theory) in water of uniform but finite depth are approximated accurately by the linearized shallow water theory only when the depth of the water is small compared with the wavelength. Therefore the assumption that, since the amplitude of the motion dies out exponentially with depth, we should expect that the hydrostatic pressure law given by (5.137) would be more accurate the deeper the depth is not true.

The perturbation approximation used here is based on Friedrich's theory for a solitary wave, which uses as the perturbation parameter the ratio of depth h to U^2/g where U is the wave speed, and this ratio is taken to be near unity.

In our usual notation we restate the three-dimensional hydrodynamic formulation for convenience.

The continuity equation for an incompressible fluid is

$$\operatorname{div} \mathbf{v} = u_x + v_y + w_z = 0. \tag{5.144}$$

Euler's equations of motion in a gravitational field

$$\begin{aligned} u_t + uu_x + vu_y + wu_z &= -\frac{p_x}{\rho_0}, \\ v_t + uv_x + vv_y + wv_z &= -\frac{p_y}{\rho_0}, \\ w_t + uw_x + vw_y + ww_z &= -\frac{p_z}{\rho_0} - g. \end{aligned} \tag{5.145}$$

The irrotationality condition

$$\mathbf{curl}\, \mathbf{v} = 0 \quad \text{or} \quad w_y - v_z = 0, \quad u_z - w_x = 0, \quad v_x - u_y = 0. \tag{5.146}$$

The BCs at the free surface

$$\frac{d\zeta}{dt} = \zeta_t + u\zeta_x + v\zeta_y + w\zeta_z = w \quad \text{at} \quad z = \zeta, \tag{5.147}$$

$$p = p_0 \quad \text{at} \quad z = \zeta. \tag{5.148}$$

The BC at the fixed surface on the ocean floor

$$uh_x + vh_y + w = 0 \qquad \text{at} \quad z = -h. \tag{5.149}$$

By letting d be a constant representative depth and k be a characteristic length such as U^2/g for a solitary wave, Stoker introduces the following dimensionless independent and dependent variables:

$$\begin{aligned}
&\bar{x} = \frac{x}{k}, \qquad \bar{y} = \frac{y}{k}, \qquad \bar{z} = \frac{z}{d}, \qquad \tau = \frac{t}{k}\sqrt{gd}, \\
&\bar{u} = \frac{u}{\sqrt{gd}}, \qquad \bar{v} = \frac{v}{\sqrt{gd}}, \qquad \bar{w} = \frac{w}{(k/d)\sqrt{gd}}, \\
&\bar{p} = \frac{p}{\rho_0 gd}, \qquad \bar{\zeta} = \frac{\zeta}{d}, \qquad \bar{h} = \frac{h}{d}.
\end{aligned} \tag{5.150}$$

To see the meaning of the new dimensionless independent variables we note that

$$\frac{x}{z} = \frac{(k/d)\bar{x}}{\bar{z}}, \qquad \frac{y}{z} = \frac{(k/d)\bar{y}}{\bar{z}}.$$

Since the ratio k/d is assumed to be very large this means that x and y are "stretched" compared to z. In particular, in the two-dimensional case (where we neglect y) this process of stretching x takes care of the approximation in shallow water theory that says we can neglect the vertical component of the particle acceleration. For a solitary wave, as mentioned above, $k = U^2/g$ which is taken to be approximately unity.

In addition Stoker introduces the perturbation parameter defined by

$$\sigma = \left(\frac{d}{k}\right)^2. \tag{5.151}$$

This stretching process of the horizontal coordinates, compared to the vertical coordinate combined with a perturbation method with respect to σ, is the characteristic feature of this approach to shallow water theory. Stoker gives the following example that illustrates the physical significance of σ: Suppose the initial elevation of a particle on the surface from its equilibrium position is $z = \zeta_0(x, y, z, 0)$. Using the transformation equations into dimensionless variables gives

$$z = \zeta_0 = d\zeta_0(\bar{x}, \bar{y}, \bar{z}, 0),$$

from which we obtain the following relationship between the initial curvature of the wave surface in physical and dimensionless variables:

$$dz_{xx} = \sigma\bar{\zeta}_{\bar{x}\bar{x}}. \tag{5.152}$$

This tells us that the product of the mean depth d and the initial curvature of the wave surface z_{xx} is very small since σ is very small and the dimensionless curvature $\bar{\zeta}_{\bar{x}\bar{x}}$ is bounded, otherwise there would be an initial discontinuity in the wave surface. We now transform the hydrodynamic formulation given by (5.144)–(5.149) into dimensionless variables by using (5.150) and obtain (omitting the overbars)

$$\sigma(u_x + v_y) + w_z = 0, \tag{5.153}$$

$$\begin{aligned} \sigma[u_t + uu_x + vu_y + p_x] + wu_z &= 0, \\ \sigma[v_t + uv_x + vv_y + p_y] + wv_z &= 0, \\ \sigma[w_t + uw_x + vw_y + p_z + 1] + ww_z &= 0, \end{aligned} \tag{5.154}$$

$$w_y - v_z = 0, \qquad u_z - w_x = 0, \qquad v_x - u_y = 0, \tag{5.155}$$

$$\sigma[\zeta_t + u\zeta_x + v\zeta_y] = w \quad \text{at} \quad z = \zeta, \tag{5.156}$$

$$p = 1 \quad \text{at} \quad z = \zeta, \tag{5.157}$$

$$\sigma[uh_x + vh_y] + w = 0 \quad \text{at} \quad z = -h. \tag{5.158}$$

Next we construct a perturbation method by expanding the dependent variables in a power series in σ.

$$\begin{aligned} \mathbf{v} &= \sum_j \sigma^j \mathbf{v}_j, \\ \zeta &= \sum_j \sigma^j \zeta_j, \\ p &= \sum_j \alpha^j p_j, \end{aligned} \tag{5.159}$$

where the expansion coefficients $\mathbf{v}_j = (u_j, v_j, w_j)$, ζ_j, and p_j are functions of the dimensionless coordinates and the ζ_j's are evaluated at $z = 0$. Inserting these power series into (5.153)–(5.158) and equating like powers of σ yields the following relations amongst the expansion coefficients of the zeroth order:

$$\begin{aligned} w_{0,z} &= 0, \\ w_0 v_{0,z} &= 0, \\ w_0 u_{0,z} &= 0, \\ w_0 w_{0,z} &= 0, \\ w_{0,y} = v_{0,z}, \qquad u_{0,z} &= w_{0,x}, \qquad v_{0,x} = u_{0,y}, \\ w_0 = 0 &\quad \text{at} \quad z = \zeta_0, \\ p_0 = 1 &\quad \text{at} \quad z = \zeta_0, \\ w_0 = 0 &\quad \text{at} \quad z = -h. \end{aligned} \tag{5.160}$$

These equations yield

$$u_0 = u_0(x, y, t), \qquad v_0 = v_0(x, y, t),$$
$$w_0 \equiv 0, \tag{5.161}$$
$$p_0(x, y, \zeta_0, t) = 0.$$

Thus we see that the zeroth-order theory (which is equivalent to setting $\sigma = 0$) yields the following approximation in dimensionless coordinates to the original hydrodynamic system and BCs given by (5.153)–(5.158):

$$u_{0,x} + v_{0,y} = 0, \tag{5.162}$$

$$u_{0,t} + u_0 u_{0,x} + v_0 u_{0,y} = -p_{0,x},$$
$$v_{0,t} + u_0 v_{0,x} + v_0 v_{0,y} = -p_{0,y}, \tag{5.163}$$
$$w_{0,t} + u_0 w_{0,x} + v_0 w_{0,y} = -p_{0,z} - 1,$$

$$w_{0,y} - v_{0,z} = 0, \qquad u_{0,z} = w_{0,x} = 0, \qquad v_{0,x} = u_{0,y}, \tag{5.164}$$

$$\zeta_{0,t} + u_0 \zeta_{0,x} + v_0 \zeta_{0,y} = 0 \quad \text{at} \quad z = \zeta_0, \tag{5.165}$$

$$p_0 = 1 \quad \text{at} \quad z = \zeta_0, \tag{5.166}$$

$$u_0 h + v_0 h_y = 0 \quad \text{at} \quad z = -h. \tag{5.167}$$

Equations (5.162)–(5.167) constitute the formulation in dimensionless coordinates of the zeroth-order approximation to the shallow water theory with the appropriate BCs.

We now write down the first-order approximation which arises from equating the coefficients of σ to the first power, after having inserted the power series expansions in the continuity equation, equations of motion, and the BCs.

$$w_{1,z} = -u_{0,x} - v_{0,y}, \tag{5.168}$$

$$u_{0,t} + u_0 u_{0,x} + v_0 u_{0,y} + p_{0,x} = 0,$$
$$v v_{0,t} + u_0 v_{0,x} + v_0 v_{0,y} + p_{0,y} = 0, \tag{5.169}$$
$$p_{0,z} + 1 = 0,$$

$$\zeta_{0,t} + u_0 \zeta_{0,x} + v_0 \zeta_{0,y} = w_1 \quad \text{at} \quad z = \zeta_0, \tag{5.170}$$

$$u_0 h_x + v_0 h_y + w_1 = 0 \quad \text{at} \quad z = -h. \tag{5.171}$$

Equation (5.168) can be integrated at once, since u_0 and v_0 are independent of z, to yield

$$w_1 = -(u_{0,x} + v_{0,y})z + f(x, y, t). \tag{5.172}$$

The function f can be determined by using (5.171), and we get

$$f(x, y, t) = -[(u_0 h)_x + (v_0 h_y)]_{z=-h}. \tag{5.173}$$

This result tells us that the vertical component of the particle velocity is linear in the depth coordinate z.

In a similar fashion the second of the equations (5.169) can be integrated and the additive arbitrary function can be evaluated; the result is

$$p_0(x, y, z, t) = \zeta_0(x, y, t) - z, \tag{5.174}$$

which is clearly the hydrostatic pressure relation in dimensionless form. In the derivation of shallow water theory in the previous section relation (5.174) arose from the a priori assumption that the vertical component of the particle acceleration can be neglected. In this perturbation method (5.174) is derived from the first-order theory (to within the first power of the perturbation parameter σ).

The values of v_1 and p_0 obtained above are now inserted in the first and third equations of (5.169) and in (5.170) to finally yield

$$u_{0,t} + u_0 u_{0,x} + v_0 u_{0,x} + \zeta_{0,x} = 0, \tag{5.175}$$

$$v_{0,t} + u_0 v_{0,x} + v_0 v_{0,y} + \zeta_{0,y} = 0, \tag{5.176}$$

$$\zeta_{0,t} + [u_0(\zeta_0 + h)_x + [v_0(\zeta_0 + h)]_y = 0, \tag{5.177}$$

which are the equations for u_0, v_0, and ζ_0 which, as mentioned, depend only on x, y, t.

We continue this process to obtain higher-order approximations. To this end, we insert the power series expansions given by (5.159) into the dimensionless form of the hydrodynamic system given by (5.153)-(5.158), and equate coefficients of like powers of σ. Equating the coefficients of σ^n in these equations yields the following between the nth and the $(n-1)$st expansion coefficients:

The continuity equation gives

$$[w_z]_n = -[u_x + v_y]_{n-1}, \tag{5.178}$$

where the notation $[\quad]_k$ means that the brackets contain the kth expansion coefficients, where $k = n$, $n-1$, respectively.

The equations of motion yield

$$\begin{aligned} [wu_z]_n &= -[u_t + uu_x + vu_y + wu_z + p_x]_{n-1}, \\ [wv_z]_n &= -[v_t + uv_x + vv_y + wv_z + p_y]_{n-1}, \\ [ww_z]_n &= -[w_t + uw_x + vw_y + ww_z + p_z + 1]_{n-1}. \end{aligned} \tag{5.179}$$

The nth expansion coefficients all satisfy the same scalar irrotationality equations.

The BCs yield the following:

$$[w]_n = -[\zeta_t + u\zeta_x + v\zeta_y]_{n-1}, \qquad [p]_n = 1 \qquad \text{at} \quad z = \zeta_n, \tag{5.180}$$

$$[w]_n = -[uh_x + vh_y]_{n-1}, \qquad \text{at} \quad z = -h, \tag{5.181}$$

where $n = 0, 1, 2, \ldots$ and $[\quad]_{n-1} = 0$ for $n - 1 < 0$. In (5.175)-(5.178) it is clear that the left-hand sides of the equations contain the nth expansion coefficients while the right-hand sides contain the $(n-1)$st expansion coefficients. This means that we have an iteration scheme whereby one can calculate the nth expansion coefficients in terms of the $(n-1)$st given the zeroth-order expansion coefficients which we need to solve. In general, all perturbation methods enjoy this iterative property: the nth expansion coefficients depend ultimately on the zeroth-order expansion coefficients.

CHAPTER 6

Sound Waves

Introduction

In Chapter 5 we studied water waves which were assumed to be waves which propagate in an incompressible fluid. Next in order of complexity is the phenomenon of sound wave propagation in air. In investigating this phenomenon we must treat the medium (air) as a compressible fluid, in contrast to the previous chapter where we were allowed to consider the medium (water) as an incompressible fluid.

In this chapter we start our investigations of wave propagation in a compressible fluid medium by considering the small amplitude oscillatory motion of a compressible fluid. The small amplitude waves that are produced are called *sound waves.* Since the amplitude is small, a linearized theory can be developed and the linearized wave equation can be thus applied.

We recall that in Chapter 4 the vibrating string was studied as a prototype for the phenomenon of sound production by stringed instruments such as the violin. Sound waves produced by such musical instruments travel in air. There is ample experimental evidence to show that sound waves are essentially small amplitude waves which satisfy an adiabatic equation of state. In subsequent chapters we shall take up the study of nonlinear wave propagation in a compressible gas where the method of characteristics is used to solve large amplitude wave propagation problems. This prepares us for the investigation of shock wave propagation in air, which will be treated in the next chapter. Indeed, we may look at the wavefront of a sound wave traveling in air as a shock front of infinitely weak strength.

6.1. Linearization of the Conservation Laws

We first state the conservation laws for a compressible adiabatic fluid in Eulerian coordinates in full nonlinear form.

Continuity equation or conservation of mass

$$\operatorname{div} \rho\mathbf{v} = -\rho_t, \tag{6.1}$$

where $\mathbf{v}$ is the particle velocity and ρ is the particle density.

Conservation of momentum

$$\frac{d\mathbf{v}}{dt} = (\mathbf{v}_t + \mathbf{v} \cdot \mathbf{grad})\mathbf{v} = -\left(\frac{1}{\rho}\right) \mathbf{grad}\, p, \tag{6.2}$$

where p is the fluid pressure. This is Euler's equation of motion. We assume zero external force.

Adiabatic equation of state

$$p = A\rho^\gamma, \qquad A = \frac{p_0}{\rho_0^\gamma}, \qquad \gamma = \frac{C_p}{C_v}, \tag{6.3}$$

where C_p and C_v are the heat capacities at constant pressure and volume, respectively. p_0, ρ_0 are defined at the equilibrium state. The wave speed c is given by

$$c^2 = \frac{dp}{d\rho} = \gamma A \rho^{\gamma-1}, \tag{6.4}$$

for an adiabatic gas, upon using (6.3). A gas obeying the adiabatic equation of state is called a *polytropic gas.*

The system (6.1)–(6.3) completely describes the propagation of waves in an extended fluid medium. The system consists of three nonlinear partial differential equations (PDEs) for the unknown $\mathbf{v}$, p, and ρ to be solved for as functions of time and space, for an adiabatic gas neglecting viscosity. For a given problem, initial and boundary conditions must be prescribed.

Linearization

Since sound waves are small amplitude waves, $\mathbf{v}$ is small, and we can thus neglect the nonlinear or convective term $(\mathbf{v} \cdot \mathbf{grad})\mathbf{v}$ of the particle acceleration in Euler's equation (6.2). For the same reason, the relative changes in pressure and density are small in the sense that we neglect second-order terms (the hypothesis of linearity). We therefore write p and in the following form:

$$p = p_0 + p', \qquad \rho = \rho_0 + \rho', \tag{6.5}$$

where p' and ρ' are the variations in pressure and density due to a traveling sound wave (perturbed pressure and density). For linearity we have the conditions

$$\frac{p - p_0}{p_0} = \frac{p'}{p_0} \ll 1, \qquad \frac{\rho'}{\rho_0} \ll 1.$$

Since the product of small terms such as $\rho'\mathbf{v}$ are second order and are therefore neglected, the linearized form of the continuity equation (6.1) becomes

$$\rho_0 \operatorname{div} \mathbf{v} = -\rho'_t. \tag{6.6}$$

Using the same approximation the linearized form of the equation of motion (6.2) becomes

$$\mathbf{v}_t = -\left(\frac{1}{\rho_0}\right) \mathbf{grad}\, p'. \tag{6.7}$$

Linearizing the adiabatic equation of state yields

$$c^2 = c_0^2 = \frac{\gamma p_0}{\rho_0}, \tag{6.8}$$

which tells us that the wave speed is constant. The system (6.6)-(6.8) represents the linearized mathematical description for wave propagation in an adiabatic gas. This is a set of three linear PDEs for $\mathbf{v}$, p', and ρ' for an unsteady three-dimensional adiabatic inviscid flow field.

Linear Wave Equations

We first eliminate p' from the three linearized equations by recognizing that $p' = p'(\rho')$, explicitly given by (6.3), so that $p'_x = (dp'/d\rho')\rho'_x = c_0^2\rho'_x$, $p'_y = c_0^2\rho'_y$, $p'_z = c_0^2\rho'_z$. This gives

$$\mathbf{grad}\, p' = c_0^2\, \mathbf{grad}\, \rho'. \tag{6.9}$$

Inserting (6.9) into (6.7) gives the linearized Euler equation for ρ'.

$$\rho_0 \mathbf{v}_t = -c_0^2\, \mathbf{grad}\, \rho'. \tag{6.10}$$

Equations (6.6) and (6.10) are the two linear PDEs for $\mathbf{v}$ and ρ'. We now eliminate $\mathbf{v}$ from these two equations by multiplying both sides of (6.10) by the div operator, which gives

$$\rho_0 \operatorname{div} \mathbf{v}_t = -c_0^2 \nabla^2 \rho'.$$

We then differentiate (6.6) with respect to t and use the result in the above expression to obtain

$$c_0^2 \nabla^2 \rho' = \rho'_{tt}, \tag{6.11}$$

which is the three-dimensional wave equation for ρ'. It is easily seen that we can eliminate ρ' from (6.6) and (6.10) to obtain

$$c_0 \nabla^2 \mathbf{v} = \mathbf{v}_{tt}, \tag{6.12}$$

which is the three-dimensional vector wave equation for $\mathbf{v}$.

And finally we can eliminate ρ' from (6.6) and (6.10) to obtain two linear PDEs for $\mathbf{v}$ and p'. We can then eliminate $\mathbf{v}$ from these resulting equations to obtain

$$c_0\nabla^2 p' = p'_{tt}, \tag{6.13}$$

which is the linear wave equation for p'.

If the fluid is irrotational a velocity potential $\phi(x, y, z, t)$ exists such that **grad** $\phi = \mathbf{v}$ or $\mathbf{v} = (\phi_x, \phi_y, \phi_z)$. From the above it follows that

$$c_0^2\nabla^2\phi = \phi_{tt}, \tag{6.14}$$

or ϕ satisfies the three-dimensional scalar wave equation.

6.2. Plane Waves

In Chapter 2 the properties of the one-dimensional wave equation were treated in detail. These properties will be used here in our discussion of plane sound wave propagation. A *plane wave* is defined as a sound wave in which all the dependent variables (pressure, particle velocity, density, etc.) depend on only one coordinate (x, say). This means that the flow field is completely homogeneous in the yz plane—at a given time all planes parallel to the yz plane have the same flow field. The velocity $\mathbf{v} = (u(x, t), 0, 0)$. The physical meaning of such a plane wave is that the direction of wave propagation is parallel to the x axis, and the wavefront is a planar surface normal to this direction. Since $\mathbf{v}$ is in the direction of wave propagation, sound waves are said to be *longitudinal waves*.

The conservation laws given by (6.1)-(6.4) reduce to the following system for a small amplitude plane wave:

Continuity equation:

$$\rho_0 u_x = -\rho_t. \tag{6.15}$$

(In our notation we shall omit the prime over the perturbed density and pressure.)

Euler's equation of motion and adiabatic condition:

$$u_t = -\left(\frac{1}{\rho_0}\right)p_x = -\left(\frac{1}{\rho_0}\right)c_0^2\rho_x, \tag{6.16}$$

where c_0 is given by (6.8).

Following Lamb [25, p. 476] we define the dimensionless variable s as the *condensation* or the relative change in density, so that

$$s = \frac{\rho - \rho_0}{\rho_0}. \tag{6.17}$$

Inserting (6.17) into (6.15) and (6.16) yields

$$u_t = -c_0 s_x, \tag{6.18}$$

$$s_t = -u_x. \tag{6.19}$$

Eliminating s from these two equations gives

$$c_0^2 u_{xx} = u_{tt}, \tag{6.20}$$

which is the one-dimensional wave equation for the particle velocity whose general solution is $u = F(x - c_0 t) + G(x + c_0 t)$, where the arbitrary functions $F(x - c_0 t)$ and $G(x + c_0 t)$ generate progressing and regressing waves, respectively; these waves travel with constant wave speed c_0. It appears from (6.19) that the corresponding value of s is given by

$$c_0 s = u = F(x - c_0 t) - G(x + c_0 t).$$

For a single waveform either G or F is zero so that $u = \pm c_0 s$, where the plus sign is taken for a progressing wave and the minus sign for a regressing wave.

The analogy with water waves of large wavelength is obvious if we identify c_0 with $\sqrt{gh}$ as given by the shallow water approximation (5.86).

6.3. Energy and Momentum

We first derive an expression for the energy of a sound wave. (In this section we revert back to using p' and ρ' for the perturbed pressure and density, since we shall display the terms involving p_0 and ρ_0 for the equilibrium state.) The energy in a unit volume of fluid is

$$\tfrac{1}{2}\rho v^2 + \rho e,$$

where v is the magnitude of $\mathbf{v}$. The first term is the kinetic energy and the second the internal energy, where e is the internal energy per unit mass. We now substitute $\rho = \rho_0 + \rho'$ in this expression and recognize that ρ_0 is the density where $v = 0$. The term $\frac{1}{2}\rho' v^2$ is a quantity of third order which we neglect, since we keep terms only up to second-order small quantities. Hence, to within second-order terms, the above expression becomes

$$\rho_0 e_0 + \rho' \frac{\partial(\rho e)}{\partial \rho} + \tfrac{1}{2}\rho'^2 \frac{\partial^2(\rho e)}{\partial \rho^2} + \tfrac{1}{2}\rho_0 v^2, \tag{6.21}$$

where the partial derivatives are evaluated at constant entropy, since sound waves propagate in an adiabatic fluid. Equation (6.21) is the approximation to the total energy compatible with the theory of small amplitude oscillations. We now transform this equation by using the first law of thermodynamics which is

$$de = T\,dS - p\,dV = T\,dS + \left(\frac{p}{\rho^2}\right) d\rho,$$

where S is the entropy per unit mass and V is the volume. From this we have

$$\left[\frac{\partial(\rho e)}{\partial \rho}\right]_S = e + \frac{p}{\rho} = w,$$

$$\left[\frac{\partial^2(\rho e)}{\partial \rho^2}\right]_S = \left(\frac{\partial w}{\partial \rho}\right)_S = \left(\frac{\partial w}{\partial p}\right)_S \left(\frac{\partial p}{\partial \rho}\right)_S = \frac{c_0^2}{\rho_0},$$

where w is the enthalpy per unit mass. Applying these equations to the approximation (6.21) for the total energy in the unit volume of fluid yields

$$\rho_0 e_0 + w_0 \rho' + \frac{\frac{1}{2} c_0^2 \rho'^2}{\rho_0} + \tfrac{1}{2}\rho_0 v^2.$$

The term $(\rho_0 e_0)$ represents the energy per unit volume when the fluid is at rest and thus does not contribute to the propagating sound wave. The next term $w_0 \rho'$ represents the change in energy due to the change in the mass of fluid per unit volume, and is zero because of the conservation of mass which states that the total mass of the fluid is unchanged, so that

$$\int \rho \, dV = \int \rho_0 \, dV \qquad \text{or} \qquad \int \rho' \, dV = 0.$$

Therefore the total change in the energy of the fluid caused by the traveling sound wave in the volume V_0 is

$$\int_{V_0} \left[\tfrac{1}{2}\rho_0 v^2 + \frac{\frac{1}{2} c_0^2 \rho'^2}{\rho_0}\right] dV,$$

where the integration is taken over the volume V_0. We shall call the integrand in this expression E which is defined as the *sound energy density*. We have

$$E = \tfrac{1}{2}\rho_0 v^2 + \frac{\frac{1}{2} c_0^2 \rho'^2}{\rho_0} = \rho_0 v^2, \tag{6.22}$$

where we have used the fact that for a sound wave

$$\rho' = \frac{\rho_0 v}{c_0}. \tag{6.23}$$

Equation (6.23) can be derived by the following argument: Consider a progressing wave. Referring to (6.16) $u = u(x - c_0 t)$ and $\rho' = \rho'(x - c_0 t)$. Substituting these functional relations for u and ρ' into (6.16) yields $du(\xi)/d\xi = (c_0/\rho_0)\, d\rho'(\xi)/d\xi$, where $\xi = x - c_0 t$. Integrating with respect to the argument ξ yields (6.23) (where u is replaced by v).

Equation (6.22) can immediately be derived from a well-known principle in mechanics: The mean (time average) potential energy of a system undergoing small amplitude oscillations is equal to its mean kinetic energy. Since the mean kinetic energy $\bar{T} = \frac{1}{2}\int \rho_0 \overline{v^2} \, dV$, where the time average is taken over the volume V_0, it follows that the mean total sound energy is $\int \bar{E} \, dV =$

$\int \rho_0 \overline{v^2}\, dV$, so that the sound energy density as given by (6.22) is the time average energy density.

We now calculate the mean flux of energy through a closed surface S_0 bounding the volume V_0 of the fluid in which a sound wave is propagated. The energy flux density is

$$\rho \mathbf{v}[\tfrac{1}{2}v^2 + w]. \tag{6.24}$$

To derive (6.24) we start by calculating the rate of change of the total energy per unit volume and write

$$[\tfrac{1}{2}\rho v^2 + \rho e]_t = \tfrac{1}{2}v^2 \rho_t + \rho \mathbf{v} \cdot \mathbf{v}_t + (\rho e)_t.$$

Using the continuity equation (6.1) and the equation of motion (6.2), the rate of change of kinetic energy becomes

$$[\tfrac{1}{2}\rho v^2]_t = -\tfrac{1}{2}v^2 \operatorname{div}(\rho \mathbf{v}) - \mathbf{v} \cdot \mathbf{grad}\, p - \rho \mathbf{v} \cdot (\mathbf{v} \cdot \mathbf{grad})\mathbf{v}.$$

We now set $\mathbf{v} \cdot (\mathbf{v} \cdot \mathbf{grad})\mathbf{v} = \frac{1}{2}\mathbf{v} \cdot \mathbf{grad}\, v^2$, and use the thermodynamic relation $dw = T\,dS + (1/\rho)\, dp$ (w is the enthalpy) to give $\mathbf{grad}\, p = \rho\, \mathbf{grad}\, w - \rho T\, \mathbf{grad}\, S$. The above equation then becomes

$$[\tfrac{1}{2}\rho v^2]_t = -\tfrac{1}{2}v^2 \operatorname{div}(\rho \mathbf{v}) - \rho \mathbf{v} \cdot \mathbf{grad}(\tfrac{1}{2}v^2 + w) + \rho T \mathbf{v} \cdot \mathbf{grad}\, S.$$

We now transform the term $(\rho e)_t$ by using the first law in the form $de = T\,dS + (p/\rho^2)\, d\rho$, the definition of enthalpy $w = e + p/\rho$, the adiabatic condition $dS/dt = \mathbf{v} \cdot \mathbf{grad}\, S + S_t = 0$, and the continuity equation. We get

$$(\rho e)_t = w\rho_t + \rho T S_t = -w \operatorname{div}(\rho \mathbf{v}) - \rho T \mathbf{v} \cdot \mathbf{grad}\, S.$$

Combining the above results for the change in the kinetic and internal energies, we finally obtain

$$[\tfrac{1}{2}\rho v^2 + \rho e]_t = -\operatorname{div}[\rho \mathbf{v}(\tfrac{1}{2}v^2 + w)].$$

We now integrate this rate of change of the total energy over the volume V_0 and use the divergence theorem to transform the volume integral to a surface integral over the surface S_0 bounding V_0. We get

$$\frac{\partial}{\partial t}\int [\tfrac{1}{2}\rho v^2 + \rho e]\, dV = -\oint [\tfrac{1}{2}v^2 + w]\rho \mathbf{v} \cdot d\mathbf{S},$$

where the volume integral is taken over V_0, $d\mathbf{S}$ is the element of surface area, and the surface integral is taken over S_0 bounding V_0. The left-hand side represents the rate of change of the total energy of fluid in the volume V_0. This is equal to the right-hand side which represents the amount of energy flowing out of V_0 across the bounding surface S_0. Hence the integrand in the surface integral is the energy flux density $\rho \mathbf{v}[\frac{1}{2}v^2 + w]$ given by (6.24).

According to the hypothesis of small oscillations we neglect the term v^2 in (6.24) which is of third order. Hence the mean energy flux density in the sound wave is $\overline{\rho w v}$. Again, using primes to represent perturbed terms, the average flux density becomes $\overline{(\rho_0 + \rho')(w_0 + w')\mathbf{v}} = \rho_0 w_0 \bar{\mathbf{v}} + \rho_0 \overline{w' \mathbf{v}}$, neglecting third-order terms. We have $w' = (\partial w/\partial p)_S p' = p'/\rho_0$, since $(\partial w/\partial p)_S = 1/\rho_0$. It follows that $\overline{\rho w \mathbf{v}} = \rho_0 w_0 \overline{\mathbf{v}} + \overline{p'\mathbf{v}}$. Therefore the total energy flux across the surface S_0 becomes

$$\oint (w_0 \overline{\rho_0 \mathbf{v}} + \overline{p'\mathbf{v}}) \cdot d\mathbf{S}.$$

Since the total quantity of fluid is unchanged in the volume V_0 (the continuity law) the time average of the mass flux through S_0 must vanish. Hence the energy flux is reduced to

$$\oint \overline{p'\mathbf{v}} \cdot d\mathbf{S}.$$

This means that the mean sound energy flux density is represented by the vector $\mathbf{q}$ where

$$\overline{\mathbf{q}} = \overline{p'\mathbf{v}}. \tag{6.25}$$

The law of *conservation of sound energy* is

$$\operatorname{div}(p'\mathbf{v}) = \operatorname{div} \mathbf{q} = -E_t. \tag{6.26}$$

In (6.26) the vector $\mathbf{q} = p'\mathbf{v}$ plays the role of the sound energy flux so that this law of conservation of sound energy is also valid at any instant, as well as for the time average.

In a traveling sound wave the pressure variation p' is related to the particle speed v by

$$p' = \rho_0 c_0 v. \tag{6.27}$$

This relationship can easily be obtained for the one-dimensional case by inserting progressing waves of the form $u = u(x - c_0 t)$ and $p' = p'(x - c_0 t)$ into (6.16) (replacing p by p'). Let the unit vector $\mathbf{n}$ be in the direction of $\mathbf{v}$ or the direction of wave propagation, since we have a longitudinal wave. We obtain

$$\mathbf{q} = c_0 \rho_0 v^2 \mathbf{n} = c_0 E \mathbf{n}. \tag{6.28}$$

Equation (6.28) tells us that the energy flux density in a plane sound wave equals the sound energy density multiplied by the velocity of sound.

Momentum Flux Density

We now calculate the momentum flux density of a sound wave by first determining the total momentum of the fluid due to the traveling wave. To this end, we consider a *wave packet* which is a sound wave occupying a finite region of space nowhere bounded by solid walls. Let $\mathbf{j}$ be the mass flux density. Then, the momentum per unit volume of fluid given by $\mathbf{j}$ equals $\rho\mathbf{v}$. Since the perturbed density is related to the perturbed pressure by

$\rho' = p'/c_0^2$ and $\mathbf{q} = \rho'\mathbf{v}$, we obtain

$$\mathbf{j} = \rho_0 \mathbf{v} + \frac{\mathbf{q}}{c_0^2} = \rho_0 \,\mathbf{grad}\, \phi + \frac{\mathbf{q}}{c_0^2}. \tag{6.29}$$

Since a sound wave involves potential flow, we wrote $\mathbf{v} = \mathbf{grad}\, \phi$. We point out the obvious fact that potential flow is valid also for nonlinear or large amplitude waves since an irrotational flow field also satisfies the exact or nonlinear Euler equation for $\mathbf{curl}\, \mathbf{v} = 0$. We again consider the fluid occupying a volume V_0 enclosed by a surface S_0. The wave packet is assumed to occupy only this volume. The total momentum of the fluid in V_0 is $\int \mathbf{j}\, dV$ which we transform to a surface integral over S_0 by using the divergence theorem. It turns out that $\int \mathbf{grad}\, \phi\, dV = \oint \phi\, d\mathbf{S} = 0$, since ϕ vanishes outside S_0 because the sound wave only occupies V_0. Therefore the total momentum of the sound wave becomes

$$\int \mathbf{j}\, dV = \frac{1}{c_0^2} \int \mathbf{q}\, dV, \tag{6.30}$$

which, in general, is not zero, thus showing a transfer of momentum across S_0. This is a second-order effect since $\mathbf{q}$ is a second-order quantity.

We now consider a sound wave propagating in an infinite medium and damped at infinity so that w', $\mathbf{v}$, etc., vanish at infinity. We calculate the time average of p'. For this we need the second-order term since p' vanishes to within the linear approximation. We start by taking the time average of Bernoulli's equation, setting $w = w_0 + w'$ and incorporating w_0 in the constant, which we then set equal to zero since it is constant throughout all space. We obtain

$$\overline{w'} + \tfrac{1}{2}\overline{v^2} = 0. \tag{6.31}$$

We now expand w' in powers of p' to within second-order terms. Using the fact that $(\partial w'/\partial p')_S = 1/\rho_0$, we obtain

$$\begin{aligned} w' &= \left(\frac{\partial w'}{\partial p'}\right)_S p' + \tfrac{1}{2}\left(\frac{\partial^2 w'}{\partial p'^2}\right)_S p'^2 = \frac{p'}{\rho_0} - \left(\frac{p'^2}{2\rho_0^2}\right)\left(\frac{\partial \rho'}{\partial p'}\right)_S \\ &= \frac{p'}{\rho_0} - \frac{p'^2}{2c_0^2\rho_0^2}. \end{aligned}$$

Substituting this equation into (6.31) yields

$$\overline{p'} = -\tfrac{1}{2}\rho_0 \overline{v^2} + \frac{\overline{p'^2}}{2\rho_0 c_0^2} = -\tfrac{1}{2}\rho_0\overline{v^2} + \frac{\overline{\rho'^2}c_0^2}{2\rho_0}. \tag{6.32}$$

Equation (6.32) shows second-order quantities and can be calculated by using the appropriate solutions of the linearized equations of motion. The expansion of the mean perturbed density in powers of the mean perturbed pressure to within second-order terms is

$$\overline{p'} = \left(\frac{\partial p'}{\partial p'}\right)_S \overline{p'} + \frac{1}{2}\left(\frac{\partial^2 \rho'}{\partial p'^2}\right)_S \overline{p'^2}. \tag{6.33}$$

If the sound wave is regarded as traveling in the volume V_0 then $v = c_0\rho'/\rho_0$ so that $\overline{p'} = 0$ from (6.32). This means that the average perturbed pressure effect in the plane wave is higher than second order. However, the second-order term for ρ' in (6.33), $\rho' = \frac{1}{2}(\partial^2\rho'/\partial p'^2)_S p'^2$, is not zero. To within the same approximation the mean value of the *momentum flux density tensor* $\bar{\mathbf{\Pi}}$ in the traveling plane sound wave is

$$\bar{\mathbf{\Pi}} = \overline{p'}\delta_{ij} + \overline{\rho' v_i v_j} = p_0\delta_{ij} + \rho_0\overline{v_i v_j}. \tag{6.34}$$

In general, the momentum flux density tensor $\mathbf{\Pi}$ is given by

$$\mathbf{\Pi} = \mathbf{n}p + \mathbf{v}(\mathbf{v}\cdot\mathbf{n}). \tag{6.35}$$

The (ij)th component of $\mathbf{\Pi}$ is

$$\Pi_{ij} = p\delta_{ij} + \rho v_i v_j. \tag{6.36}$$

Equation (6.35) or (6.36) can be derived as follows: We shall use tensor notation. The continuity equation can be written as

$$(\rho v_j)_{,j} = -\rho_t,$$

where $(\rho v_j)_{,j} \equiv (\partial v_1)/\partial x_1 + (\partial v_2)/\partial x_2 + (\partial v_3)/\partial x_3$.

Euler's equations of motion is

$$v_{i,t} = -v_j v_{i,j} - \left(\frac{1}{\rho}\right)p_x \quad \text{(summed over } j\text{)}, \qquad i, j = 1, 2, 3.$$

Using the continuity equation and Euler's equations we obtain

$$\begin{aligned}(\rho v_i)_{,t} &= \rho v_{i,t} + \rho_t v_i \\ &= -\rho v_j v_{i,j} - p_x - v_i(\rho v_j)_{,j} \\ &= -p_{x_i} - (\rho v_i v_j)_{,j} \\ &= -\delta_{ij} p_{x_i} - (\rho v_i v_j)_{,j} \\ &= -\Pi_{ij,j},\end{aligned}$$

where Π_{ij} is the (ij)th component of the momentum flux density tensor given by (6.36).

We now go back to the expression (6.34) for the mean value of the momentum flux density tensor $\bar{\mathbf{\Pi}}$. The equilibrium pressure p_0 does not relate to the sound wave. For the second term, which is quadratic in the particle velocity, we again introduce the unit vector $\mathbf{n}$ in the $\mathbf{v}$ direction or the direction of wave propagation. Using (6.22) the mean value of the *momentum flux density tensor* in a sound wave becomes (using tensor notation)

$$\bar{\Pi}_{ij} = \bar{E} n_i n_j. \tag{6.37}$$

For example, if the wave is propagated in the x direction the only nonzero component of $\bar{\Pi}_{ij}$ is $\bar{\Pi}_{xx} = E$. Thus, according to the approximation used the mean momentum flux density has magnitude E and is in the direction of the wave propagation.

6.4. Reflection and Refraction of Sound Waves

Consider a monochromatic (single frequency) planar sound wave propagating from one fluid to another fluid of different density such that there is a planar interface between the two fluids (i.e., air and water). If the incident sound wave traveling in medium (1) approaches the interface at an angle of incidence different from the normal, a refracted wave propagates into medium (2) with a refracted angle, in general, different from the angle of incidence. The *angle of incidence* is defined as the angle that the incident sound wave makes with the normal to the surface (interface between regions (1) and (2)). The *angle of refraction* is the angle the refracted wave makes with the normal. Another wave arising from the interface surface is at least partly reflecting. This is a *reflected wave* that propagates back into region (1) with an *angle of reflection* (the angle between the reflected wave and the normal) equal to the angle of incidence. Consequently, the motion in medium (1) is due to a combination of the incident and reflected waves, while the motion in medium (2) is due only to the refracted wave.

The relationship amongst these three waves is determined by the boundary conditions BCs that prevail on the interface or surface of separation. These BCs require that the normal components of the particle velocity be equal, and the pressure be equal on either side of the interface.

Before we continue our discussion of reflection and refraction of sound waves we interpose some brief remarks on the nature of the solution of traveling sound waves *traveling in an arbitrary direction.* To this end, we shall introduce a quantity **k** called the wave vector, which is defined below. Since a sound wave is longitudinal, it is therefore irrotational so that the vorticity vector $\boldsymbol{\psi}$ is identically zero. Therefore, instead of dealing with the velocity vector **v** we may focus our attention on the velocity potential ϕ. Solving for ϕ as a function of space and time allows us to determine the unsteady velocity field. Suppose a sound wave travels in an arbitrary direction **n**, the unit normal to the wave surface. The current discussion briefly treats such sound waves propagating in an arbitrary direction in three-dimensional space by representing time-harmonic solutions for the velocity potential. For simplicity we consider monochromatic waves (having a single frequency ω). The wave speed c is c_i in medium (i) $(i = 1, 2)$. We now define the *wave vector* **k**† by

$$\mathbf{k}=\left(\frac{\omega}{c}\right)\mathbf{n}=\left(\frac{2\pi}{\lambda}\right)\mathbf{n}. \tag{6.38}$$

We note that even though ω is the same in mediums (1) and (2), since we may have $c_2 \neq c_2$, it follows that the wavelengths $\lambda_1 \neq \lambda_2$, so that the wavelengths are different in the two media if the wave speeds are different. Introducing the vector $\mathbf{r}=(x, y, z)$ as the radius vector to a point on the wave surface whose coordinates are (x, y, z), it follows that the scalar product $\mathbf{k}\cdot\mathbf{r}=(\omega/c)\mathbf{n}\cdot\mathbf{r}$ is the projection of $\mathbf{r}$ in the direction normal to the surface or in the direction of the propagating wave. The velocity potential for a monochromatic sound wave traveling in the $\mathbf{n}$ direction has a time-harmonic solution of the form

$$\phi(x, y, z, t)=\text{Re}[A \exp(i(\mathbf{k}\cdot\mathbf{r}\mp\omega t))], \tag{6.39}$$

where A is a constant called the *complex amplitude.* We may set $A=ae^{i\alpha}$ where a is the real amplitude and α is the real phase angle. The minus sign corresponds to a progressing wave and the plus sign to a regressing wave. Clearly A may take on different values for each of these waves. We may then rewrite (6.39) as

$$\phi(x, y, z, t)=a\cos(\mathbf{k}\cdot\mathbf{r}\mp\omega t+\alpha). \tag{6.40}$$

Let $\mathbf{k}=(k_x, k_y, k_z)$ and the direction cosines of $\mathbf{n}$ be (l, m, n). Then (6.39) or (6.40) become

$$\begin{aligned}\phi(x, y, z, t)&=A\exp(i(lk_x x+mk_y y+nk_z z\mp\omega t))\\&=a\cos(lk_x x+mk_y y+nk_z z\mp\omega t+\alpha),\end{aligned}$$

where $l^2+m^2+n^2=1$.

For a progressing sound wave propagating in the positive x direction, these expressions reduce to

$$\begin{aligned}\phi(x, t)&=A\exp(i(kx-\omega t)\\&=a\cos(kx-t+\alpha),\end{aligned}$$

where $k_x\equiv k=\omega/c$.

Monochromatic waves are very important in wave propagation problems since any wave whatsoever can be represented as a linear combination of monochromatic waves with various wave vectors, amplitudes, and frequencies. This decomposition of an arbitrary waveform into monochromatic waves is an example of an expansion in a Fourier series or Fourier transform,

† Note that **k** has the dimension of L^{-1}. The magnitude of **k** is called the *wave number.*

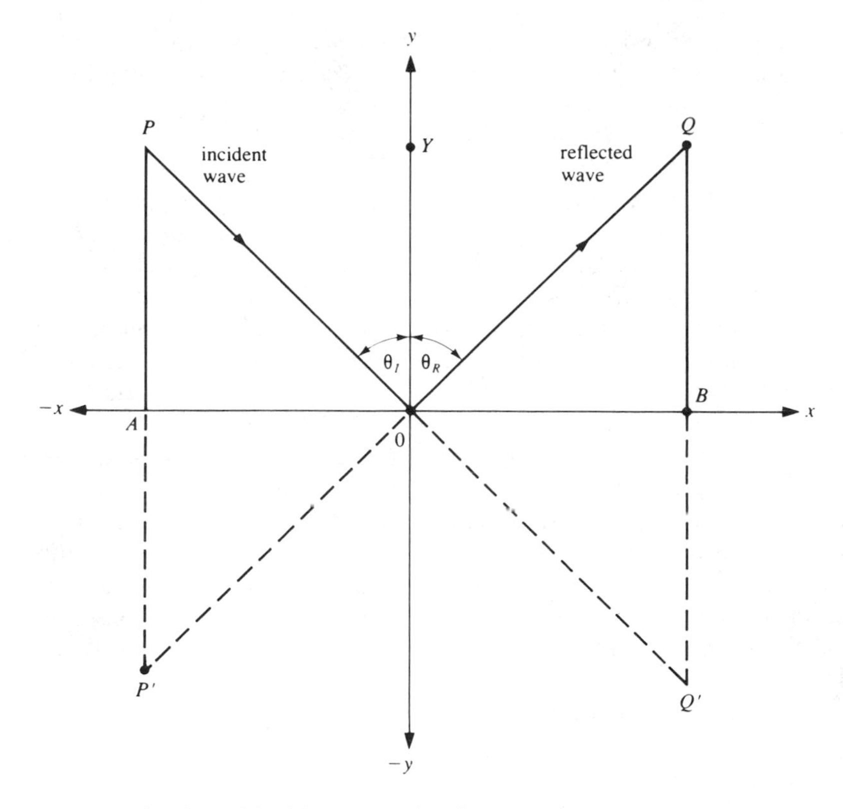

Fig. 6.1. Reflection of incident wave showing that $PO + OQ$ is minimum path.

and is called the *spectral resolution*. The terms of the expansion are called the *monochromatic components* or *Fourier components* of the wave.

We now return to the reflection and refraction of a sound wave from medium (1) into medium (2).† We take the (x, y) plane as the plane in which the waves propagate. We let the $y = 0$ plane (the x axis) be the interface surface, and let the region $y > 0$ represent medium (1) and region $y < 0$ represent medium (2).

We first consider the simpler case of a purely reflected sound wave at the x axis which acts as a mirror. The mathematics of reflection is the same as for a light wave except that the physics is different in that a *light wave is a transverse wave.* Figure 6.1 shows an incident wave PO reflected at O

---

† The reflection of stress waves from a free surface and the refraction of these waves into a medium of different refractive index is discussed in detail in [13, Chap. 4, p. 107]. In this work it was pointed out that stress waves can be transverse as well as longitudinal and hence more than one reflected and refracted wave can result from a given incident wave, in contradistinction to sound waves where a single reflected and refracted wave ensue.

(on the x or horizontal axis) into a reflected wave OQ. Drop perpendicular lines from P and Q to the x axis and extend them. Let P' be the image of P, and Q' the image of Q so that $PA = AP'$ and $QB = BQ'$. From the figure it follows that $\measuredangle YOP = \theta_I = \measuredangle YOQ = \theta_R$, so that the angle of incidence equals the angle of reflection. Also, from the figure, we have $OQ' = OQ$. This fact tells us that the path POQ' equals the path POQ. Now POQ' is a straight line by construction and is thus the shortest path from P to Q' or P to O to Q. We thus have the following important minimum principle: Consider the given points P and Q where P is the source of the sound wave and Q is the sound receptor. Let O be a variable point on the x axis. The particular position of O which minimizes the sum of the lengths PO and OQ is the point O on the x axis where the incident wave starting at P is reflected into the reflected wave that passes through Q.†

We now investigate the refraction into medium (2) of an incident wave propagating in medium (1). We first prove the assumption used above that the reflected wave lies in the same plane as the incident wave, and also prove the same statement for the refracted wave. The incident wave lies in the (x, y) plane. Then $k_z = 0$ for the incident wave. Since the BCs on the x axis do not depend on the z coordinate we must have $k_z = 0$ for the reflected and the refracted waves so that they also lie in the (x, y) plane. It also follows from the BCs that ω and k_x are, respectively, the same in media (1) and (2). From the definition of $\mathbf{k}$ given by (6.38) we have

$$k_{x,\text{inc}} = k_{x,\text{refl}} = \left(\frac{\omega}{c}\right) \sin \theta,$$

for the incident and reflected waves, thus giving another proof that the angle of incidence equals the angle of reflection.

For an incident wave propagating in medium (1) with a wave speed c_1 and a refracted wave in medium (2) with a wave speed c_2, let k_{x1} be the x component of the wave number of the incident wave and k_{x2} be the corresponding component of the wave number of the refracted wave. Then, since $k_{x1} = k_{x2}$, we have from (6.38)

$$k_{x1} = \left(\frac{\omega}{c_1}\right) \sin \theta_1 = k_{x2} = \left(\frac{\omega}{c_2}\right) \sin \theta_2,$$

† This is essentially the same principle of the equality of the angle of incidence and reflection of a light ray assuring the shortest possible path of the ray in going from its source to a reflected point. This principle goes back to antiquity; it was discovered by Heron of Alexandria. Incidentally, this minimization principle can be extended to the game of billiards by constructing multiple reflecting waves, as long as we assume perfectly elastic collisions and neglect rotational effects such as English.

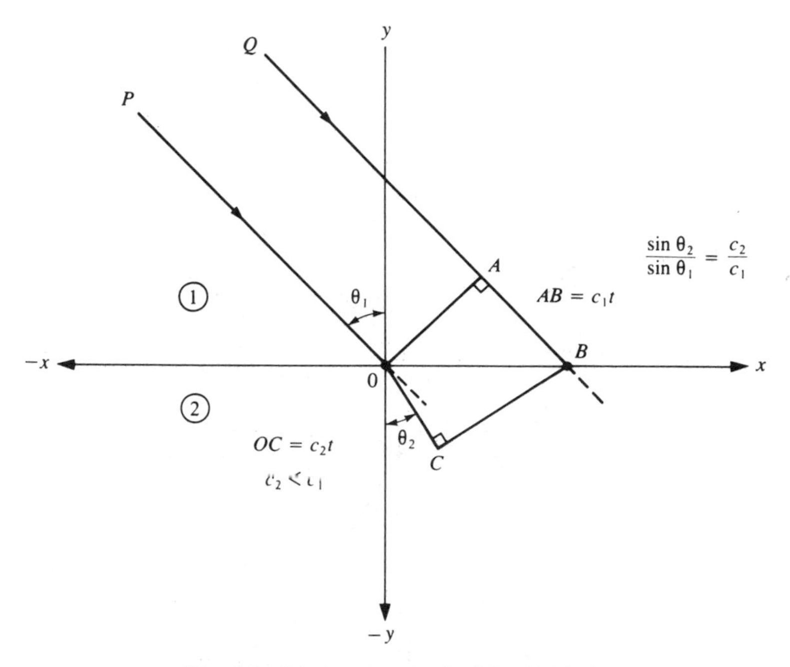

Fig. 6.2. Geometric proof of Snells' law.

yielding *Snell's Law*† for sound waves which is expressed as

$$\frac{\sin \theta_1}{\sin \theta_2} = \frac{c_1}{c_2}. \tag{6.41}$$

We now give a geometric proof of (6.38). Figure 6.2 shows two incident waves in medium (1), *PO* and *QAB*, making an incident angle θ_1 with the y axis. Let *OA* be the common wavefronts of these parallel waves so that *OA* is perpendicular to *PO* and *QB*. In a time t the incident wavefront *OA* travels a distance $AB = c_1 t$ so that the refracted wavefront after a time t is *CB* which is perpendicular to *OC* (the portion of the refracted wave from O to C). *OC* makes the refracted angle θ_2 with the negative y axis. Since the refracted wave travels with the wave speed c_2 we must have $OC = c_2 t$. From the figure we have

$$\sin \theta_1 = \frac{AB}{OB} = \frac{c_1 t}{OB}, \qquad \sin \theta_2 = \frac{OC}{OB} = \frac{c_2 t}{OB},$$

which yields (6.41).

We now obtain a quantitative relationship amongst the intensities of these three waves. We consider Fig. 6.3 which shows the same coordinate system

† In Chapter 8 we again take up Snell's law from the point of view of stress waves.

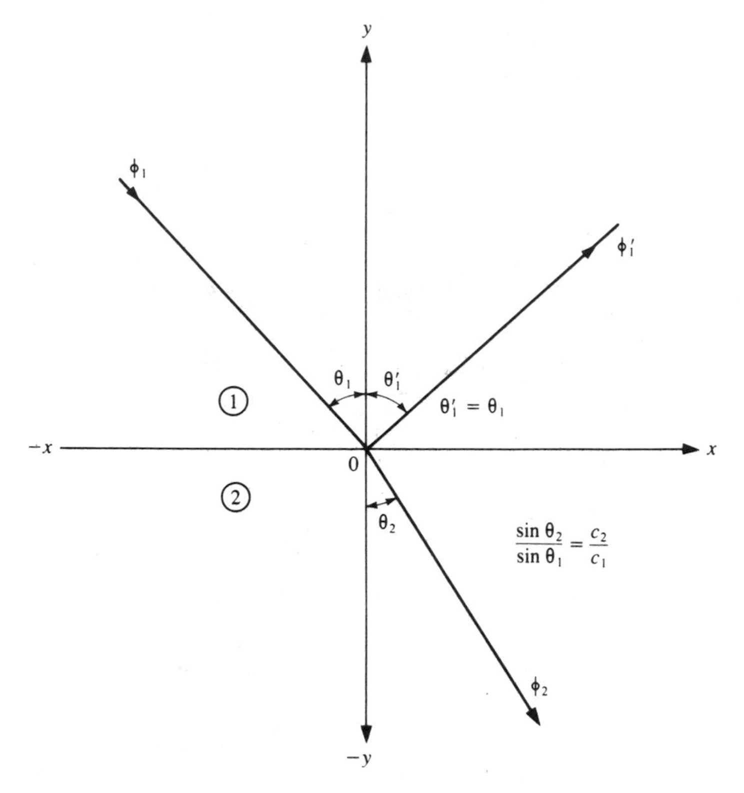

Fig. 6.3. Incident, reflected, and refracted waves.

as Figs. 6.1 and 6.2. The incident wave is represented by the potential ϕ_1, the reflected wave by ϕ_1', and the refracted wave by ϕ_2. According to the geometry of Fig. 6.3 we write the various potentials as

$$\begin{aligned}
\phi_1 &= A_1 \exp\left(i\left[\left(\frac{\omega}{c_1}\right)x \sin\theta_1 - \left(\frac{\omega}{c_1}\right)y\cos(\theta_1 - \omega t)\right]\right),\\
\phi_1' &= A_1' \exp\left(i\left[\left(\frac{\omega}{c_1}\right)x \sin\theta_1' + \left(\frac{\omega}{c_1}\right)y\cos(\theta_1' - \omega t)\right]\right),\\
\phi_2 &= A_2 \exp\left(i\left[\left(\frac{\omega}{c_2}\right)x \sin\theta_2 - \left(\frac{\omega}{c_2}\right)y\sin(\theta_2 - \omega t)\right]\right).
\end{aligned} \tag{6.42}$$

The BCs on $x=0$ are: (1) the continuity of pressure, and (2) the continuity of the normal component of the particle velocity. These become

$$\text{BC (1):} \quad [p]_{y=0} \quad \text{or} \quad -\rho_1(\phi_1 + \phi_1')_t = -\rho_2\phi_{2t}, \tag{6.43}$$

$$\text{BC (2):} \quad [v]_{y=0} \quad \text{or} \quad (\phi_1 + \phi_1')_y = \phi_{2y}, \tag{6.44}$$

where $[\quad]_{y=0}$ means the jump in the quantity in brackets as the boundary $y=0$ vanishes. These are the appropriate continuity conditions at the interface. Using (6.42), the BCs (6.43) and (6.44) yield

$$(1): \qquad \rho_1(A_1+A_1')=\rho_2 A_2, \tag{6.45}$$

$$(2): \qquad \left(\frac{\cos\theta_1}{c_1}\right)(A_1-A_1')=\left(\frac{\cos\theta_2}{c_2}\right)A_2. \tag{6.46}$$

We define the *reflection coefficient* R as the ratio of the time average of the energy flux densities in the reflected to the incident wave. From (6.28) we see that the energy flux density in a plane wave is $\rho c v^2$. From this expression and the definition of R we have

$$R=\frac{\rho_1 c_1 \overline{v_1'^2}}{\rho_1 c_1 \overline{v_1^2}}=\frac{|A_1'|^2}{|A_1'|^2}. \tag{6.47}$$

Dividing (6.45) and (6.46) by A_1, eliminating the ratio A_2/A_1 from the resulting equations, and solving for R in (6.47) yields

$$R=\left[\frac{\rho_2 c_2 \cos\theta_1-\rho_1 c_1 \cos\theta_2}{\rho_2 c_2 \cos\theta_1+\rho_1 c_1 \cos\theta_2}\right]^2.$$

Eliminating θ_2 in this expression by using Snell's law (6.41) gives the reflection coefficient in the form

$$R=\left[\frac{\rho_2 c_2 \cos\theta_1-\rho_1\sqrt{(c_1^2-c_2^2\sin^2\theta_1)}}{\rho_2 c_2 \cos\theta_1+\rho_1\sqrt{(c_1^2-c_2^2\sin^2\theta_1)}}\right]^2. \tag{6.48}$$

If the incident sound wave is normal to the interface then $\theta_1=0$ so that (6.48) becomes

$$R=\left[\frac{\rho_2 c_2-\rho_1 c_1}{\rho_2 c_2+\rho_1 c_1}\right]^2. \tag{6.49}$$

The term $\rho_i c_i$ is called the *acoustic impedance* of the sound wave in medium (i). We digress a moment to describe its physical significance vis-à-vis an interesting electrical analogy. The acoustic impedance represents the ratio of the force of a sound wave on a unit surface area (normal to the propagating wave) to the area times the strain rate. For simplicity, we consider a uniaxial stress wave propagating normal to a planar surface. Let σ be the stress on the surface element, ε the strain, D the operator $d/dt=i\omega$ (ω is the angular frequency) for a time-harmonic stress, and Z the impedance defined by $Z=\sigma/D\varepsilon$. We then see that the acoustic impedance ρc has the dimensions of Z. The reason for Z being defined as the *ratio of stress to strain rate* is by analogy with electrical circuit theory where the electrical impedance is defined as the ratio of voltage to current. There is an interesting analogy between the mechanical variables of stress and strain and the electrical variables of voltage and charge. Stress is analogous to voltage and strain is analogous to electrical charge. Since current is the time rate of change of charge it follows that strain rate $D\varepsilon$ is analogous to current; hence the electrical impedance is analogous to the mechanical impedance as defined above.

We are now in a position to explain the physical significance of (6.49). It tells us that for normal incidence of a sound wave the reflection coefficient is equal to the square of the ratio of the difference of acoustic impedances to the sum of the acoustic impedances. In a sense, the acoustic impedance invokes the properties of the medium in which the wave travels, since it is defined as the density of the medium times the wave speed in the medium. If the two acoustic impedances of a normal incident wave are equal then the reflection coefficient vanishes; this is clear since the properties of medium (2) (described by ρc) are the same as those of medium (1) so that the wave travels through medium (2) unchanged—there is no reflection.

We next ask: What is the relationship between θ_1 and the parameters ρ and c in the two media when we set $R=0$? This is the case of no reflection which means the incident wave is totally refracted. To this end, we set $R=0$ in (6.48). After a little manipulation we obtain

$$\tan^2\theta_1 = \frac{\rho_2^2 c_2^2 - \rho_1^2 c_1^2}{\rho_1^2(c_1^2 - c_2^2)}. \tag{6.50}$$

From (6.50) we obtain either of the two following inequalities which insure total refraction:

$$\begin{aligned} &(1) \quad \rho_2 c_2 > \rho_1 c_1 \quad \text{and} \quad c_1 > c_2 \\ \text{or} \quad &(2) \quad \rho_1 c_1 > \rho_2 c_2 \quad \text{and} \quad c_2 > c_1. \end{aligned} \tag{6.51}$$

6.5. Sound Wave Propagation in a Moving Medium

Qualitative Ideas

We consider a ringing bell as a source of spherical sound waves. Let the bell be at a point O in a three-dimensional medium of uniform density, temperature, and atmospheric pressure so that the wave speed c is constant. If the bell moves with a constant velocity q in the positive x or horizontal direction, with respect to a fixed or laboratory coordinate system, we have three cases:

(1) $q=0$—the bell is stationary.
(2) $q<c$—the bell moves subsonically.
(3) $q>c$—the bell moves supersonically.

For case (1) the bell emits spherical sound waves whose wavefronts are spherical surfaces which expand uniformly with O as the fixed origin. A sound wave emanating from the bell will have its wavefront expand with

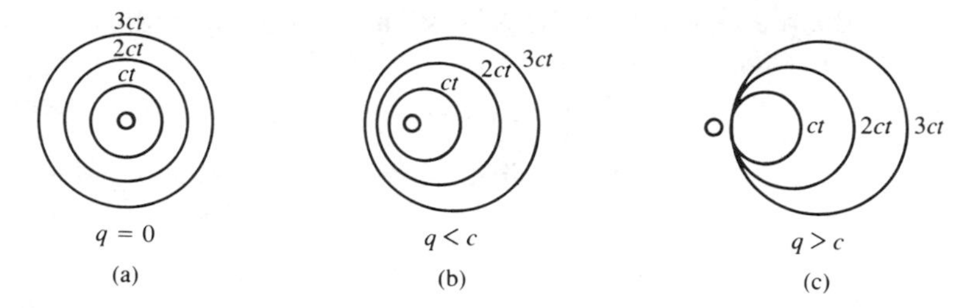

Fig. 6.4. Planar projection of spherical wavefronts for (a) stationary, (b) subsonic, and (c) supersonic bell.

the constant speed of sound c such that at times t, $2t$, $3t, \ldots,$ the sound disturbance will have reached points which lie on the concentric spherical surfaces of center O and radii ct, $2ct$, $3ct, \ldots$. For case (2) the spherical surfaces will still have radii ct, $2ct$, $3ct, \ldots,$ but their centers will be displaced to the right by amounts qt, $2qt$, $3qt, \ldots,$ respectively. Since $q < c$, these surfaces will never intersect or touch each other so that the sound disturbance will eventually permeate all space. For case (3) the sound waves will still lie on the surfaces of radii ct, $2ct$, $3ct, \ldots,$ but, since $q > c$, the disturbances will never reach points which lie outside a cone whose vertex is at O. It is clear that O is the common point of tangency of all the surfaces. This cone is called the *Mach cone*; it is the envelope of the expanding spherical sound waves. Let A be the semicone angle. A is called the *Mach angle*, and the geometry gives $\sin A = c/q = 1/M$ where M is called the *Mach number.* Figure 6.4 illustrates the situation by representing projections in the (x, y) plane for the three cases. Chapter 7 on fluid dynamics treats the supersonic case in some detail from the point of view of quasilinear PDEs.

Quantitative Treatment

The relationship given by (6.38) between wave number and frequency $(k = \omega/c)$ is only valid for a monochromatic sound wave propagated *in a medium at rest.* It is not difficult to generalize this expression for a sound wave propagating in a moving medium—the observer being in a fixed coordinate system.

Let $X = (x, y, z)$ be a fixed coordinate system and let $X' = (x', y', z')$ be a coordinate system moving with velocity $\mathbf{u}$ with respect to X. For simplicity we assume X' does not rotate. We now consider a homogeneous fluid medium (air) and fix the X' coordinate system in the fluid medium. This clearly means the fluid is at rest in the X' coordinate system. The velocity potential for a monochromatic progressing sound wave in the system X' has the usual time-harmonic form

$$\phi(x', y', z', t) = A \exp[i(\mathbf{k} \cdot \mathbf{r}' - kct)], \qquad \omega = kc,$$

where $\mathbf{r}'$ is the radius vector in the X' system. Let $\mathbf{r}$ be the radius vector in the X system. Then $\mathbf{r}'$ is related to $\mathbf{r}$ by

$$\mathbf{r}' = \mathbf{r} - \mathbf{u}t.$$

This means that in the fixed coordinate system X the velocity potential is given by

$$(x, y, z, t) = A \exp[i[\mathbf{k} \cdot \mathbf{r} - (kc + \mathbf{k} \cdot \mathbf{u})t]].$$

The factor multiplying t in the exponent of this expression tells us that the frequency ω in the moving medium is related to the wave vector $\mathbf{k}$ by

$$\omega = kc + \mathbf{k} \cdot \mathbf{u}. \tag{6.52}$$

Clearly if the fluid is at rest, $\mathbf{u} = 0$ so that (6.54) reduces to

$$\omega = kc \qquad \text{for} \qquad \mathbf{u} = 0.$$

Equation (6.52) allows us to investigate the *Doeppler effect* which gives the relationship between the frequency ω_0 of oscillations of a stationary source of sound waves and the frequency ω of the sound heard by a moving observer. Consider a source of sound waves at rest relative to the medium (at rest in the X' system). Let the sound waves be heard by an observer moving with a velocity $\mathbf{u}$ relative to the source. In system X' we have $k = \omega_0/c$, where ω_0 is the frequency of oscillations of the source. In the system X moving with the observer (the observer is fixed in X), the X' system moves with velocity $-\mathbf{u}$, and the frequency of the sound waves as heard by the observer is $\omega = ck - \mathbf{u} \cdot \mathbf{k}$, upon using (6.52). Let θ be the angle between the direction of the velocity $\mathbf{u}$ and the wave vector $\mathbf{k}$. Upon setting $k = \omega_0/c$ and using the fact that $\mathbf{u} \cdot \mathbf{r} = ur \cos\theta$, we find that the frequency of the sound waves received by the moving observer is

$$\omega = \omega_0\left[1 - \left(\frac{u}{c}\right) \cos\theta\right]. \tag{6.53}$$

In investigating the Doeppler effect we impose the following restriction on (6.53): $|u|/c < 1$, or *the speed of the moving observer with respect to the source must be subsonic.* Since $\mathbf{u}$ is the relative velocity between the source and object, the same restriction is valid for a moving source and a stationary object. We showed above, in analyzing Fig. 6.4, that for the case of the bell moving supersonically the sound field is not eventually heard everywhere in space, but is restricted to the Mach cone. $u = c$ is the special case of transonic flow where the Mach angle $A = \pi/2$, which is the transition between subsonic and supersonic flow. The supersonic case will be reserved for the chapter on fluid dynamics.

Let us now consider the case of a stationary observer and a moving source. We assume the source (fixed in the X' coordinate system) moves subsonically with a velocity $-\mathbf{u}$ with respect to the observer fixed in the X

coordinate system. Since the source is at rest in X', the frequency of the emitted sound waves must equal the frequency ω_0 of the oscillating source. Changing the sign of **u** in (6.52) and again introducing the angle θ between **u** and **k** we have

$$\omega_0 = kc\left[1 - \left(\frac{u}{c}\right)\cos\theta\right], \qquad \omega = kc.$$

Comparing this equation to (6.53) we see that ω and ω_0 are interchanged. Solving for ω yields

$$\omega = \frac{\omega_0}{[1 - (u/c)\cos\theta]}. \tag{6.54}$$

Equation (6.54) gives the relationship between the frequency of oscillations of the moving source and the frequency of sound heard by the observer at rest. Equation (6.54) expresses the Doeppler effect for the case of the sound source moving with respect to the observer.

We now give a physical interpretation of the Doeppler effect vis-à-vis (6.54). Recall that, since the source sends out spherical sound waves, **k** is in the direction of the outgoing normal to the wavefronts or in the radial direction to the spherical surfaces, so that θ is the angle between the line of sight of the source and the observer and the direction of the velocity **u** of the source. Figure 6.5 shows a stationary observer at O and a source at S moving with a velocity **u** with respect to O. If $\theta = \pm\pi/2$, then $\cos\theta = 0$ so that $\omega = \omega_0$. This means that if the source travels in the normal direction

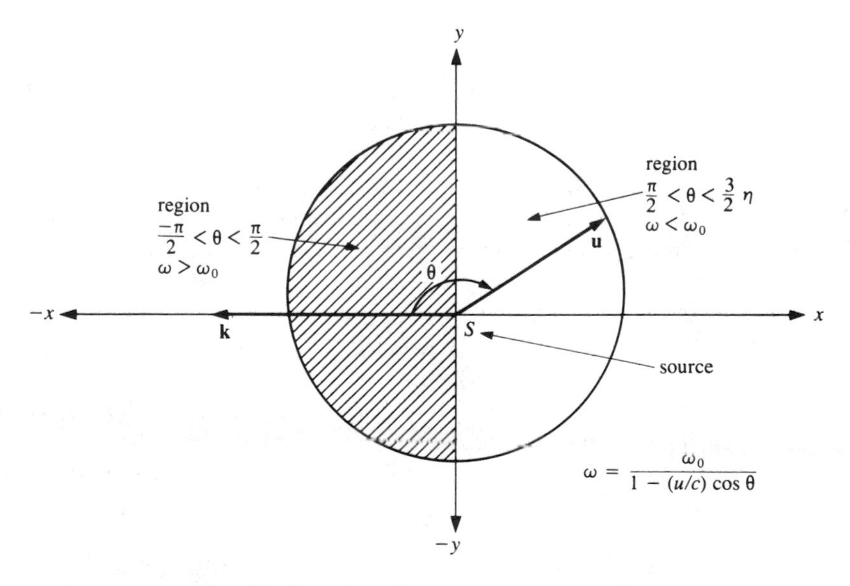

Fig. 6.5. Doppler effect showing range of **u**.

to the line of sight then the frequency of sound that the observer hears is equal to the frequency of oscillations of the source. From the figure we see that if θ has the range $\pi/2 < \theta < \frac{3}{2}\pi$ then $\mathbf{u}$ is directed away from S so that the observer moves away from the source. $\cos\theta < 0$ and (6.54) tell us that the frequency the observer hears is less than that of the oscillating source. Conversely, if the direction of $\mathbf{u}$ lies in the region $-\pi/2 < \theta < \pi/2$ then the observer moves toward the source so that $\cos\theta > 0$, and the frequency the observer hears is greater than that of the oscillating source.

6.6. Spherical Sound Waves

In discussing the Doeppler effect above it was tacitly assumed that the source emitted spherical sound waves. We now investigate the properties of these spherical waves quantitatively. To this end, we consider a sound wave in which the distribution of density, particle velocity, and pressure depends only on the distance from the sound source which is idealized as a point source. Such a wave is called a *spherical sound wave.* As usual, in order to deal with the linear wave equation with constant coefficients, we assume a homogeneous medium so that c is constant.

We first write down the wave equation for the velocity potential $\phi(r, t)$ which represents a spherical wave. Since ϕ is a function only of the distance r from the source and time t, the wave equation becomes, using the Laplacian in spherical coordinates assuming spherical symmetry,

$$\left(\frac{c^2}{r^2}\right)(r^2\phi_r)_r - \phi_{tt} = 0. \tag{6.55}$$

We seek a general solution of (6.55) as the sum of progressing and regressing waves. For a progressing wave the function $f(r-ct)$ does not satisfy the wave equation in spherical coordinates, (6.55), but satisfies the one-dimensional wave equation $c^2 f_{rr} - f_{tt} = 0$ with r as the spatial coordinate. Therefore, in order to obtain a solution to (6.55) we shall try a functional form such as $r^n f(r-ct)$ for a progressing wave and attempt to solve for the integer n. The reason for assuming such a form is that for a progressing spherical waveform the function $f(r-ct)$ must be a factor in the solution. Physically speaking, we also know that the amplitude of this waveform must decrease with increasing r. The simplest assumption is to attempt to construct a solution where $f(r-ct)$ is multiplied by r^n for a negative integer n. The answer turns out to be $n=-1$ so that the amplitude of the spherical waveform varies inversely as the distance from the sound source. We see this by substituting $\phi(r, t) = r^n f(r-ct)$ into (6.55) and using the equation $c^2 f_{rr} - f_{tt} = 0$. We finally obtain the following PDE for

$$f(r-ct)\colon\ 2(n+1)f_r + n(n+1)f = 0,$$

which is satisfied if we set $n=-1$. This gives us the required result.

We may now write the general solution of (6.55) in terms of progressing and regressing waves

$$\phi(r, t) = \left(\frac{1}{r}\right) f(r - ct) + \left(\frac{1}{r}\right) g(r + ct). \tag{6.56}$$

The first term of the right-hand side represents a progressing wave, and the second term a regressing wave. The functions $f(r-ct)$ and $g(r+ct)$ are arbitrary functions of their respective arguments and are completely determined by prescribing the appropriate initial and boundary conditions for a given problem. A physical interpretation of (6.56) is: The first term is an *outgoing wave* from the point sound source whose amplitude decreases inversely as the distance from the source. The wavefronts propagate as a continuous system of expanding spherical surfaces whose common centers are at the point source of sound. The intensity of this outgoing wave is given by the square of the amplitude, and therefore falls off inversely as the square of the distance from the source. This is clearly so since the total energy flux in the wave is distributed over the spherical surface (wavefront) whose area increases as r^2. A similar statement is valid for the second term of the right-hand side of (6.56), except that this term represents an *incoming wave*, i.e., a wave traveling in toward the source.

Standing Waves

We recall that in Chapter 5 standing wave solutions were defined as being of the general form $\phi(r, t) = \bar{\phi}(r) f(t)$ which means that a standing waveform is one where we can separate out the spatial from the temporal parts of the solution. As pointed out in Chapter 5, a standing wave has a wave shape which is fixed in space within a multiplying factor that depends on time. This tells us that the points of maxima, minima, etc. are independent of time. We are interested in time-harmonic monochromatic waveforms. Consequently, we may write such a typical spherical standing wave in the form

$$\phi(r, t) = A[\exp(-i\omega t)]\left(\frac{1}{r}\right) \sin kr, \qquad k = \frac{\omega}{c}, \tag{6.57}$$

where the amplitude of the wave is A/r (A is a constant). Note that (6.57) tells us that ϕ is bounded at $r=0$. Indeed, we have $\phi(0, t) = A \exp(-i\omega t)$. This means that the standing wave expressed by (6.57) is independent of the sound source. This is clear; for if we consider outgoing or incoming spherical waves, their amplitudes would become infinite at $r=0$ (at the source). Another way of looking at standing wave solutions for spherical waves is as follows: Since a standing wave involves no sound source in the field (infinite space) $\phi(r, t)$ must remain finite at $r=0$. This can be so by setting $\phi(0, t) = 0$ in (6.56), yielding $f(-ct) = -g(ct)$. From this expression we obtain the general expression for a standing wave in the form

$$\phi(r, t) = \left(\frac{1}{r}\right)[f(r - ct) - f(r + ct)]. \tag{6.58}$$

We may interpret (6.58) as follows: The spherical standing wave solution for $\phi(r, t)$ is the difference between an outgoing and an incoming spherical wave. The analogy with one-dimensional standing waves was given in Chapter 2 where two lateral sinusoidal waves of equal amplitude approach each other from either side of a finite string fixed at each end. The resulting wave pattern is a standing wave, which is a sinusoidal function of time whose amplitude is spatially modulated sinusoidally.

Consider an outgoing time-harmonic monochromatic spherical wave of the form

$$\phi(r, t) = \left(\frac{A}{r}\right) \exp[i(kr - \omega t)], \tag{6.59}$$

where k is the wave number. An interesting fact is that (6.59) satisfies the following PDE:

$$\left(\frac{1}{r^2}\right)(r^2\phi_r)_r + k^2 = -4\pi A e^{-i\omega t}\delta(\mathbf{r}), \tag{6.60}$$

where the Dirac delta function $\delta(\mathbf{r}) = \delta(x)\delta(y)\delta(z)$. This means that the right-hand side of (6.60) is zero everywhere except at the origin where it is infinite. Now we surround the origin by a small sphere and integrate (6.60) over the volume of this sphere. In the limit, as the radius of the sphere tends to zero, we obtain $-4\pi A e^{-i\omega t}$ on each side of the equation.

The properties of $\delta(\mathbf{r})$ are: (1) $\delta(\mathbf{r}) = 0$ everywhere except at $\mathbf{r} = 0$ where $\delta(0) = \infty$, and (2) $\int\int\int_{-\infty}^{\infty} \delta(\mathbf{r})\, dx\, dy\, dz = 1$. The delta function is not really a function but is the limit of a sequence of functions. For example, in one-dimensional x space we can construct a sequence of pulse functions $\delta_\varepsilon(x)$ depending on the small positive parameter ε, defined as follows:

$$\delta(x) = \begin{cases} 1/\varepsilon & \text{for} \quad -\varepsilon/2 < x < \varepsilon/2, \\ 0 & \text{for} \quad |x| > \varepsilon. \end{cases} \tag{6.61}$$

This tells us that we have a sequence of pulses centered at $x = 0$ whose pulse width decreases and whose amplitude increases as ε decreases in such a manner that the area under each pulse is unity. The limit function as $\varepsilon \to 0$ is the one-dimensional delta function which has the properties stated above.

Another example is given by a sequence of Gaussian probability functions in three-dimensional space normalized to a unit volume. As the standard deviation tends to zero this sequence tends to the delta function. Again, suppose we have a function $f(x, y, z)$ such that $f = f_0$ at the origin. Integrating over all space gives $\int\int\int_{-\infty}^{\infty} \delta(x)\delta(y)\delta(z)f(x, y, z)\, dx\, dy\, dz = f_0$, which means that $\delta(x)\delta(y)\delta(z)$ picks out the value of $f(x, y, z)$) at the origin when integrated over all space.

6.7. Cylindrical Sound Waves

We consider a cylindrical coordinate system with x being the cylindrical axis. Let R be the perpendicular distance from x to a point on the cylindrical surface. Let r be the radius of a spherical surface tangent to this cylindrical surface whose center is on the x axis a distance x from the point of intersection of R with x. We then have $r^2 = R^2 + x^2$. The velocity potential $\phi(R, t)$ represents a cylindrical wave if it exhibits cylindrical symmetry, i.e., it is spatially dependent on R. $\phi(R, t)$ is then said to be an axisymmetric solution of the wave equation. Before we write down the wave equation for ϕ in cylindrical coordinates we attempt to determine the general form of such solutions by appealing to the general spherically symmetric solution (6.56). The function $\phi(R, t)$ can be obtained from (6.56) by integrating this equation over x from 0 to ∞. Since $x = \sqrt{r^2 - R^2}$, we have $dx = r\,dr/\sqrt{(r^2 - R^2)}$. We therefore find that the general axisymmetric solution is

$$\phi(R, t) = \int_R^\infty \frac{f(r-ct)}{\sqrt{(r^2 - R^2)}}\,dr + \int_R^\infty \frac{g(r+ct)}{\sqrt{(r^2 - R^2)}}\,dr. \tag{6.62}$$

The first term on the right-hand side represents an outgoing cylindrical wave, and the second term an incoming one. If we introduce the variables of integration $z = \pi \mp ct$ (corresponding to the outgoing, incoming cylindrical wave, respectively), (6.62) becomes

$$\phi(R, t) = \int_{R-ct}^\infty \frac{f(z)}{\sqrt{[(z+ct)^2 - R^2]}}\,dz + \int_{R+ct}^\infty \frac{g(z)}{\sqrt{[(z-ct)^2 - R^2]}}\,dz. \tag{6.63}$$

Similarly to the spherical wave case, standing waves are obtained when we set $f(r-ct) = -g(r+ct)$. It is easily seen that a standing cylindrical wave can be represented by the following form for the velocity potential:

$$\phi(R, t) = \int_{R-ct}^{R+ct} \frac{F(z)}{\sqrt{[(R^2 - (z+ct)^2]}}\,dz, \tag{6.64}$$

where $F(z)$ is an arbitrary function which can be determined by prescribing appropriate initial and boundary conditions.

We now write down the wave equation for the velocity potential in cylindrical coordinates with axial symmetry

$$\left(\frac{c^2}{R}\right)(R\phi_R)_R - \phi_{tt} = 0. \tag{6.65}$$

To obtain standing monochromatic, axially symmetric, cylindrical waves we attempt to solve (6.65) by separating variables and hence write ϕ in the form

$$\phi(R, t) = e^{-i\omega t} f(R),$$

where the differential equation for $f(R)$ is found by inserting this expression for $\phi(R, t)$ into (6.65). We obtain

$$f'' + \frac{f'}{R} + k^2 f = 0, \qquad \frac{\omega^2}{c^2} = k^2. \tag{6.66}$$

Equation (6.66) is Bessel's equation of zeroth order. As is well known in differential equations, the general solution of (6.66) is a linear combination of Bessel's function of zeroth order of the first kind $J_0(kR)$, and the second kind $Y_0(kR)$ which has a singularity at $R=0$. For a stationary cylindrical wave ϕ must be bounded at $R=0$ so that we neglect the Y_0 part of the solution. Therefore, for a stationary cylindrical wave we obtain

$$\phi(R, t) = Ae^{-i\omega t} J_0(kR). \tag{6.67}$$

Since $J_0(0)=1$ we have $\phi(0, t)A = e^{-i\omega t}$, where the constant A is the amplitude. If R is large J_0 can be replaced by its asymptotic expansion for large values of the argument, so that $\phi(R, t)$ takes the asymptotic form

$$\phi(R, t) \sim A\left(\frac{2}{\pi}\right)^{1/2}\left(\frac{1}{(kR)^{1/2}\cos(kR-(\pi/4))}\right), \tag{6.68}$$

see [41, p. 368].

The solution for the potential for a monochromatic *outgoing wave* has a logarithmic singularity at the source $R=0$, located on the x axis. It is given by

$$\phi(R, t) = Ae^{-i\omega t} H_0^1(kR),\dagger \tag{6.69}$$

where H_0^1 is the Hankel function of the first kind of order zero. As $R \to 0$ this function has a logarithmic singularity, so that the expression for the potential for the outgoing wave in the neighborhood of the source has the approximation

$$\phi(R, t) \sim \left(\frac{2iA}{\pi}\right) \log(kR) e^{-i\omega t}. \tag{6.70}$$

At large distances from the axis the potential is given by the asymptotic expansion

$$\phi(R, t) \sim A\left(\frac{2}{\pi kR}\right)^{1/2} \exp\left[i\left(kR - \omega t - \frac{\pi}{4}\right)\right]. \tag{6.71}$$

† See [11, Vol. II, p. 194].

Equation (6.71) tells us that at large distances from the source the amplitude of the velocity potential for a cylindrical wave diminishes inversely as the square root of the distance from the axis, so that the intensity decreases inversely as $1/R$. Clearly this result arises from the fact that the total energy flux is distributed over a cylindrical surface whose area increases proportionally to R as the wave is propagated.

Another approach to the solution for the velocity potential for an axially symmetric cylindrical wave is as follows: Consider the (x, y) plane and let the z axis be the axis of a right circular cylinder of radius R. $\phi(x, y, t)$ satisfies the two-dimensional wave equation in the (x, y) plane

$$c^2(\phi_{xx} + \phi_{yy}) - \phi_{tt} = 0. \tag{6.72}$$

The transformation into polar coordinates R, given by $x = R\cos\theta$, $y = R\sin\theta$, performed on this wave equation and invoking axial symmetry (ϕ is independent of θ) yields (6.65). However, we now consider (6.72) and attempt to write down plane wave solutions in rectangular coordinates. We start with a plane wave solution for an arbitrary direction angle α

$$\exp[ik(x\cos\alpha + y\sin\alpha)]\exp[-i\omega t].$$

Integrating this expression with respect to the direction angle α and invoking the polar coordinates R, θ yields the axisymmetric standing wave solution

$$\phi(x, y, t) = e^{-i\omega t}\int_0^{2\pi} \exp[ikR\cos(\theta - \alpha)]\, d\theta = 2\pi e^{-i\omega t} J_0(kR),$$

which is proportional to the right-hand side of (6.67).

A cylindrical outgoing wave differs from a spherical or plane wave in one important aspect: It has a forward wavefront but no backward front. Once the cylindrical sound wave has reached a given point in space the amplitude then diminishes rather slowly for large time. This is seen by setting $g(z) = 0$ in (6.63), and letting $\xi_1 = R - ct_1 < z < R + ct_1 = \xi_2$. The potential then becomes

$$\phi(R, t) = \int_{R-ct_1}^{R+ct_1} \frac{f(z)}{\sqrt{[(z+ct)^2 - R^2]}}\, dz \qquad \text{for} \quad \xi_1 < t < \xi_2.$$

As $t_1 \to \infty$ this expression tends to zero as

$$\phi(R, t) \sim \frac{1}{ct}\int_{R-ct_1}^{R+ct_1} f(z)\, dz, \qquad t \text{ large}.$$

This means that the potential diminishes inversely as the time for large time. Thus the velocity potential in an outgoing cylindrical wave with axial symmetry vanishes rather slowly as time becomes infinite.

We note a property common to both spherical and cylindrical waves. Before these outgoing waves arrive at a field point the velocity potential must be identically zero. From the nature of these outgoing waves, the potential must return to zero after these waves pass the field point. Thus the potential must vanish before and after the waves pass any point in space. From this fact we can deduce an important conclusion concerning the distribution of condensations and rarefactions in a spherical and cylindrical wave by investigating the pressure variations p'. We recall that $p' = -\rho\phi_t$. It is then clear that if we integrate p' over all time for a given r or R the result is zero. Thus

$$\int_{-\infty}^{\infty} p' \, dt = 0.$$

This tells us that as the spherical or cylindrical wave passes a given field point both condensations ($p' > 0$) and rarefactions ($p' < 0$) will be observed at the field point. In this respect, both spherical and cylindrical waves differ markedly from plane waves, which may consist of condensations or rarefactions only.

6.8. General Solution of the Wave Equation

We shall now derive an expression for the solution of the three-dimensional wave equation for the velocity potential. This ultimately allows us to obtain the velocity and pressure fields in terms of the prescribed initial velocity and pressure distribution.

Let $\phi(x, y, z, t)$ and $\psi(x, y, z, t)$ be two solutions of the three-dimensional wave equation which vanish at infinity, so that

$$c^2\nabla^2\phi - \phi_{tt} = 0, \qquad c^2\nabla^2\psi - \psi_{tt} = 0, \tag{6.73}$$

where ∇^2 is the Laplacian operator in three-dimensional space. We take $\phi(x, y, z, t)$ to be the velocity potential and $\psi(x, y, z, t)$ to be an arbitrary *test function* which we ultimately define in such a manner as to make the calculations easier. Finally, we define the integral I by

$$I = \int (\phi\psi_t - \psi\phi_t) \, dV, \tag{6.74}$$

where the volume integral is taken over all space. Taking the partial derivative of I with respect to t, and using (6.73) gives

$$I_t = c^2 \int (\phi\nabla^2\psi - \psi\nabla^2\phi) \, dV = c^2 \int \operatorname{div}(\phi \, \mathbf{grad} \, \psi - \psi \, \mathbf{grad} \, \phi) \, dV.$$

We transform the last integral to a surface integral by use of the divergence theorem. Since the surface is at infinity, and by invoking the boundary conditions at infinity, we conclude that $I_t = 0$. This yields the fact that I is independent of time. Since I involves a volume integration over all space we find that

$$I=\int(\phi\psi_t-\psi\phi_t)\,dV=\text{const.} \tag{6.75}$$

Let us now choose the following solution of the wave equation for the test function $\psi(x, y, z, t)$:

$$\psi(r, t; t_0)=\delta[r-c(t_0-t)] \qquad \text{for} \quad t_0-t>0. \tag{6.76}$$

where δ is the Dirac delta function, r is the magnitude of the radius vector from some point O whose coordinates are (x, y, z) to a point O' whose coordinates are (x', y', z'), and t_0 is some instant of time. We interpret (6.76) as follows: The argument of the delta function is zero for $r=c(t_0-t)$. All points O' are on this spherical surface of radius r which depends on t_0 and t. For $t=0$ the spherical surface has the radius $r=ct_0$. We shall attempt to obtain a general solution for ϕ which depends on the distribution of ϕ and ϕ_t on this spherical surface (the initial distributions). Note that as t increase from zero to t_0 the surface shrinks to zero at $t=t_0$. The test function defined by (6.76) allows us, in a straightforward way, to obtain an expression for the velocity potential in terms of the initial distribution of the potential and its time derivative on the surface of a sphere of radius $r=ct_0$, centered at O.

We now calculate the volume integral of ψ over all space. We have, using (6.76)

$$\int\psi\,dV=\int_0^\infty 4\pi r^2\psi\,dr=4\pi\int_0^\infty r\delta[r-c(t_0-t)]\,dr.$$

The argument of the delta function is zero for $r=c(t_0-t)$, so that this equation becomes

$$\int\psi\,dV=4\pi c(t_0-t). \tag{6.77}$$

Differentiating this equation with respect to t, keeping t_0 fixed, yields

$$\int\psi_t\,dV=-4\pi c. \tag{6.78}$$

We now make use of the fact that I is constant, as shown by (6.75), by evaluating I both for $t=t_0$ and for $t=0$ and equating the results. Setting $t=t_0$ in (6.76), inserting the resulting expression in (6.75), and using the properties of the delta function, allows us to take the values of ϕ and ϕ_t at the point O and hence take them outside the integral. We obtain

$$I(x, y, z, t_0)=\phi(x, y, z, t_0)\int\psi_t\,dV-\phi_t(x, y, z, t_0)\int\psi\,dV.$$

According to (6.76) and (6.78) the second integral vanishes at $t=t_0$, and we obtain

$$I(x, y, z, t_0)=-4\pi c\phi(x, y, z, t_0). \tag{6.79}$$

now calculate I for $t=0$. First we observe that $\psi_t = -\psi_{t_0}$. We then set $\phi(x, y, z, 0) = \phi_0$ and $\psi(x, y, z, 0) = \psi_0$. I at $t=0$ then becomes (using extended notation for partial derivatives)

$$I(x, y, z, 0) = -\frac{\partial}{\partial t_0} \int \phi_0 \psi_0 \, dV - \int \psi_0 \frac{\partial \phi_0}{\partial t} \, dV.$$

We set the volume element $dV = r^2 dr \, d\Omega$ where $d\Omega$ is an element of solid angle, use the properties of the delta function, and obtain

$$(x, y, z, 0) = -\frac{\partial}{\partial t_0}\left(ct_0 \int \phi_0(r = ct_0) \, d\Omega \right) - ct_0 \int \frac{\partial \phi_0(r = ct_0)}{\partial t} \, d\Omega. \quad (6.80)$$

Finally, we set $I(x, y, z, t_0) = I(x, y, z, 0)$ by using (6.79) and (6.80), and solve for $\phi(x, y, z, t)$ (omitting the subscript in t_0)

$$\phi(x, y, z, t) = \frac{1}{4\pi}\left[\left(\frac{\partial}{\partial t}\right) t \int \phi_0(r = ct) \, d\Omega + t \int \frac{\partial \phi_o(r = ct)}{\partial t} \, d\Omega \right]. \quad (6.81)$$

Equation (6.81) is called *Poisson's formula* for the three-dimensional velocity potential. It gives the spatial distribution of the potential at any instant in terms of the initial distribution of the potential and its time derivative. It is clear that we may obtain the initial velocity and pressure distributions on the spherical surface of radius ct from the initial ϕ and ϕ_t (and conversely). We see that the value of the potential at time t is determined by the values of ϕ and ϕ_t at time $t=0$ that are on the surface of a sphere of radius $r = ct$ centered at O, whose coordinates are (x, y, z).

6.9. Huyghens' Principle

We now examine the physical nature of the solution for $\phi(x, y, z, t)$ as given by Poisson's formula (6.81) vis-à-vis *Huyghens' principle.* The Dutch physicist, mathematician, and astronomer, Christian Huyghens (1629–1695) enunciated the following pivotal principle in wave propagation:

Every point on an advancing wavefront can be considered as a source of secondary waves which, in a homogeneous medium, spread out as spherical wavelets. A later position of the wavefront is given by the envelope of the secondary wavelets.

We first give a qualitative interpretation of Huyghens' principle by appealing to Fig. 6.6 which shows a point sound source at S and a part of a spherical wavefront W_1 at time t_1 moving with a speed c so that the radius of the wavefront is $r_1 = ct_1$. We now construct the spherical wavefront W_2 at a later time t_2 using Huyghens' principle which states that each point on W_1 is the source of a secondary wave. Therefore we erect a continuum of spherical surfaces whose centers are on the surface W_1 and whose radii are

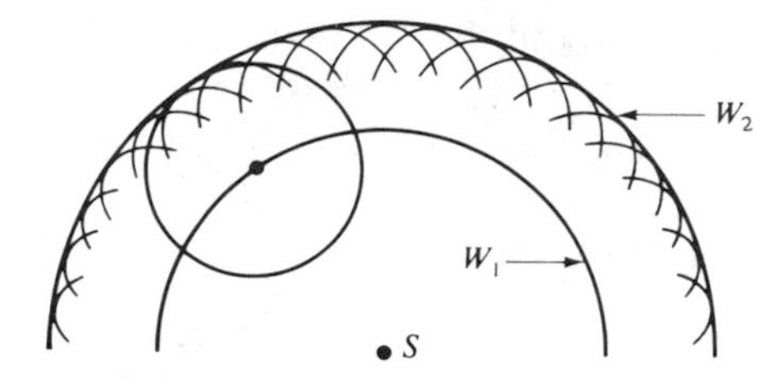

Fig. 6.6. Huyghen's principle for a spherical wavefront.

$c(t_2 - t_1)$. The envelope of these surfaces is the wavefront W_2 which is a spherical surface of radius $r_2 = ct_2$.

Referring to Poissons' formula (6.81), we can interpret the integrals on the right-hand side as representing envelopes of the secondary waves emanating from the surface $r = ct$, which involve both ϕ and ϕ_t. It is clear that we are dealing with three-dimensional wave propagation which exhibit spherical wavefronts. Huyghens' principle asserts that a sharply localized initial state (where ϕ and ϕ_t are defined) is observed later at a different place as an effect that is equally sharply delineated. In the limiting case where the source of sound is a point source, the effect at another point (a field point) will be felt only at a definite instant of time t which depends on the distance from the source to the field point. In general, for three-dimensional wave propagation, the value of the potential $\phi(x, y, z, t)$ at a field point P: (x, y, z) at time t depends on the initial values of ϕ and ϕ_t in a certain domain of space, called the *domain of dependence* belonging to P. For spherical wave propagation this domain of dependence is the surface of the sphere of radius ct whose center is P. The disturbance at P at time t therefore does not depend on the initial data inside or outside the spherical surface of radius ct. This means that in three-dimensional space, in which waves are propagated according to the three-dimensional wave equation, sharp signals are transmitted and can be recorded as sharp signals, according to Huyghens' principle. This is clearly seen from Poisson's formula (6.81) where the integrals can be interpreted as representing the mean values of the ICs over the spherical surface $r = ct$ of center at the field point (x, y, z).

For the case of two-dimensional wave propagation (involving solutions of the two-dimensional wave equation for the potential) Huyghens' principle is not valid. The reason is that, in the case of two-dimensional space, the domain of dependence consists of the whole interior and the circumference of the circle of radius $r = ct$ around the point P: (x, y). This will be proved below.

To understand this basic difference in the character of the solutions for the three- and two-dimensional wave equations, we investigate these solutions in more detail. For the three-dimensional wave equation, we can put Poisson's formula (6.81) in the following form:

$$\phi(x, y, z, t) = \left(\frac{\partial}{\partial t}\right) M(f) + tM(g), \tag{6.82}$$

where f and g are the ICs

$$\phi(x, y, z, 0) = f(x, y, z), \qquad \phi_t(x, y, z, 0) = g(x, y, z), \tag{6.83}$$

and $M(h)$ (where h stands for f or g, respectively) is defined by

$$M(h) = \tfrac{1}{4} \int \int_\Omega h(x+\xi, y+\eta, z+\zeta)\, d\Omega, \tag{6.84}$$

where $d\Omega$ is the element of surface area on the sphere Ω of radius $\xi^2 + \eta^2 + \zeta^2 = c^2 t^2$. It is clear that the solution for $\phi(x, y, z, t)$ as given by (6.81) is the same as (6.82), (6.83), and (6.84). The definition of the function M as given by (6.84) tells us that *$M(f)$ and $M(g)$ denote the mean value of the* ICs *over the surface of the sphere of radius ct with center at* (x, y, z). This concept of the solution for the potential, being given as the sum of the mean values of the two ICs over the spherical surface, is an example of *spherical means* leading to Darboux's equation for mean values which is related to the three-dimensional wave equation. This method is treated in detail in [11, Vol. II, p. 699]. Again, the point is brought out sharply from (6.82), (6.83), and (6.84), that the solution for the potential in three dimensions at (x, y, z) depends only on the average values of the ICs on the surface of the sphere of radius ct centered at the field point (x, y, z).

Method of Descent

We pointed out above that the character of the solution for the two-dimensional wave equation is quite different from that of the three-dimensional wave equation. It was stated above that, for the two-dimensional wave equation, the solution at a field point does not depend only on the ICs on a circle but on the domain of dependence which consists of the interior and circumference of a circle. To prove this statement we shall now derive an expression for the potential $\phi(x, y, t)$ in two-dimensional space. Instead of deriving the solution for the potential from the two-dimensional wave equation in (x, y) space, we apply a method used by the French mathematician J. Hadamard called the *method of descent.* We consider the two-dimensional wave equation $c^2(\phi_{xx} + \phi_{yy}) - \phi_{tt} = 0$ to be a special case of the three-dimensional wave equation, in which both the initial data and the solution itself are independent of the z coordinate. We thus "descend" from three to two dimensions. To this end we assume the ICs (6.83) are given by $f = f(x, y)$ and $g = g(x, y)$. Then (6.84) becomes

$$M(h) = \frac{1}{4\pi} \int \int_\Omega h(x+\xi, y+\eta)\, d\Omega.$$

Expressing the element of surface area of the sphere $\xi^2+\eta^2+\zeta^2=r^2=c^2t^2$ as

$$c^2t^2\,d\Omega=\left(\frac{ct}{\zeta}\right)d\xi\,d\eta=\frac{ct}{\zeta\sqrt{(c^2t^2-\xi^2-\eta^2)}}\,d\xi\,d\eta,$$

the above expression for $M(h)$ becomes

$$M(h)=\left(\frac{1}{2\pi ct}\right)\int\int_{\sqrt{\xi^2+\eta^2}\le ct}\frac{h(x+\xi,\,y+\eta)}{\sqrt{(c^2t^2-\xi^2-\eta^2)}}\,d\xi\,d\eta. \tag{6.85}$$

The integration is taken over the area $\sqrt{\xi^2+\eta^2}\le ct$. Then the solution of the wave equation in two dimensions for the velocity potential becomes

$$\phi(x,y,t)=\left(\frac{\partial}{\partial t}\right)tM(f)+tM(g), \tag{6.86}$$

where $M(f)$ and $M(g)$ are given by (6.85) for $h=(f,g)$, respectively. As mentioned above, the integrations for M, as shown by (6.85), are taken over the interior and circumference of the circle whose region is $\xi^2+\eta^2\le ct$, which is the domain of dependence of the solution at the field point (x, y). The solution for the two-dimensional potential, as expressed by (6.85) and (6.86), is a proof of the assertion given above that in two-dimensional space the effect observed at the field point (x, y) at time t depends on the entire sound radiation which has occurred at the point source up to the time $t-r/c$.

CHAPTER 7

Fluid Dynamics

Introduction

In Chapter 6 we investigated sound wave propagation in a compressible fluid (air) under the hypothesis that sound waves are considered to be due to small amplitude oscillatory motion of the medium. Therefore Chapter 6, which was devoted to sound waves in air, involved a linearized theory whereby the linear wave equation was invoked. In this chapter we shall extend the theory of wave propagation in a compressible fluid to a more general treatment, in which we shall take into account large amplitude wave propagation of supersonic flow (which involves nonlinear phenomena), and shock waves (which involve discontinuities in some of the dynamic and thermodynamic variables). We shall show, for example, that the wave front for a sound wave is the limiting case of the shock front for supersonic flow where the shock strength becomes infinitely weak and the flow field becomes linearized. It also appears that, for subsonic flow, there can be no wave propagation because of the different character of the partial differential equations (PDEs) that govern the flow. Transonic flow involves a transition between subsonic and supersonic flow. Steady flows in which this transition occurs are called mixed or transonic flows, and the surface where the transition occurs is called the transitional or sonic surface. Chaplygin's equation is particularly useful in investigating the flow near this transition, where we obtain the Euler–Tricomi equation whose mathematical properties are important in studying this transitional region. One of its interesting properties is that it is hyperbolic in a certain region (insuring supersonic flow) and elliptic in another region (subsonic flow). This equation will be investigated in some detail.

The mathematical structure for supersonic wave propagation was presented in Chapter 3. In that chapter, it was shown that either a system of first-order quasilinear PDEs or an equivalent second-order PDE is obtained from the three conservation laws: mass or continuity, momentum, and energy. These are the field equations of mathematical physics whose solution yields the supersonic flow field for the particle velocity, pressure, etc. In that chapter it was also pointed out that the character of these PDEs can be determined by investigating the principal part. It was shown that a second-order PDE in two dimensions can be classified into three categories: hyperbolic, parabolic, elliptic. It will be shown in this chapter, when we investigate two-dimensional steady flow, that the hyperbolic equation represents supersonic flow, the parabolic, transonic flow, and the elliptic, subsonic flow. The pivotal feature of the hyperbolic PDE is the method of characteristics. This was described in Chapter 3, starting with a single first-order PDE, then extending the theory to a system of two first-order PDEs vis-à-vis the nonlinear wave equation, and finally developing the method of characteristics for a pair of first-order quasilinear PDEs, and then an equivalent second-order quasilinear PDE. It was pointed out that the key physical problem is the Cauchy initial value (IV) Problem. The geometric structure underpinning the method of characteristics was also presented. This involved the concepts of Monge axis, pencil and cone, directional derivatives, characteristic directions, etc. These concepts will be applied in this chapter where we explore the nature of the solutions of these quasilinear PDEs for supersonic flow fields.

As implied above, the physical basis for the system of PDEs governing the various flow fields are the conservation laws: continuity, momentum, and energy. These conservation laws were described in various places in this work, wherever necessary. In Chapter 2 they were introduced as physical principles. The Lagrangian and Eulerian representations were described, and the physics of these laws were discussed in terms of these representations in a somewhat qualitative manner. In Chapter 3 these conservation laws again appear, but from the viewpoint of weak solutions of quasilinear PDEs which emphasize the global nature of the solutions. These weak solutions account for discontinuities such as shocks that occur across wave fronts. In Chapter 5 the conservation laws are again treated from the point of view of an incompressible fluid such as water; but a deeper mathematical description was given. And finally, in Chapter 6, the conservation laws again came up; this time from the point of view of propagation of sound waves as small amplitude waves in a compressible fluid (air). The conservation laws were used in the Eulerian representation and were linearized in accordance with the hypothesis of small amplitude sound wave propagation.

In this chapter we shall again make use of this Eulerian representation of the conservation laws as a starting point for a thorough discussion of the resulting system of quasilinear PDEs. However, we shall extend the treatment of Chapter 6 to include the nonlinear terms in the conservation equations. And finally, in the second part of this chapter which deals with a viscous fluid medium, we shall extend the conservations laws to include viscosity effects; this is done by generalizing the energy equation. We finally arrive at the Navier–Stokes equations which describe a viscous fluid. It was pointed out that the Eulerian representation of the conservation laws is more convenient for fluid flow where moving boundaries are considered.

In contrast, the Lagrangian representation of the conservation laws are used for the investigation of wave propagation in solids where there is comparatively little particle motion compared to fluids.

This chapter will treat wave propagation both in inviscid fluids where viscosity effects are neglected, and in viscous fluids where we cannot neglect these effects. It will therefore be convenient to divide the chapter into two parts: Part I—Inviscid Fluids; Part II—Viscous Fluids.

PART I. INVISCID FLUIDS

Introduction

In this part of Chapter 7 we shall assume the fluid medium (air) to be *inviscid, adiabatic,* and *isentropic.* An inviscid fluid means one where we do not take into account viscosity effects so that no dissipation of thermal energy is considered. An adiabatic fluid means one where there is no heat exchange between the fluid medium and its surroundings. This latter statement means that, for a thermodynamically reversible process, the medium is considered to be isentropic—the entropy of the fluid system is constant. We shall first investigate the mathematical and physical properties of one-dimensional compressible flow, then two-dimensional steady supersonic, transonic and subsonic flows, three-dimensional flow, and shock wave propagation. In keeping with the purpose of this work the emphasis will, as usual, be on wave propagation.

7.1. One-Dimensional Compressible Inviscid Flow

As pointed out [12, Chap. III], "Isentropic flow of compressible fluids admits of a fairly exhaustive mathematical treatment if the state of the medium depends only on the time t and a single Cartesian coordinate." It was also pointed out [12, Chap. 5] that, "One-dimensional gas dynamics is an excellent example of the interplay of the mathematics of quasilinear PDEs and the physics of continuous media." We therefore start our investigations of nonlinear wave propagation in inviscid compressible fluids with the one-dimensional unsteady isentropic case.

To fix our ideas we start with the physical model of a long thin tube of gas extending along the x axis from 0 to ∞. The fact that the tube is thin makes the geometry one dimensional, which means we neglect gradients of velocity, etc. normal to the axis. The initial conditions (ICs) are that the gas (air) in the tube is in a quiescent state of uniform particle velocity u_0, pressure p_0, and density ρ_0. At the front end ($x = 0$) we have a piston which starts to move with a constant piston velocity U_P. If it moves to the right

($U_P > 0$) then the gas in the neighborhood of the piston is compressed so that a compression wave is propagated into the gas. If $U_P < 0$ then the neighboring gas suffers a negative pressure so that a rarefactions wave is initiated at the piston and propagates into the tube. Thus we see that, in either case, the motion of the gas is then caused by the action of the piston. The reason why we restrict the motion of the piston to be constant is that if its velocity changes, then a series of waves of continuously varying intensity peel off from the piston making the flow unnecessarily complicated. As long as the piston maintains a constant velocity, a single wave peels off from the piston at $t=0$ where the motion of the piston must be idealized to have an infinite acceleration in order to suddenly have a constant nonzero velocity for $t>0$.

We shall represent the motion of the gas, piston, etc., by trajectories in the (x, t) plane where, by convention, the x axis is the abscissa. The $C_\pm$ characteristics will also be represented as curves on which disturbances must be propagated. Also various regions such as the domain of dependence and zone of influence will be represented in this plane. Obviously, the ICs are given on the x axis.

The *conservation laws* are nonlinear, one-dimensional, and unsteady. The energy equation is represented by the adiabatic equation of state so that the pressure is a unique function of the density only. For the one-dimensional case the particle velocity $\mathbf{v} = (u, 0, 0)$ where $u = u(x, t)$, so that the *continuity equation* (5.2) becomes

$$\rho_t + \operatorname{div}(\rho u) = \rho_t + u\rho_x + \rho u_x = 0. \tag{7.1}$$

The *conservation of momentum or Euler's equation* (5.6) (for $\mathbf{F}=0$) becomes

$$\rho u_t + \rho u u_x + p_x = 0. \tag{7.2}$$

The first term of (7.2) represents the linear portion of the inertia, the second term the nonlinear part of the inertia or convective term, and the third term the external force due to the pressure gradient.

The *energy equation or equation of state* for an adiabatic medium (air) is given by

$$p = A\rho^\gamma, \qquad A = \frac{p_0}{\rho_0^\gamma}, \qquad \gamma = \frac{C_p}{C_v}, \tag{7.3}$$

where C_p and C_v are the specific heats at constant pressure and volume, respectively, and p_0 and ρ_0 are defined at the undisturbed state. For air at normal temperatures we have $\gamma = 1.4$. Let c be the speed of the propagating compression wave. This wave speed is given by

$$c^2 = \frac{dp}{d\rho}. \tag{7.4}$$

For an adiabatic gas (7.4) becomes

$$c^2 = \gamma A \rho^{\gamma-1}, \tag{7.5}$$

so that c as well as p are prescribed functions of ρ.

Characteristic Equations and Curves

We now investigate the properties of (7.1) and (7.2) vis-à-vis characteristic theory developed in Chapter 3, Section 3.3. In particular, we put these equations in the form given by (3.19) and (3.20); we therefore rewrite (7.1) and (7.2) as the following system:

$$L_1 = \rho_t + u\rho_x + \rho u_x = 0, \tag{7.6}$$

$$L_2 = \rho u_t + \rho u u_x + c^2 \rho_x = 0, \tag{7.7}$$

where L_1 and L_2 stand for linear combinations of the partial derivatives ρ_x, ρ_t, u_x, u_t. We manipulated (7.2) as follows: Since $p = p(\rho)$ we have, upon using (7.4), $p_x = (dp/d\rho)\rho_x = c^2\rho_x$, where c^2 is a prescribed function of ρ given by (7.5). This accounts for the $c^2\rho_x$ term of L_2 in (7.7). We note that in the system (3.19) and (3.20), (3.20) is a special case of (7.7). The correspondence between this system and (7.6), (7.7) is obvious. Following Chapter 3, we form a linear combination $L = L_1 + \lambda L_2$ that depends on the parameter λ. Multiplying (7.7) by λ and adding (7.6) yields

$$L = \rho(1 + \lambda u)u_x + \lambda \rho u_t + (u + \lambda c^2)\rho_x + \rho_t = 0. \tag{7.8}$$

Equation (7.8) gives a linear combination of u_x, u_t, ρ_x, ρ_t which depends on the parameter λ. In general, λ depends on x and t, but at a given field point in the (x, t) plane λ is fixed. The crux of the method of characteristics, as described in Chapter 3, is that, at each field point, we want the *directional derivatives* du/dt and $d\rho/dt$ to be in the same direction. This direction is the *characteristic direction.* In other words, we want that linear combination L or that value of λ which yields this characteristic direction. (We point out again that this direction, in general, depends on each field point.) It turns out that we shall obtain two values of λ corresponding to two different directions, since the system (7.6) and (7.7) represents two first-order quasilinear PDEs. This means we shall obtain two distinct families of characteristic curves represented by $C_\pm$; each field point has two distinct characteristic line elements passing through it which represent two different characteristic directions $C_\pm$. C_+ will have a positive slope and C_- a negative slope. To obtain the characteristic directions, we apply the rule given by (3.21) to (7.8) which becomes

$$\text{coef.}(u_x) : \text{coef.}(u_t) = \text{coef.}(\rho_x) : \text{coef.}(\rho_t) = \frac{dx}{dt},$$

or

$$\frac{1+\lambda u}{\lambda \rho}=u+\lambda c^2=\frac{dx}{dt}. \tag{7.9}$$

Solving (7.9) for λ and dx/dt yields the following system:

$$\begin{aligned} &\text{for} \quad \lambda=\frac{1}{c}, \qquad \frac{dx}{dt}=c+u \qquad C_+, \\ &\text{for} \quad \lambda=-\frac{1}{c}, \qquad \frac{dx}{dt}=-(c-u) \quad C_-. \end{aligned} \tag{7.10}$$

The important result given by the system (7.10) is the crux of characteristic theory for the one-dimensional, unsteady, adiabatic flow case. It tells us the following: For the linear combination L given by $\lambda=1/c$, the *characteristic ordinary differential equation* (ODE) $dx/dt=c+u$ defines a positive *characteristic direction* dx/dt at each field point, since the right-hand side is positive. This ODE is, in general, nonlinear. In order to solve it we must know the right-hand side as a function of (x, t). Suppose we have solved for the flow field at a particular field point whose coordinates are (x, t). (We shall indicate below how characteristic theory, as described here, can allow us to obtain an approximation to the solution for $u(x, t)$ and $\rho(x, t)$.) Knowing $u(x, t)$ and $\rho(x, t)$ means we know the right-hand side as a function of (x, t) since c is a prescribed function of ρ. Since the characteristic ODE is first order, a single integration yields a one-parameter family of characteristic curves called the C_+ family. We now apply the same reasoning to the second equation in the system (7.10). The other linear combination L, which is given by setting $\lambda=-1/c$, yields the characteristic ODE $dx/dt=-(c-u)$ of negative slope, since the particle velocity is always much less than the wave speed. Its solution yields a one-parameter family of characteristic curves of negative slope dx/dt called the C_- family. We thus have the following situation: If we know the flow field in a given region in the (x, t) plane, then at each field point we have two different characteristic directions given by the slopes of the $C_\pm$ characteristics passing through the field point. Globally speaking, we have then constructed two families of curvilinear characteristics (the $C_\pm$ characteristics) in the flow field called a *characteristic net.* The importance of this characteristic net is that each element of the net (each characteristic curve) picks up initial data from the x axis. This will be described subsequently.

Parametric Form of the Characteristic Equations

Another way of looking at these two families of characteristic equations is to put them into parametric form according to the following scheme: as we know, the system (7.10) consists of the two characteristic ODEs, each of which is first order. Since we require two integrations, we can introduce two parameters (r, s) (r is a constant of integration defining a C_+, and s is a constant of integration defining a C_- characteristic). This means that, if we fix r and vary s we generate a C_+ characteristic, and conversely for a C_-. The slope of a C_+ characteristic and a C_- characteristic can then be written in the following parametric form:

$$C_+: \quad \frac{dx}{dt}=\frac{x_s}{t_s}, \qquad C_-: \quad \frac{dx}{dt}=\frac{x_r}{t_r}. \tag{7.11}$$

Then the characteristic ODEs (7.10) can be written in the parametric form

$$\begin{aligned} x_s &= (c+u)t_s, \\ x_r &= -(c-u)t_r. \end{aligned} \tag{7.12}$$

The solution of the system (7.12) yields the two families of characteristics in the parametric form

$$\begin{aligned} C_+: &\quad x=x(s, r_i), \qquad t=t(s, r_i), \\ C_-: &\quad x=x(r, s_j), \qquad t=t(r, s_j). \end{aligned} \tag{7.13}$$

From this notation it is clear that, at each field point P: (x, t) in the region of solution, there exists a unique pair of the parameters (r, s), namely, (r_i, s_j) which defines a C_+ and a C_- characteristic, respectively, passing through P. Since, in general, the characteristic ODEs (7.10) or (7.12) are nonlinear, the families of characteristics depend on the solution $u(x, t)$ at each field point. Nevertheless, finite difference methods have been developed based on this exposition of characteristic theory which allow us to obtain an approximation to the solution at each field point in terms of data at field points at previous times, so that the solution field eventually depends on data on the x axis (the Cauchy IV Problem). The reader is referred to [12] for a finite difference treatment based on characteristic theory. The parameters r, s have important physical significance which will be discussed below.

We are now in a position to discuss the significance of the two linear combinations of the partial derivatives L given by $\lambda = \pm 1/c$ which correspond to the characteristic directions given by the slopes of $C_\pm$, respectively. Inserting $\lambda = \pm 1/c$ into (7.8) and using the characteristic ODEs (7.10), we obtain the following system that holds on the C_+ and C_- characteristics, respectively:

$$\begin{aligned} \frac{dx}{dt}\left(ux+\frac{c\rho_x}{\rho}\right)+u_t+\frac{c\rho_t}{\rho}=0 \quad &\text{on} \quad C_+: \quad \frac{dx}{dt}=c+u, \\ \frac{dx}{dt}\left(u_x-\frac{c\rho_x}{\rho}\right)+u_t-\frac{c\rho_t}{\rho}=0 \quad &\text{on} \quad C_-: \quad \frac{dx}{dt}=-(c-u). \end{aligned} \tag{7.14}$$

The system (7.14) is a set of first-order quasilinear PDEs for u and ρ (recall that c is a prescribed function of ρ). The first equation is valid on the C_+

family, and the second on the C_- family. They are expressed as linear combinations of u_x, u_t, ρ_x, and ρ_t; where dx/dt is the slope of the C_+ or C_- characteristic, as shown. The system (7.14) must therefore be solved simultaneously at a field point P: (x, t) for the solution u_P, and ρ_P. This is done by integrating the first equation of (7.14) along a C_+ characteristic QP from a field point Q at a previous time step (on which the data u_Q, and ρ_Q are known) to the field point P. (Knowing u_Q, ρ_Q, and the adiabatic equation of state allows us to construct the characteristic segment QP.) Note that r (the constant of integration along C_+) is obtained in terms of given data u_Q, ρ_Q. Simultaneously, we integrate the second equation of (7.14) along a C_- characteristic RP from another field point R at a previous time step to P. At R we know the data u_R, ρ_R, and hence can determine the constant of integration s and the C_- line segment RP. The result of these two integrations yields two equations from which we can solve for u_P and ρ_P at P. Clearly we can find two such field points at previous time steps (the time steps might not be the same), such that a C_+ segment from one of these points and a C_- segment from the other drawn forward in time intersect at the field point P. This is shown in Fig. 7.1 where the C_+ line segment QP is drawn from Q to P and the C_- line segment RP is drawn from R to P. Knowing u_Q, ρ_Q we can determine r from these data at Q. Similarly, we can determine s from the data u_R, ρ_R at R. It is clear that we must take the previous field points Q and R to be close enough to P so that the respective characteristic line segments are straight lines, meaning that the time steps must be small. Knowing these straight line characteristic segments QP and RP greatly simplifies the calculations for a specific problem. Of course, refinements can be used to obtain greater accuracy.

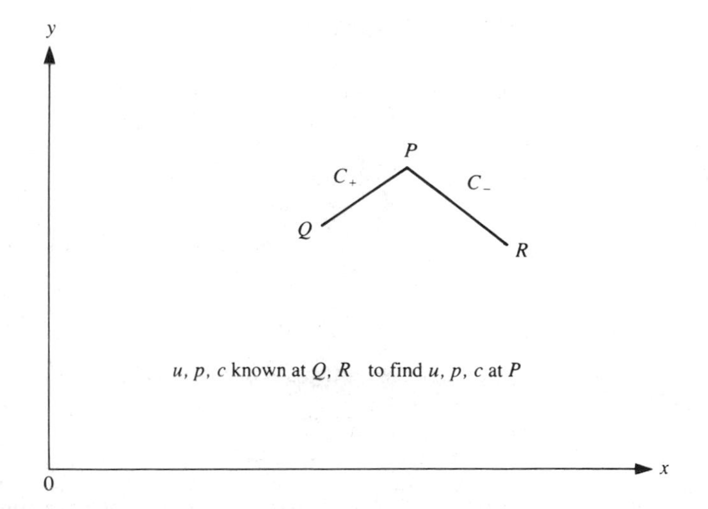

Fig. 7.1. Characteristic line segments from Q and R to field point P.

Our task, therefore, is to devise a method of integrating the first equation of (7.14) along a C_+ characteristic segment, and the second along a C_- segment. This means that we want to find a function of u and ρ, say $f(u, \rho)$, that is constant along a C_+ characteristic, and a function $g(u, \rho)$ that is constant along a C_- characteristic. Suppose at a field point P: (x, t) we have found a function f such that $f(u, \rho) = \text{const.} = f_i$ along a C_+, and $g(u, \rho) = \text{const.} = g_j$ along a C_-. Suppose we also know these constants f_i and g_j (from known data at previous time intervals). Then the set of equations, $f(u, \rho) = f_i$, $g(u, \rho) = g_j$, allows us to solve for u_P and ρ_P and thus find the flow field at the field point P, if we can find the value of the constants f_i, g_j. Suppose we have a particular pair of C_+ and C_- characteristics. Then the following equations hold: $df/dt = 0$ on C_+, $dg/dt = 0$ on C_-. Expanding these total derivatives gives

$$\begin{aligned} \left(\frac{dx}{dt}\right) f_x + f_t &= 0 \quad \text{on} \quad C_+: \quad \frac{dx}{dt} = c + u, \\ \left(\frac{dx}{dt}\right) g_x + g_t &= 0 \quad \text{on} \quad C_-: \quad \frac{dx}{dt} = -(c - u). \end{aligned} \tag{7.15}$$

Comparing (7.15) to (7.14) we cannot quite get a correlation between these two systems because, for instance, the factor multiplying dx/dt in the first equation of (7.14) is $u_x + c\rho_x/\rho$ while the corresponding factor of the first equation of (7.15) is f_x (a single dependent variable). However, Riemann got around this by the ingenious trick of defining a function l such that

$$l_x = \frac{c\rho_x}{\rho}, \qquad l_t = \frac{c\rho_t}{\rho}, \qquad dl = \left(\frac{c}{\rho}\right) d\rho, \qquad l = \int_{\rho_0}^{\rho} \frac{c(\rho)}{\rho}\, d\rho. \tag{7.16}$$

Equation (7.16) allows us to write (7.14) as

$$\begin{aligned} \left(\frac{dx}{dt}\right)(u_x + l_x) + (u_t + l_t) &= \left(\frac{dx}{dt}\right)(u + l)_x + (u + l)_t = 0 \quad \text{on } C_+, \\ \left(\frac{dx}{dt}\right)(u_x - l_x) + (u_t - l_t) &= \left(\frac{dx}{dt}\right)(u - l)_x + (u - l)_t = 0 \quad \text{on } C_-. \end{aligned} \tag{7.17}$$

It is clear that, defining l by (7.16), we can then correlate (7.17) with (7.15) by setting

$$f = u + l = \text{const.} \quad \text{on } C_+, \qquad g = u - l = \text{const.} \quad \text{on } C_-.$$

Since $d\rho = c^{-2}\, dp$, by using (7.16) we may also write l as

$$l(p) = \int_{p_0}^{p} \frac{dp}{\rho c}.$$

We recognize the term ρc as the *acoustic impedance*, whose physical significance was described in Chapter 6.

We now identify the constants of integration (r, s) with f and g by setting

$$f = u + l = r, \qquad g = u - l = s. \tag{7.18}$$

We note that f_i is identified with r, and g_j with s. This means that r is constant on a C_+ (picks out a particular C_+) and s is constant on a C_-. The parameters (r, s) are called the *Riemann invariants.*

Before we go any further let us calculate r, s for our adiabatic or polytropic gas. This is easily done by inserting (7.3) into (7.16) and integrating. For an adiabatic gas we then obtain the following expression for $l(\rho)$:

$$l(\rho) = \frac{2c}{\gamma - 1}. \tag{7.19}$$

Inserting (7.19) into (7.18) yields the following situation for an adiabatic gas:

$$\begin{aligned} &\text{slope of } C_+ = \frac{dx}{dt} = u + c, \qquad u + \frac{2c}{\gamma - 1} = r = \text{const. on } C_+,\\ &\text{slope of } C_- = \frac{dx}{dt} = u - c, \qquad u - \frac{2c}{\gamma - 1} = s = \text{const. on } C_-. \end{aligned} \tag{7.20}$$

Consider the special case $\gamma = 3$, which is unrealistic physically, since $\gamma = 1.4$ for air. We then obtain

$$\begin{aligned} &u + c = r = \text{const.} \quad \text{on } C_+: \quad \frac{dx}{dt} = u + c \quad \text{or} \quad \frac{dx}{dt} = r \quad \text{on } C_+,\\ &u - c = s = \text{const.} \quad \text{on } C_-: \quad \frac{dx}{dt} = u - c \quad \text{or} \quad \frac{dx}{dt} = s \quad \text{on } C_-. \end{aligned}$$

This tells us that the characteristics in the (x, t) plane are straight lines for the special case $\gamma = 3$.

Calculation of the Flow Field for an Adiabatic Gas

The system (7.20) allows us to calculate the flow field in the (x, t) plane for an adiabatic gas. We shall calculate this field in an approximate way by showing how to obtain the particle velocity and density at a finite number of points. This is an example of a *finite difference method*, since we discretize the problem by performing calculations that depend on a finite set of field points.

We shall calculate (u_P, ρ_P) at the nth time step in terms of data previously obtained at the neighboring field points Q, R at the $(n-1)$st time step. Figure 7.2 is a more complete representation of Fig. 7.1. It shows a *characteristic net* constructed as follows: We shall call each field point a *nodal*

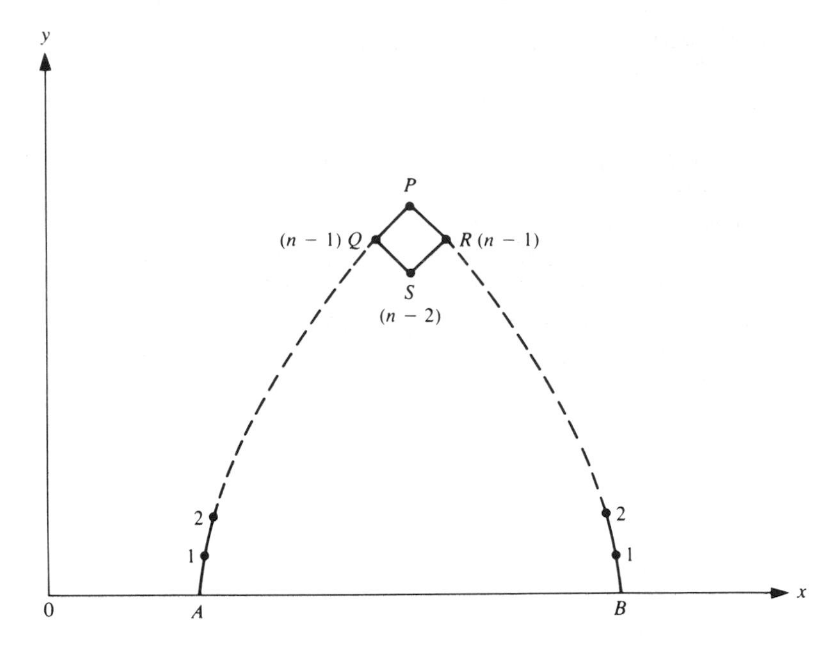

Fig. 7.2. Characteristic net showing n steps.

point or *node*. Each node has a C_+ and a C_- characteristic line segment projected back one step in time. The node P, at the nth time step, is reached by the characteristic segments C_+: $\overline{QP}$ and C_-: $\overline{RP}$, where Q and R are the nodes at the $(n-1)$st time step so that two nodes are involved. (Recall that it is not necessary that $t_Q = t_R$.) Similarly, projecting the characteristic segments from Q and R back one step in time we see that, at the $(n-2)$nd time step, the three nodes S, T, U are involved. In general, at the $(n-k)$th time step, $k+1$ nodes are involved $(k=0, 1, 2, \ldots, n)$. This means that, for this process of "marching backwards in time", at each previous time step one more node is involved. $k=n$ corresponds to the zeroth time step which is on the x axis, so that we must prescribe initial data (u, ρ) at these $n+1$ nodes in order to obtain solutions at the nodes making up this characteristic net. This clearly means that the solution for (u_P, ρ_P) at node P depends eventually only on the initial data.

Coordinates of P

To calculate the solution (u_P, ρ_P) at the node P we must first obtain the coordinates of P, knowing those of the nodes Q and R at the $(n-1)$st time step. To this end, we apply the characteristic ODEs (7.20). According to our approximation we assume that the C_+ characteristic segment $\overline{QP}$ and the C_- segment $\overline{RP}$ are straight lines. Therefore integrating the characteristic

ODEs $dx/dt = u + c$ along $\overline{QP}$, and $dx/dt = u - c$ along $\overline{RP}$ yields

$$\begin{aligned} &\text{on } \overline{QP}: \quad x_P - x_Q = z_Q(t_P - t_Q), \qquad z_Q = \left(\frac{dx}{dt}\right)_Q = u_Q + c_Q, \\ &\text{on } \overline{RP}: \quad x_P - x_R = z_R(t_P - t_R), \qquad z_R = \left(\frac{dx}{dt}\right)_R = u_R - c_R. \end{aligned} \tag{7.21}$$

In deriving (7.21) we calculated the slope z_Q or QP from the data u_Q and c_Q at the node Q, and similarly for the negative slope z_R of RP (where we used the data at R). The solution of (7.21) is

$$\begin{aligned} x_P &= \left(\frac{1}{c_Q + u_Q + c_R - u_R}\right)[(c_Q + u_Q)x_R + (c_R - u_R)x_Q \\ &\quad - (c_Q c_R - u_Q u_R)(t_Q - t_R)], \\ t_P &= \left(\frac{1}{c_Q + u_Q + c_R - u_R}\right)[x_R - x_Q + (c_R - u_R)t_R + (c_Q + u_Q)t_Q]. \end{aligned} \tag{7.22}$$

For the special case $u_Q = u_R = 0$, $c_Q = c_R = c_0$, $t_Q = t_R$, (7.22) becomes

$$x_P = \tfrac{1}{2}(x_R - x_Q), \qquad t_P = \left(\frac{1}{2c_0}\right)(x_R - x_Q) + t_R.$$

Obviously, this means that the triangle PQR in Fig. 7.2 is isosceles with the base $\overline{QR}$ parallel to the x axis.

Calculation of (u_P, c_P)

Knowing the coordinates (x_P, t_P) of the node P in terms of those of Q and R, we now attempt to calculate (u_P, c_P) in terms of data at Q and R by using the definition of the Riemann invariants r and s defined for an adiabatic gas by (7.20). Referring again to Fig. 7.2, on C_+: $\overline{QP}$ we have $r = \text{const.}$, and on C_-: $\overline{RP}$ we have $s = \text{const.}$ We therefore obtain from (7.20)

$$\begin{aligned} &\text{on } \overline{QP}: \quad u_P + \mu c_P = r = u_Q + \mu c_Q, \qquad \mu = \frac{2}{\gamma - 1}, \\ &\text{on } \overline{RP}: \quad u_P - \mu c_P = s = u_R - \mu c_R. \end{aligned} \tag{7.23}$$

The solution of (7.23) is

$$\begin{aligned} u_P &= \tfrac{1}{2}[u_Q + u_R + \mu(c_Q - c_R)], \\ c_P &= \tfrac{1}{2}[u_Q - u_R + \mu(c_Q + c_R)]. \end{aligned} \tag{7.24}$$

For the case $c_Q = c_R = c_0$, u_P is the average value of u_Q and u_R, whereas $c_P = c_0 + \tfrac{1}{2}(u_Q - u_R)$.

Since ρ_P can be obtained from c_P from (7.5) (having obtained the constant A from the initial density and pressure), it is therefore clear that we have presented a procedure for obtaining an approximation for the solution for (u, ρ) at a finite number of field points, in terms of the prescribed initial data on the x axis, for an adiabatic gas.

We now take a closer look at Fig. 7.2. The polygonal curve AP is made up of n C_+ line segments, each one joining two neighboring nodes. Similarly, the polygonal curve BP is composed of n C_- line segments. We shall call the curve AP the frontier C_+ characteristic (even though it is composed of n characteristics), and we shall call the curve BP the C_- frontier characteristic. Then the triangle APB, composed of these two frontier characteristics and the interval $[A, B]$, contains $\frac{1}{2}n(n+1)$ nodes including the points A, B, and P. Suppose the ICs are:

(1) The gas is quiescent, meaning that $\rho = \rho_0$, $c = c_0$, $u = u_0 = 0$, on the portion of the x axis outside the interval $[A, B]$.
(2) u is prescribed and is not zero and $c = c_0$ for the nodes on the interval $[A, B]$.

If we attempt to build up solutions for u and c, in the region off the x axis outside the triangle APB by using (7.24), we find that $c = c_0$ and $u = 0$ so that the gas in this region is quiescent. This means that the only nodes or field points whose solutions are affected by the ICs on the interval $[A, B]$ are those points in the triangle APB and on the frontier characteristics AP and BP. This important concept of the initial data on $[A, B]$, effecting only the field points in the region bounded by the frontier characteristics from P intersecting A and B, is one of the key features of wave propagation phenomena. We shall discuss this below.

Domain of Dependence and Range of Influence†

As pointed out above, the solution at P depends only on the initial data on the interval $[A, B]$ of the x axis intercepted by the frontier characteristics AP and BP from P. This interval is called the *domain of dependence* of P, since no initial data outside $[A, B]$ effects the solution at P. This is shown in Fig. 7.3(a). There also exists a *range of influence* which is defined as the region influenced by the initial data in $[A, B]$. This is the region bounded by the C_- characteristic from A and the C_+ characteristic from B, as shown in Fig. 7.3(b). These important concepts of domain of dependence and range of influence occur in all wave propagation problems, not only for an adiabatic gas. It is a general property of the hyperbolic PDEs which govern wave propagation, that characteristic curves stemming from the IV curve demark domains of dependence and ranges of influence. The reason for this is that disturbances in the initial data can only be propagated along characteristics. We recall that in Chapter 3 there is a discussion of these

† Refer also to Chapter 3, Section 3.5 on propagation of discontinuities.

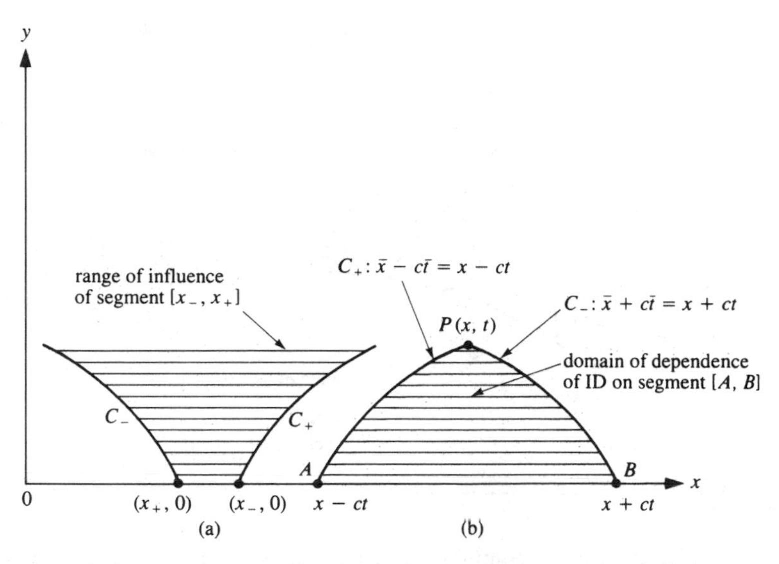

Fig. 7.3. (a) Domain of dependence and (b) range of influence of field point P in the (x, t) plane.

concepts in connection with Riemann's method of solving the Cauchy IV Problem.

Propagation of Discontinuities Along Characteristics

We shall use these concepts to show that any discontinuity in the initial data is propagated only along characteristic curves. We now discuss Fig. 7.4 which will demonstrate this assertion. Let A and B be two points on the IV curve (curve Γ). Draw a C_+ characteristic from A, a C_- characteristic from B, and let them intersect at the field point P. Suppose there is a discontinuity in the initial data at point A. Let C be a point on Γ to the left of A. Let Q be any field point in the curvilinear strip $CAPD$. From the definitions of domain of dependence, the solution at every point Q inside this strip is determined by the initial data on the interval CB. Clearly the initial data on the interval AB is different from that on CB (because of the discontinuity at A). Now fix A and let C approach A along Γ. Then the curve CD tends to the characteristic AP and Q tends to a point on AP, thus proving that a discontinuity in the initial data at A propagates along the characteristic AP (since Q is an arbitrary point on AP). Clearly the same proof holds for the C_- characteristic BP.

Linear Case

In Chapter 6 we discussed sound wave propagation in a compressible fluid. It was pointed out that sound waves have small amplitude oscillations so

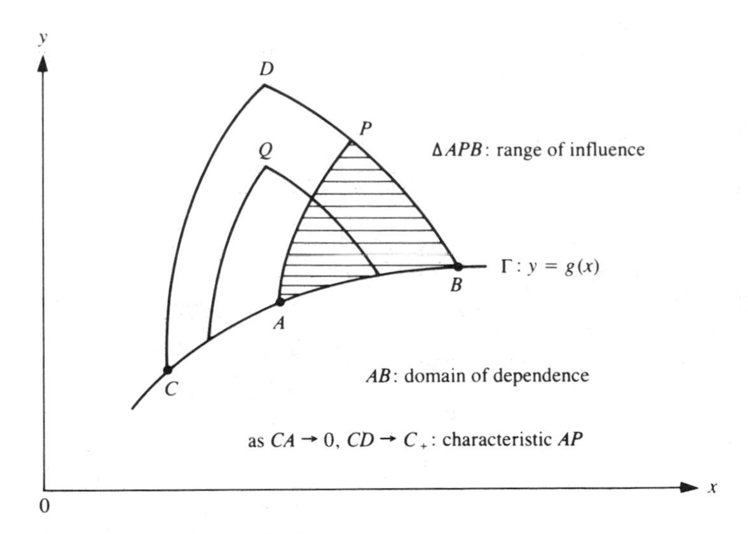

Fig. 7.4. Propagation of discontinuities along characteristics showing domain of dependence and range of influence.

that the equations of motion for a one-dimensional unsteady gas are linearized. The method of characteristics developed above was general enough to encompass large amplitude wave propagation in such a gas. It is instructive here to again linearize the flow equations given here by (7.6) and (7.7) and apply the method of characteristics to sound wave propagation in an adiabatic gas.

Linearization means neglecting products of small terms (quadratic terms). Therefore the linearized continuity equation (7.6) becomes

$$\rho_0 u_x + \rho'_t = 0, \tag{7.25}$$

where ρ' is the perturbed density. The linearization of the equation of motion (7.7) is

$$\rho_0 u_t + c_0^2 \rho'_x = 0. \tag{7.26}$$

The linear analogue of (7.8) is obtained by taking a linear combination of (7.25) and (7.26) yielding

$$\rho_0 u_x + \lambda \rho_0 u_t + \lambda c_0^2 \rho'_x + \rho'_t = 0.$$

The conditions for the du/dt and $d\rho'/dt$ to be in the same direction are obtained in the usual way, yielding the characteristic ODEs

$$\frac{dx}{dt} = \begin{cases} c_0, & C_+: \quad \lambda = 1/c_0, \\ -c_0, & C_-: \quad \lambda = -1/c_0. \end{cases} \tag{7.27}$$

The solution of the characteristic ODEs (7.27) yields a C_+ family of straight line characteristics of slope $1/c_0$ (in the (x, t) plane), and a family of C_-

characteristics of slope $-1/c_0$. An important point is that these characteristics are independent of the solution. This principle of the characteristics being independent of the solution is a property of linear hyperbolic PDEs in general; in this case, the straight line characteristics are due to the constant coefficients in the PDEs. Comparing (7.27) with (7.10) we observe that for the linear case $c = c_0 = \text{const.}$, and u is neglected in the slopes dx/dt.

For the linear case the proper linear combinations of partial derivatives are obtained by setting $\lambda = \pm 1/c_0$, so that the linear analogue of (7.14) becomes

$$\begin{aligned} \left(\frac{dx}{dt}\right)\left(u + \frac{c_0\rho'}{\rho_0}\right)_x + \left(u + \frac{c_0\rho'}{\rho_0}\right)_t &= 0 \quad \text{on } C_+, \\ \left(\frac{dx}{dt}\right)\left(u - \frac{c_0\rho'}{\rho_0}\right)_x + \left(u - \frac{c_0\rho'}{\rho_0}\right)_t &= 0 \quad \text{on } C_-. \end{aligned} \tag{7.28}$$

From (7.28) we obtain the linearized Riemann invariants (r, s).

$$\begin{aligned} u + \frac{c_0\rho'}{\rho_0} &= r = \text{const.} \quad \text{on } C_+: \quad \frac{dx}{dt} = c_0, \\ u - \frac{c_0\rho'}{\rho_0} &= s = \text{const.} \quad \text{on } C_-: \quad \frac{dx}{dt} = -c_0. \end{aligned} \tag{7.29}$$

The system (7.29) can easily be obtained from (7.16) and (7.18). Using the approximation $c = c_0$ we get

$$l(\rho') = \int_{\rho_0}^{\rho_0+\rho'} \left(\frac{c_0}{\rho}\right) d\rho = c_0 \log\left(1 + \frac{\rho'}{\rho_0}\right) \sim \frac{c_0\rho'}{\rho_0}.$$

Inserting this approximation for $l(\rho')$ into (7.18) yields (7.29).

Solution of the Linearized Initial Value Problem

We shall now apply the system (7.29) to obtain the solution for u_P and ρ'_P at a field point in terms of the initial conditions. This is related to the *Cauchy initial value* (IV) *Problem* that was treated in Chapter 4 where we obtained D'Alembert's solution. There is a difference in the physics of the problem as well as the initial data: The problem treated in Chapter 4 concerned the vibrating string and dealt with the wave equation for the lateral displacement of the string—the initial data were the prescribed displacement and velocity on the x axis. For the case considered here, we have a first-order linear system to be solved for the particle velocity and perturbed density. The ICs are the prescribed velocity and density on the x axis.

Consider Fig. 7.5. From the field point P draw a C_+ characteristic which intersects the x axis at point A, and a C_- characteristic which intersects the x axis at point B. From (7.29) we obtain the equations for the

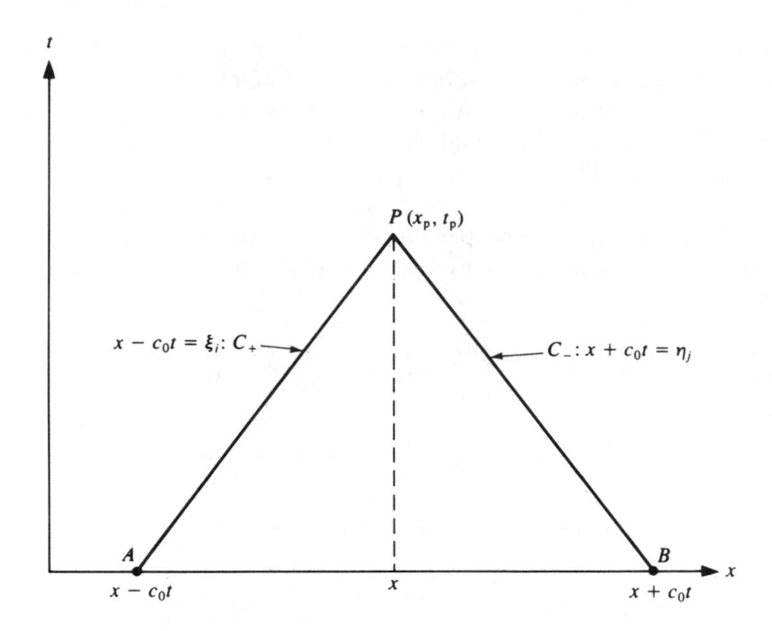

Fig. 7.5. Linear characteristics from field point P showing characteristic coordinates.

characteristics AP and BP.

$$C_{+}:\ AP\quad x-c_0t=\xi_i, \qquad C_{-}:\ BP\quad x+c_0t=\eta_j. \tag{7.30}$$

Note that (ξ, η) are the characteristic coordinates defined by (4.2) for the linear wave equation. Equations (7.30) tell us that the characteristic coordinate $\xi=\xi_i$ is constant along AP while η varies generating the C_+ characteristic, and $\eta=\eta_j$ is constant along BP while ξ varies. (ξ_i, η_j) are determined by the coordinates of A, B, respectively. We have

$$\text{at}\quad A:\quad \xi_i=x_P-c_0t_P, \qquad \text{at}\quad B:\quad \eta_j=x_P+c_0t_P,$$

where (x_P, t_P) are the coordinates of P. Applying (7.29) and calculating r from the initial data at A and s from the initial data at B, yields

$$\begin{aligned} &\text{on}\quad AP:\quad u_P+\frac{c_0\rho'_P}{\rho_0}=r=u_A+\frac{c_0\rho'_A}{\rho_0},\\ &\text{on}\quad BP:\quad u_P-\frac{c_0\rho'_P}{\rho_0}=s=u_B-\frac{c_0\rho'_B}{\rho_0}. \end{aligned} \tag{7.31}$$

The solution of (7.31) for u_P is very simply expressed in terms of the average of the IVs of u and ρ' at A and B, since the geometry of Fig. 7.5 is symmetric (triangle *APB* is isosceles). We obtain

$$u_P = \tfrac{1}{2}(u_A + u_B) + \left(\frac{c_0}{2\rho_0}\right)(\rho'_A + \rho'_B),$$
$$\rho'_P = \left(\frac{\rho_0}{2c_0}\right)(u_A - u_B) + \tfrac{1}{2}(\rho'_A - \rho'_B). \tag{7.32}$$

Note that the second equation of (7.32) involves differences of the initial data at A and B. Clearly (7.32) must reduce to $u_P = u_B = u_A$ for the trivial case where the initial data is uniform, i.e., where $u_A = u_B$, $\rho'_A = \rho'_B$.

And finally, for the linear case the domain of dependence and range of influence shown in Fig. 7.3(a), (b) are particularly simple: For Fig. 7.3(a) the characteristic AP, which is composed of polygonal line segments for the quasilinear case, is reduced to a single straight line characteristic AP for the linear case. A similar statement holds for the characteristic BP. Similarly, for Fig. 7.3(b) the C_- segments emanating from A are replaced by a single C_- straight line characteristic from A for the linear case, and C_+ at B are replaced by a single C_+ straight line emanating from B.

7.2. Two-Dimensional Steady Flow

We now investigate the two-dimensional steady, isentropic, irrotational flow in the (x, y) plane. In Chapter 5 the continuity equations and Euler's equation of motion were developed and then adapted to the needs of our investigations of wave propagation of waves in water which was considered to be an incompressible fluid. In this section we relax the restriction of incompressibility, since in this chapter we are concerned with wave propagation in a compressible fluid. We maintain the condition of irrotationality, which means $\mathbf{curl}\,\mathbf{q} = 0$ where $\mathbf{q}$ is the particle velocity. The restriction to a two-dimensional fluid in a steady state condition means that $\mathbf{q} = (u, v, 0)$, where u, v are x, y components of the velocity, respectively, and are functions of (x, y) only since the flow is time-independent.

The *irrotational condition* in two dimensions becomes

$$v_x - u_y = 0. \tag{7.33}$$

The *conservation laws* are given in the Eulerian representation in two dimensions. The *conservation of mass* or continuity equation becomes

$$(\rho u)_x + (\rho v)_y = 0, \tag{7.34}$$

where the density ρ is also a function of (x, y). The flux of fluid in two dimensions is $\rho\mathbf{q} = (\rho u, \rho v, 0)$. This represents the mass of fluid flowing through a unit cross-sectional area per unit time. Expanding the left-hand side of (7.34) and using the fact that

$$\frac{d\rho}{dt} = u\rho_x + v\rho_y, \qquad \frac{dx}{dt} = u, \qquad \frac{dy}{dt} = v,$$

yields

$$\frac{d\rho}{dt} = -\rho(\mathbf{div}\,\mathbf{q}) = -\rho(u_x + v_y), \tag{7.34'}$$

which say that the time rate of change of the density equals the negative of the divergence of the velocity times the density.

The *conservation of linear momentum* or *equations of motion* become

$$\begin{aligned} \frac{du}{dt} &= uu_x + vu_y = -\frac{p_x}{\rho} \qquad x \text{ direction}, \\ \frac{dv}{dt} &= uv_x + vv_y = -\frac{p_y}{\rho} \qquad y \text{ direction}. \end{aligned} \tag{7.35}$$

Since we have a steady state condition, $u_t = v_t = 0$, so that the x and y components of the particle acceleration, du/dt, dv/dt, are given as the convective terms which are quadratic terms representing the spatial distribution of the acceleration.

Bernoulli's Law

We shall make use of the fact that an adiabatic gas is defined by $dp/d\rho = c^2$ where c^2 is a prescribed function of $q^2 = u^2 + v^2$. This is given by Bernoulli's law which can be put in the form

$$\mu^2 q^2 + (1 - \mu^2)c^2 = c_*^2, \qquad \mu^2 = \frac{\gamma - 1}{\gamma + 1}, \tag{7.36}$$

where c_* is the *critical sound speed.* The critical sound speed allows us to ascertain the supersonic or subsonic character of the flow by comparing the flow speed q with c_*. This is seen by casting (7.36) in the following form:

$$q^2 - c_* = (1 - \mu^2)(q^2 - c^2). \tag{7.36'}$$

From (7.36') we have the conditions

$$\begin{aligned} q > c_* &\quad \text{if and only if} \quad M > 1, \qquad M = q/c, \\ q < c_* &\quad \text{if and only if} \quad M < 1, \end{aligned}$$

where M is the Mach number. These conditions prove the above assertion that c_* determines the supersonic or subsonic character of the flow. In other words, if $M^* = q/c_*$ is the Mach number corresponding to the critical sound speed, then $M^* \gtrless 1$ if and only if $M \gtrless 1$.

In order to derive Bernoulli's law (7.36), which is an energy equation, we present the necessary thermodynamic relations that enter into the energy equation for an adiabatic gas. The first law of thermodynamics can be written as

$$di = \frac{dp}{\rho} + T\,dS,$$

where i is the *specific enthalpy* (enthalpy per unit mass) defined by

$$i = e + \frac{p}{\rho},$$

where e is the specific internal energy, S is the specific entropy, and T is the temperature. For an isentropic fluid

$$\frac{dS}{dt} = \mathbf{q} \cdot \mathbf{grad}\, S + S_t = 0.$$

Euler's equation of motion is

$$\frac{d\mathbf{q}}{dt} = -\left(\frac{1}{\rho}\right) \mathbf{grad}\, p \qquad \text{or} \qquad \left(\frac{d}{dt}\right)\left(\tfrac{1}{2}q^2 + \frac{p}{\rho}\right) = 0.$$

From these expressions we immediately deduce that on each streamline of a steady flow we have

$$\left(\frac{d}{dt}\right)(\tfrac{1}{2}q^2 + i) = \mathbf{q} \cdot \frac{d\mathbf{q}}{dt} + \mathbf{q} \cdot \mathbf{grad}\, i = \mathbf{q} \cdot T\, \mathbf{grad}\, S = 0.$$

From this we obtain the following form of Bernoulli's law for steady state flow:

$$\tfrac{1}{2}q^2 + i = \tfrac{1}{2}\hat{q}^2, \tag{7.37}$$

where $\hat{q}$ is called the *limit speed* and is constant along a streamline. (It may, of course, have different values along different streamlines.) Equation (7.37) tells us that the particle speed q cannot exceed the limit speed $\hat{q}$, so that $q < \hat{q}$ and q approaches $\hat{q}$ as i tends to zero.

The isentropic condition is $dS = 0$ so that

$$di = \frac{dp}{\rho} = \left(\frac{c^2}{\rho}\right) d\rho.$$

For an adiabatic gas (7.5) gives the relation between c and ρ, so that integrating i with respect to ρ yields

$$i = \frac{c^2}{\gamma - 1}.$$

Substituting this expression for i into (7.37) gives Bernoulli's law in the form (7.36), which is what we set out to prove.

Characteristics for Two-Dimensional Steady State Flow

We shall now obtain the $C_\pm$ families of characteristics in the (x, y) plane for a two-dimensional steady state adiabatic flow field. To this end, we eliminate ρ from the equations of motion by using the continuity equation and obtain two first-order quasilinear PDEs (one representing the irrotationality condition) for the components of the particle velocity u, v. The equations of motion (7.35) can be put in the form

$$uu_x + vu_y = -\frac{c^2\rho_x}{\rho},$$

$$uv_x + vv_y = -\frac{c^2\rho_y}{\rho}.$$

Multiplying the first equation by u, the second by v, and adding gives

$$u^2u_x + uv(u_y + v_x) + v^2v_y = -\left(\frac{c^2}{\rho}\right)(u\rho_x + v\rho_y).$$

Using the continuity equation (7.34), the right-hand side of the above expression becomes $c^2(u_x + v_y)$. We thereby obtain

$$(c^2 - u^2)u_x - uv(u_y + v_x) + (c^2 - v^2)v_y = 0. \tag{7.38}$$

The irrotational condition (7.33) and equation (7.38) are the required pair of first-order quasilinear PDEs for u and v, whose solution gives the velocity field in the (x, y) plane. We shall call them the flow equations. (Recall that c^2 is a known function of q^2 for our adiabatic gas.) According to the method of characteristics, we want to find the differential equations for the slopes of the characteristic curves (in the (x, y) plane), which have the property that along these slopes the directional derivatives du/dx and dv/dx are to be in the same direction. Since the system (7.33) and (7.38) involve the four derivatives $u_x, \ldots, v_y$, we attempt to find that linear combination of (7.33) and (7.38) such that these directional derivatives are in the same direction. Therefore, in the usual manner we construct a linear combination of (7.33) and (7.38) by multiplying (7.33) by a parameter λ and adding (7.38). We obtain

$$(c^2 - u^2)u_x - (uv + \lambda)u_y - (uv - \lambda)v_x + (c^2 - v^2)v_y = 0. \tag{7.39}$$

We recall the rule that the directional derivatives du/dx and dv/dx be in the same direction

$$\text{coef.}(u_y) : \text{coef.}(u_x) = \text{coef.}(v_y) : \text{coef.}(v_x) = \frac{dy}{dx} \equiv z, \tag{7.40}$$

where z is the slope of the characteristics, which is to be determined from the solution of the required characteristic differential equation.

Applying this rule (given by (7.40)) to (7.39) yields

$$z = -\frac{uv + \lambda}{c^2 - u^2} = \frac{c^2 - v^2}{-uv + \lambda}. \tag{7.41}$$

Solving this system for λ and z gives

$$\lambda_{\pm} = \pm c\sqrt{q^2 - c^2}, \tag{7.42}$$

$$z_{\pm} = \frac{-uv \pm c\sqrt{q^2 - c^2}}{c^2 - u^2}. \tag{7.43}$$

It is easily seen from (7.43) that $z_{\pm}$ are the roots of the following quadratic equation:

$$(c^2 - u^2)z^2 + 2uvz + (c^2 - v^2) = 0. \tag{7.44}$$

Equation (7.44) is the required characteristic ODE for the slope z of the characteristic curves. It is a first-order, second degree differential equation whose coefficients are known functions of u and v. The roots $z_{\pm}$ of (7.44) are given by (7.43). They are classified into three types according to the nature of the discriminant which is $q^2 - c^2$.

(1) If $q^2 > c^2$ the flow is supersonic.
(2) If $q^2 = c^2$ the flow is transonic.
(3) If $q^2 < c^2$ the flow is subsonic.

Thus the nature of the discriminant determines the type of flow. It also determines the type of PDE. We have seen that quasilinear PDEs can be classified into three types:

(1) Hyperbolic, for supersonic flow where there are two families of characteristics (characterized by two real unequal roots).
(2) Parabolic, for transonic flow where there is one family of characteristics characterized by two real and equal roots.
(3) Elliptic, for subsonic flow, where there are no real characteristics since the roots are complex conjugates.

Thus the classification of the roots $z_{\pm}$ of (7.44) characterizes the type of flow given by the quasilinear system (7.33) and (7.38). These results are summarized in Table 7.1.

Table 7.1. Classification of roots z for two-dimensional steady flow.

Nature of roots z	Type of PDE	$M = q/c$	Type of flow
Real unequal	Hyperbolic	>1	Supersonic
Real equal	Parabolic	$=1$	Transonic
Complex conjugates	Elliptic	<1	Subsonic

Velocity Potential

Since the flow is irrotational we have potential flow, meaning that a velocity potential $\phi(x, y)$ exists such that

$$\mathbf{grad}\ \phi = \mathbf{q} \qquad \text{or} \qquad u = \phi_x, \qquad v = \phi_y, \qquad w = 0. \tag{7.45}$$

Clearly (7.45) satisfies (7.33) identically. Inserting (7.45) into (7.38) gives

$$(c^2 - \phi_x^2)\phi_{xx} - 2(\phi_x\phi_y)\phi_{xy} + (c^2 - \phi_y^2)\phi_{yy} = 0. \tag{7.46}$$

Equation (7.46) is a second-order quasilinear PDE for $\phi(x, y)$, and is an example of the generic form given by (3.2). The correspondence among the coefficients is $a = c^2 - \phi_x^2$, $b = -2\phi_x\phi_y$, $c = c^2 - \phi_y^2$, $d = 0$. We may therefore apply the method of characteristics given in Section 3.4 of Chapter 3 to (7.46). The characteristic equation given by (3.24) becomes (7.44) where u and v are given by (7.45). Therefore Table 7.1 is a special case of Table 3.1. Thus we see that the two-dimensional isentropic, steady state, irrotational flow is equivalent to the Cauchy IV Problem for a quasilinear PDE for the velocity potential.

Hodograph Transformation

This is a mapping of the flow equations from the (x, y) coordinates or physical space to (u, v) coordinates or velocity space, which is sometimes called the *hodograph plane*. We shall develop this transformation and apply it to the flow equations (the first-order system) given by (7.33) and (7.38).

Jacobian or Mapping Function

In order to develop the hodograph transformation we shall construct the Jacobian or mapping function that allows us to go from physical to hodograph space. This allows us to obtain a nonsingular mapping $(x, y) \rightleftarrows (u, v)$, meaning that the mapping has a unique inverse. Suppose we have found the hodograph transformation. Then we can solve for the derivatives $u_x, \ldots, v_y$ in terms of the derivatives $x_u, \ldots, y_v$, and thereby transform the system (7.33) and (7.38) into a system of first-order PDEs in the hodograph or (u, v) plane whose coefficients are known functions of u and v. Therefore the transformed system of PDEs is linear—this gives us a great advantage. This system is then solved for $x = x(u, v)$, $y = y(u, v)$. The inverse mapping gives the flow field $u = u(x, y)$, $v = v(x, y)$.

We start by expanding the differentials dx and dy to obtain

$$dx = x_u\, du + x_v\, dv, \qquad dy = y_u\, du + y_v\, dv. \tag{7.47}$$

We first divide these equations by dx (keeping y constant) and obtain the following pair of algebraic equations for u_x and u_y:

$$1 = x_u u_x + x_v v_x,$$
$$0 = y_u u_x + y_v v_x.$$

The solution of this system is

$$u_x = \frac{y_v}{\det(J)}, \qquad v_x = -\frac{y_u}{\det(J)}, \tag{7.48}$$

where the Jacobian of the mapping from (x, y) to (u, v) is

$$J = \begin{pmatrix} x_u & x_v \\ y_u & y_v \end{pmatrix} \quad \text{so that} \quad \det(J) = x_u y_v - x_v y_u. \tag{7.49}$$

We now divide (7.47) by dy (keeping x constant) and obtain

$$0 = x_u u_y + x_v v_y,$$
$$1 = y_u u_y + y_v v_y.$$

The solution of this system is

$$u_y = -\frac{x_v}{\det(J)}, \qquad v_y = \frac{x_u}{\det(J)}. \tag{7.50}$$

We also expand du and dv by the same procedure of first keeping v constant, then u constant, and obtain algebraic equations for x_u, x_v, y_u, y_v whose solutions are

$$x_u = \frac{v_y}{\det(j)}, \qquad y_u = -\frac{v_x}{\det(j)},$$
$$x_v = -\frac{u_y}{\det(j)}, \quad y_x = \frac{u_x}{\det(j)}, \tag{7.51}$$

j is the Jacobian of the mapping from (u, v) to (x, y) space, and is given by

$$j = \begin{pmatrix} u_x & u_y \\ v_x & v_y \end{pmatrix}, \qquad \text{where} \quad \det(j) = u_x v_y - u_y v_x. \tag{7.52}$$

It is easily seen that $Jj = I$, the 2×2 identity matrix.

We summarize these results as follows:

$$\begin{pmatrix} u_x & v_x \\ u_y & v_y \end{pmatrix} = \det(j) \begin{pmatrix} y_v & -y_u \\ -x_v & x_u \end{pmatrix}. \tag{7.53}$$

We assume $\det(j) \neq 0$. Equation (7.53) is the required hodograph transformation from the (x, y) to the (u, v) plane. It gives the solution for $u_x, \ldots, v_y$ in terms of $x_u, \ldots, y_v$. Applying (7.53) to the flow equations (7.33) and

(7.38) yields

$$-y_u + x_v = 0, \tag{7.54}$$

$$(c^2 - u^2)y_v + uv(x_v + y_u) + (c^2 - v^2)x_u = 0. \tag{7.55}$$

The assertion that the hodograph transformation yields a system of linear PDEs is correct, as seen by inspecting (7.54) and (7.55). The importance of this simplification into a linear system is that the two families of characteristics in the hodograph or (u, v) plane are independent of the solution.

Characteristics in the Hodograph Plane

We shall now obtain the two families of characteristics in the hodograph plane of the system (7.54) and (7.55). We do this by determining that linear combination of the derivatives x_u, x_v, y_u, y_v, which puts the directional derivatives dx/du and dy/du in the characteristic direction dv/du. We shall call the characteristic curves in the (u, v) plane corresponding to the slopes $(dv/du)_\pm$, $\Gamma_\pm$. We determine these characteristic slopes or the characteristic ODEs in the usual way by determining the proper linear combination of (7.54) and (7.55). To this end we multiply (7.54) by the parameter λ, add the result to (7.55), and obtain

$$(c^2 - u^2)y_v + (uv + \lambda)x_v + (uv - \lambda)y_u + (c^2 - v^2)x_u = 0. \tag{7.56}$$

To obtain the proper linear combination we apply the following law to (7.56):

$$\text{coef.}(x_v) : \text{coef.}(x_u) = \text{coef.}(y_v) : \text{coef.}(y_v).$$

We obtain

$$Z_\pm = \frac{uv + \lambda_\pm}{c^2 - v^2} = \frac{c^2 - u^2}{uv - \lambda_\pm}, \qquad \text{where} \quad Z_\pm = \left(\frac{dv}{du}\right)_\pm. \tag{7.57}$$

From (7.57) we obtain

$$Z_\pm = \left(\frac{dv}{du}\right)_\pm = \frac{uv \pm c\sqrt{(q^2 - c^2)}}{c^2 - v^2}, \tag{7.58}$$

$$\lambda_\pm = \pm c\sqrt{(q^2 - c^2)}. \tag{7.59}$$

Equation (7.59) gives the proper linear combination; $\Gamma_\pm$ corresponds to λ_+ and Γ_- to λ_-. Equation (7.58) gives the roots $Z_\pm$ of the characteristic ODE which is easily seen to be

$$(c^2 - v^2)Z^2 - 2uvZ + (c^2 - u^2) = 0. \tag{7.60}$$

Equation (7.60) is a first-order second degree ODE being a quadratic in the characteristic slope Z in the hodograph plane. The solution of the characteristic ODE yields the curves $\Gamma_\pm$ corresponding to the plus or minus directions. By comparing the roots $Z_{\pm}$ given by (7.58) with the roots $z_{\pm}$ given by (7.43) we get the following result:

$$Z_{\pm} = -\frac{1}{z_{\mp}} \quad \text{or} \quad \left(\frac{dv}{du}\right)_{\pm} = -1 \bigg/ \left(\frac{dy}{dx}\right)_{\mp}. \tag{7.61}$$

From equation (7.61) we get the following interesting result: If a point, whose coordinates are (x, y) and its image (u, v), is represented in the same coordinate plane, then the slopes of C_+ and Γ_- and C_- and Γ_+ (through the corresponding points (x, y) and (u, v)) are orthogonal, respectively.

Geometric Interpretation of Characteristics

We now give a geometric interpretation of the $C_{\pm}$ and $\Gamma_{\pm}$ characteristics. Figure 7.6 shows a C_+ and a C_- characteristic stemming from a given field point P in the (x, y) plane. The particle velocity $\mathbf{q}$ at P is shown in the figure, as is the angle α between $\mathbf{q}$ and C_+. α is also the angle between $\mathbf{q}$ and C_-. θ is the angle that $\mathbf{q}$ makes with the positive x axis. Figure 7.7 shows the geometry of the $\Gamma_{\pm}$ characteristics. From the geometry in the (x, y) plane we obtain

$$u = q\cos\theta, \qquad v = q\sin\theta, \qquad z = \tan(\theta \pm \alpha). \tag{7.62}$$

Inserting (7.62) into the characteristic ODE for z, given by (7.44) yields, after some trigonometric transformations,

$$\sin\alpha = \frac{c}{q} = \frac{1}{M}, \qquad \text{where the Mach number} \quad M = \frac{q}{c}. \tag{7.63}$$

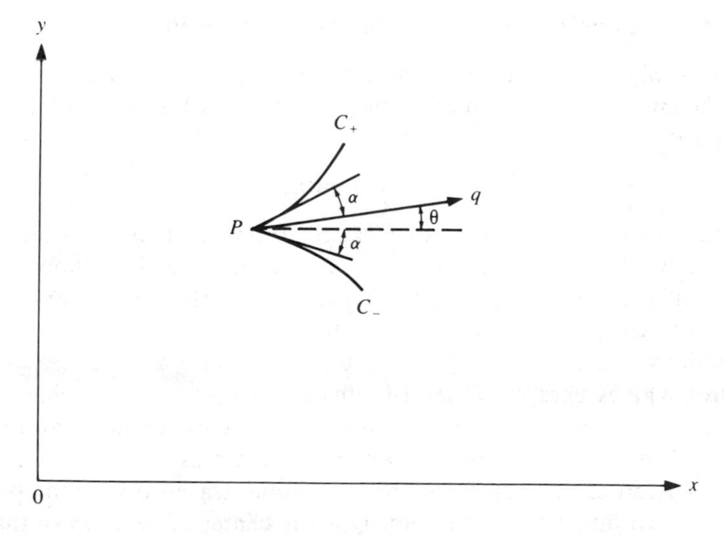

Fig. 7.6. Characteristics and particle velocity at field point P.

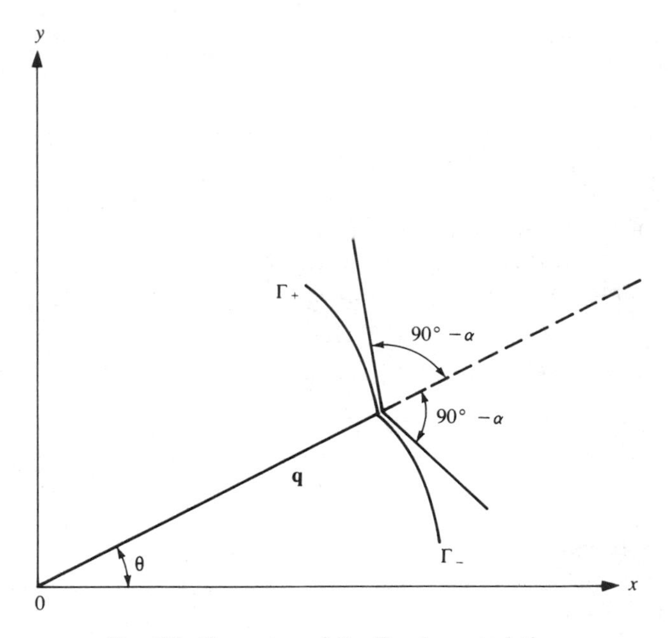

Fig. 7.7. Geometry of the $\Gamma_\pm$ characteristics.

Also, in the hodograph plane for the $\Gamma_\pm$ characteristics we have

$$M=\frac{q}{c}=\frac{1}{\cos\alpha'},\qquad \text{where}\quad \alpha'=\frac{\pi}{2}-\alpha. \tag{7.64}$$

Characteristics as Epicycloids in the Hodograph Plane

We now apply the above results to an adiabatic gas. We use Bernoulli's law or the energy equation in the form given by (7.36). Inserting (7.63) into (7.36) gives

$$q^2[\mu^2+(1-\mu^2)\sin^2\alpha]=c_*^2. \tag{7.65}$$

We now use (7.65) to give an elegant graphical interpretation of the characteristics in the hodograph plane. Specifically, we shall show that for an adiabatic gas the Γ characteristics in the hodograph plane are epicycloids generated by the points on a circle of diameter $c_*[(1/\mu)-1]=\hat{q}-c_*$ which rolls on the sonic circle of radius $u^2+v^2=c_*$.

Figure 7.8 gives a graphical description of the trajectory of a Γ characteristic curve in the hodograph plane. It shows a circle of center O' and radius $QO'=r$ rolling without slipping on a circle of radius $OQ=c_*$. The curve AP is a portion of the trajectory, the epicycloid, traced out by the point P fixed on the rolling circle. The point Q is the center of rotation so that QP is perpendicular to the tangent to the trajectory at P. $PO=\mathbf{q}$. Angle QPO

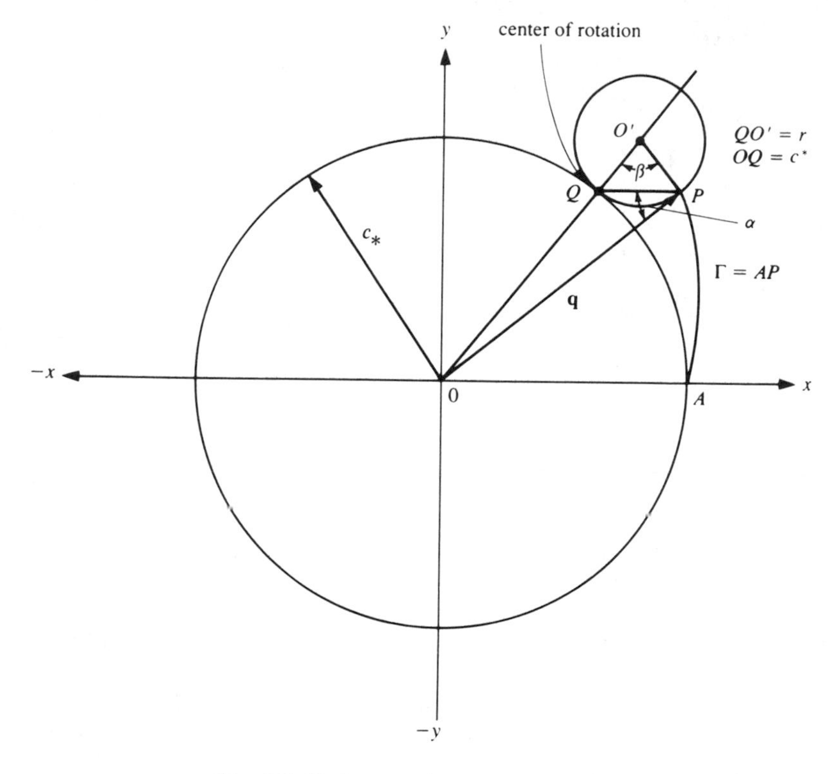

Fig. 7.8. Γ characteristic as epicycloid.

turns out to be α which is given by (7.63). It also turns out that $QO' = \frac{1}{2}(\hat{q} - c_*)$. This will now be proved from Fig. 7.8 and (7.65). To identify the trajectory AP with an epicycloid, it is sufficient to prove that the angle $QPO = \alpha$. From the figure we have angle $QO'P = \beta$ and angle $O'QP = (\pi - \beta)/2$ since triangle $QO'P$ is isosceles. Then angle $OQP = (\pi + \beta)/2$. We obtain from triangle $OO'P$ the relation

$$q^2 = (c_* + r)^2 + r^2 - 2r(c_* + r)\cos\beta.$$

Applying the law of sines to triangle OQP yields

$$\left(\frac{1}{c_*}\right)\sin\alpha = \left(\frac{1}{q}\right)\sin(\tfrac{1}{2}(\pi + \beta)) \qquad \text{or} \qquad \cos\frac{\beta}{2} = \left(\frac{q}{c_*}\right)\sin\alpha.$$

This gives

$$\cos\beta = 2\left(\frac{q}{c_*}\right)^2 \sin^2\alpha - 1.$$

Then q^2 becomes

$$q^2=(c_*+r)^2+r^2-2r(c_*+r)\left[2\left(\frac{q}{c_*}\right)^2\sin^2\alpha-1\right].$$

This simplifies to

$$q^2\left[1+4\left(\frac{r}{c_*^2}\right)(c_*+r)\sin^2\alpha\right]=(2r+c_*)^2. \tag{7.66}$$

Comparing (7.66) with (7.36) gives the result

$$r=\left(\frac{c_*}{2}\right)(\mu^{-1}-1)=\tfrac{1}{2}(\hat{q}-c_*). \tag{7.67}$$

This was the expression we set out to prove. The components of $\mathbf{q}$ are

$$\begin{aligned} u&=(c_*+r)\cos\theta-r\cos\left((c_*+r)\left(\frac{\theta}{r}\right)\right),\\ v&=(c_*+r)\sin\theta-r\sin\left((c_*+r)\left(\frac{\theta}{r}\right)\right). \end{aligned} \tag{7.68}$$

The system (7.68) gives the trajectory of a Γ characteristic if we plot v versus u. Note that the starting position of the rolling circle corresponds to $\theta=0$ where $u=c_*$, $v=0$. This corresponds to the point A in Fig. 7.8. Again referring to the figure, through each point of the annular ring $c_*<q^2<\hat{q}^2$ there pass two epicycloids (one for Γ_+ and one for Γ_-) so that this ring is covered with a net of two families of Γ characteristics in the hodograph plane.

Steady state plane flow offers a rich variety of special problems such as flow in nozzles and jets and around aerodynamic bodies. This is beyond the scope of this book, since the emphasis here is on wave propagation phenomena. The reader is referred to such works as [10] for a treatment of these topics.

7.3. Shock Wave Phenomena

Introduction

Up to now we have been investigating wave propagation in a one-dimensional unsteady isentropic fluid and a two-dimensional steady state, irrotational, isentropic fluid without shocks. We saw that the flow equations arose from the three conservation laws governing a continuous medium: mass, momentum, and energy. The energy equation assumed an adiabatic equation of state where the pressure is a unique function of the density. We have assumed that p, ρ, T, and S, are continuous across the wave fronts which are the characteristics. Any disturbance in the initial data can only be propagated along characteristic curves with the speed of sound c. These families of characteristics represent wave fronts for a very weak disturbance (a sound wave) across which the dynamic and thermodynamic variables are continuous. This means that we are dealing with an adiabatic isentropic flow, such that there is a thermodynamically reversible transition of flow across the wave front.

These assumptions are valid only if the gradients of velocity and temperature across the wave front are small. Otherwise, a mathematical description of the physical situation must take into account the effect of irreversible thermodynamic processes caused by friction and heat conduction that occur across the wave front. Experience from experimental data shows that these irreversible processes occur in gases only in the narrow neighborhood of the wave front, the so-called transition zone where the gradients are large, while outside this zone the flow is essentially adiabatic and isentropic. If this transition zone is assumed to be infinitely thin, then the gradients are replaced by discontinuities in velocity, temperature, pressure, etc., and lead to the so-called "jump conditions" across the wave front. We assume that a flow involving such discontinuities at the wave front is governed by the same conservation laws, but we must take into account these discontinuities. However, an additional condition must prevail, namely, that the entropy must increase across the wave front, which is now called a shock wave. This increase of entropy is the result of an irreversible thermodynamic process. This was shown by Lord Rayleigh who demonstrated that an adiabatic reversible transition across the shock wave would violate the principle of conservation of energy. The proof by Rayleigh that the entropy increases across a shock front means that the shock wave is a compression wave (the pressure increases across the shock) and not a rarefaction wave. Hugoniot also showed this by proving that in the absence of heat conduction and viscosity outside the transition zone, conservation of energy implies conservation of entropy except in the transition zone where the entropy increases.

Qualitative Ideas on Shock Waves

It was mentioned above that the shock front is a discontinuity surface across which there is a sudden transition or jump in the dynamical and thermodynamic variables. There is flow of gas across this shock surface. We mention in passing that there is another type of surface discontinuity called a *contact surface*. This surface separates two parts of the medium without any flow across the surface. Thus the contact surface moves with the fluid and separates two zones of different densities and entropies, for example, but the pressures and normal components of the velocity are continuous across the surface.

Shock Tube

We shall use the shock tube as a physical model for one-dimensional shock wave propagation. We consider a long thin tube of gas initially in a quiescent

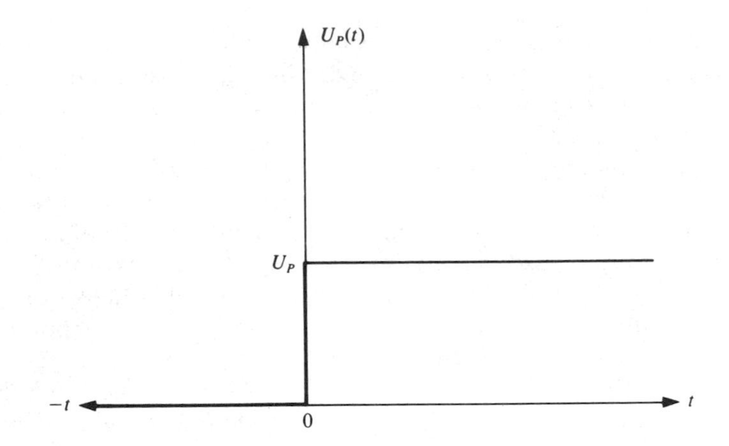

Fig. 7.9. Piston velocity versus time.

state: at rest with constant density and pressure so that $\rho = \rho_0$, $p = p_0$, $\mathbf{q} = 0$, where $\mathbf{q} = (u, 0, 0)$. We assume one-dimensional unsteady flow. The tube is represented by the positive x axis, and a piston is placed at the origin $(x = 0)$ at $t < 0$. At $t = 0$ the piston moves to the right with a constant velocity U_P. Clearly this is an idealization, for at $t = 0$ we assume a finite jump in the piston velocity giving an infinite acceleration no matter how small the piston velocity is. Figure 7.9 shows a plot of piston velocity versus time.

The motion of the gas which results from suddenly moving the piston cannot be continuous in the neighborhood of $x = t = 0$. What happens is that initially a shock wave forms at the right-hand side of the piston, and eats into the quiet gas with a constant shock wave velocity U which we shall subsequently prove to be supersonic $(U > c)$. The situation is shown in Fig. 7.10. If the piston velocity were not constant then a continuous family of shock waves would peal off from the piston. Figure 7.11 shows the trajectories of the piston, shock front, the particle velocity (which has zero velocity until disturbed by the shock whereupon it takes on the value of the piston velocity), and the C_+ characteristic stemming from the origin. The situation depicted by this figure is an approximation which holds in the neighborhood of the piston. A more exact solution can only be obtained

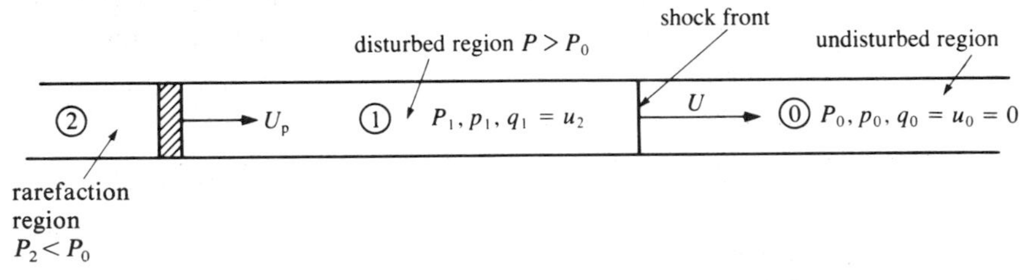

Fig. 7.10. Shock tube showing advancing shock front.

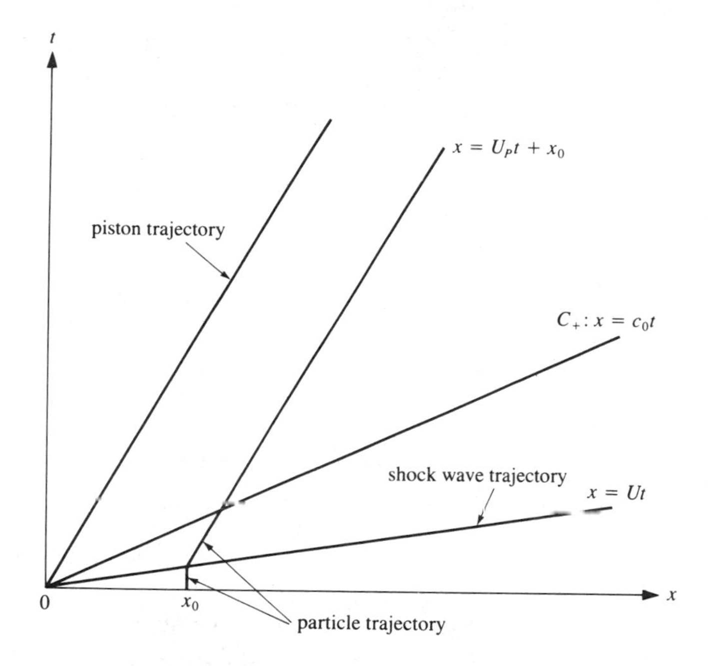

Fig. 7.11. Various trajectories in the (x, t) plane.

by solving the quasilinear flow equations which hold in the wedge region bounded by the piston and shock wave trajectories. The solution will, in general, give a nonconstant c which will yield curved characteristics.

Suppose we stepwise decrease the piston velocity U_P. For each value of U_P we obtain a shock wave with a corresponding velocity U. As U_P decreases, U decreases; and as $U_P \to 0$, $U \to c_0$ the velocity of a sound wave. We thus get a sequence of shock waves traveling at slower velocities until, in the limit, the shock wave approaches a sound wave. In addition, as U decreases to c_0 the shock strength as measured by $(p - p_0)/p_0$ becomes infinitely weak. p is the disturbed pressure and p_0 the quiescent pressure. We thus get a sequence of shocks eating into the quiet gas of slower velocity and weaker strength which, in the limit, approaches a sound wave.

In describing qualitatively what happens across the shock wave we shall make use of certain definitions. The *front side* is defined as the side of the shock front across which the gas enters. The other side is called the *back side*. The gas particles cross the shock wave from front to back. The undisturbed region is designated by the subscript (0), and the disturbed region by the subscript (1). The direction in which the shock front moves is given by U. There are three possibilities:

1. *Advancing shock front.* Suppose the velocity u_0 on the front side is zero. The shock wave advances into the quiescent region (0) with velocity U when observed from the front side, which turns out to be supersonic.

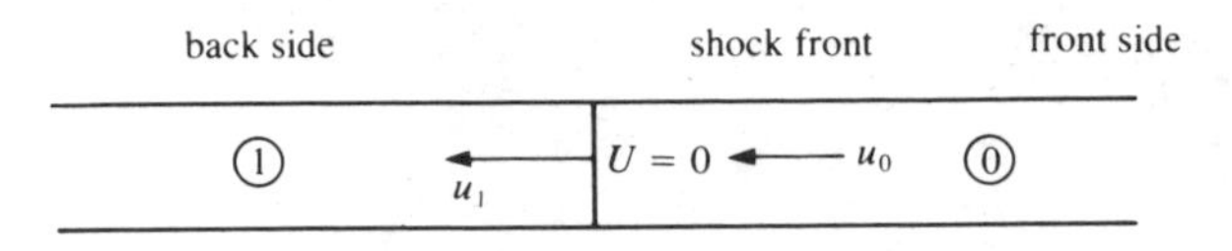

Fig. 7.12. Stationary shock front.

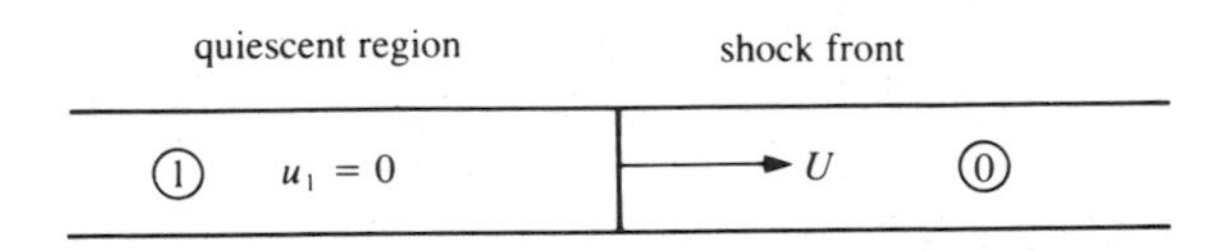

Fig. 7.13. Receding shock front.

The velocity of the shock wave when observed from region (1), the high pressure side, is $U - u_1$. This turns out to be subsonic when observed from the back side. The shock wave moves supersonically into the quiescent zone quickly enveloping more and more of the gas, which follows at a speed less than that of the shock front. This was described above for the moving piston in a quiet gas and is the situation described in Fig. 7.10.

2. *Stationary shock front.* Any shock front is stationary (velocity of the shock front is zero) when observed from a coordinate system fixed in the shock wave. This is seen in Fig. 7.12, where $U = 0$, u_0 is supersonic and u_1 is subsonic.
3. *Receding shock front.* This is interpreted by Fig. 7.13 where the shock front recedes into region (0) with velocity U, leaving behind a high pressure zone (1) where $u_1 = 0$. Such a receding shock is encountered as a shock wave reflecting from a wall.

Analogy with Particle Motion

Suppose a column of cars is going along a single lane highway at a low speed. Then any sudden deceleration of one car will give the next car behind time to react (if he is far enough behind), so that he and also the other cars behind will not need to suddenly decelerate. These cars are said to be moving less than the "sound speed" or critical speed. Suppose the column speeds up so that if a car in front suddenly slows down then the one behind him is forced to suddenly slow down, and similarly for the next behind that, and so on. The critical speed of the column at which this process occurs is the sound speed. Suppose the column moves faster than this sound speed and the cars are too close. Then if the first car suddenly stops, the second will crash into him, the third into the second, etc. A receding shock wave will then be produced which travels back with a characteristic speed if all the cars of the column are an equidistance apart and are traveling with the same speed. Clearly the trajectory of the shock wave is the locus of points of intersection of neighboring cars as they hit. We can easily visualize a forward-facing shock wave impinging on a quiescent zone by considering a car moving with a speed greater than the sound speed and crashing into a column of stationary cars parked in neutral with their brakes off. The first car that is hit will hit and set into motion the second car, etc. The locus of point of intersection of these successively hit cars is the trajectory of a shock wave advancing into a state at rest. We can represent each car by a mass coupled to neighboring masses by nonlinear springs. Then a system of coupled ODEs can be derived which describe the onset of motion of the successively struck masses. We assume two masses collide when the spring between them undergoes maximum compression. We alluded to this approach previously (Chapter 2) in developing the one-dimensional wave equation by a limiting process from a system of linear coupled ODEs. A more complete description of this process is contained in [12]. The difference here is that the spring force no longer obeys Hooke's law but satisfies a more complicated nonlinear law of repulsion.

Quantitative Treatment

Jump Conditions Across the Shock Front. We restate the conservation laws in the Eulerian representation for a one-dimensional gas from the point of view of shock wave propagation, where we shall incorporate jumps in the appropriate quantities. Let x be the Eulerian coordinate and $a(x, t)$ be the Lagrangian coordinate. Suppose a portion of gas occupies the interval (a_0, a_1) at time t, where $a_0 = a_0(t)$, $a_1 = a_1(t)$. We state the conservation laws for the gas occupying this interval.

Conservation of Mass (Continuity Equation)

$$\frac{d}{dt}\int_{a_0}^{a_1} \rho(x, t)\, dx = 0. \tag{7.69}$$

$\rho\, dx$ is the element of mass occupying the interval $(x, x + dx)$ at time t, where $a_0 \leq x \leq a_1$. In this form the integration is with respect to x over the interval (a_0, a_1) for a fixed t. a_0, a_1, of course, depend on t.

Conservation of Momentum. The rate of increase of linear momentum of a column of gas occupying the interval (a_0, a_1) equals the net force on the column, which is assumed to be due only to the pressure gradient acting at the ends of the column. We get

$$\frac{d}{dt}\int_{a_0}^{a_1} \rho u\, dx = p_0 - p_1, \tag{7.70}$$

where the velocity $\mathbf{q} = (u, 0, 0)$ and p_0, p_1 is the pressure at a_0, a_1, respectively.

Conservation of Energy. To formulate the energy equation for the column of gas in the interval (a_0, a_1), we equate the rate of increase of the total energy to the power input due to the pressure gradient at the ends of the column. The energy equation for the column therefore becomes

$$\frac{d}{dt}\int_{a_0}^{a_1} \rho(e+\tfrac{1}{2}u^2)\,dx = p_0u_0 - p_1u_1, \tag{7.71}$$

where e is the specific internal energy, $\frac{1}{2}u^2$ is the kinetic energy per unit mass, and p_iu_i is the power input due to the pressure at a_i $(i=0, 1)$.
Entropy Change. The total entropy of the column is

$$\int_{a_1}^{a_0} \rho S\,dx,$$

where S is the specific entropy. The statement that the column of gas does not lose its entropy is given by

$$\frac{d}{dt}\int_{a_0}^{a_1} \rho S\,dx > 0. \tag{7.72}$$

Jump Conditions. The integrals in the left-hand sides of (7.69)-(7.71) are of the form

$$I = \int_{a_0}^{a_1} F(x, t)\,dx. \tag{7.73}$$

We now assume that there exists an interior point $x = \bar{x}$ in the column $(a_0 < \bar{x} < a_1)$ where the integrand F is discontinuous. Specifically, at a fixed t there is a finite jump discontinuity in F at x. Let $F_\pm$ be the value of F as it approaches $\bar{x}$ from the right, left, respectively. Then $F_+ - F_- \neq 0$. We now calculate dI/dt using the formula for differentiating an integral with respect to the parameter t. We obtain

$$\begin{aligned}\frac{dI}{dt} &= \left(\frac{d}{dt}\right)\left[\int_{a_0}^{\bar{x}_-} + \int_{\bar{x}_+}^{a_1}\right] F(x, t)\,dx \\ &= \left(\frac{dx}{dt}\right)_- F(\bar{x}, t) - \left(\frac{da}{dt}\right)_0 F(a_0, t) + \left(\frac{da}{dt}\right)_1 F(a_1, t) \\ &\quad - \left(\frac{d\bar{x}}{dt}\right)_+ F(x_+, t) + \int_{a_0}^{a_1} \frac{\partial F(x, t)}{\partial t}\,dx.\end{aligned} \tag{7.74}$$

$\partial F/\partial t$ is assumed to be continuous across $\bar{x}$ so that

$$\frac{\partial F_+}{\partial t} - \frac{\partial F_-}{\partial t} = 0,$$

using obvious notation. We now shrink the column by the limiting process of letting $a_1 - a_0 \to 0$ such that $a_0 \leq x \leq a_1$. Then $F_- \to F(a_0, t) \equiv F_0$ and

$F_+ \to F(a_1, t) \equiv F_1$. We also have $da_i/dt = u_i$ $(i = 0, 1)$, where u_i is the velocity at the ends of the column, and $d\bar{x}/dt = U$ is the velocity of the point of discontinuity $\bar{x}$. $\bar{x}$ is the location of the shock front so that $d\bar{x}/dt$ is the shock wave velocity. Applying this limiting process to (7.74) gives

$$\lim_{a_1 - a_0 \to 0} \left(\frac{dI}{dt}\right) = (u_1 - U)F_1 - (u_0 - U)F_0.$$

Let $v_i = u_i - U$ $(i = 0, 1)$, so that v_i is the velocity of each end of the column relative to the shock wave velocity. Then the jump conditions become

$$\frac{dI}{dt} = v_1 F_1 - v_0 F_0, \quad F = \begin{cases} \rho & \text{(conservation of mass),} \\ \rho u & \text{(conservation of momentum),} \\ \rho(e + \frac{1}{2}u^2) & \text{(conservation of energy).} \end{cases} \quad (7.75)$$

Shock Conditions. Applying the jump conditions (7.75) to (7.69)-(7.72) yields the following jump conditions or "shock conditions" that prevail across the shock front:

mass: $$\rho_1 v_1 = \rho_0 v_0 = m$$
$$= \text{mass flux across shock wave at } \bar{x}; \quad (7.76)$$

momentum: $$(\rho_1 u_1)v_1 - (\rho_0 u_0)v_0 = p_0 - p_1; \quad (7.77)$$

energy: $$\rho_1 v_1(e_1 + \tfrac{1}{2}u_1^2) - \rho_0 v_0(e_0 + \tfrac{1}{2}u_0^2) = p_0 u_0 - p_1 u_1; \quad (7.78)$$

increase of entropy: $$\rho_1 v_1 S_1 - \rho_0 v_0 S_0 > 0. \quad (7.79)$$

Recall that the subscripts "0" and "1" refer to the undisturbed and disturbed region, respectively, in the neighborhood of the shock front whose trajectory is given by $x = \bar{x}(t)$. Note that the position $\bar{x}$ represents a discontinuity "surface" in one dimension. Equations (7.76)-(7.78) are the shock conditions, and (7.79) is an inequality representing the increase of entropy across the shock wave for an irreversible thermodynamic process. For a shock front there is flow across this discontinuity surface so that $m \neq 0$. For a contact discontinuity surface there is no mass flow across the surface so that $m = 0$. We consider only the shock front here. Each of the above three shock conditions can be written in a form involving only the relative velocities v_0, v_1. This means that the shock conditions are "invariant" with respect to a translation of the coordinate system when the shock wave velocity U is constant. This means that the shock conditions have the same form for an observer fixed in space or fixed on the shock wave (case of a stationary shock). To see this, we write the momentum equation (7.77) in the following form using (7.76) and $u_i = v_i + U$:

$$mu_1 - mu_0 = m(v_1 + U) - m(v_0 + U) = m(v_1 - v_0) = p_0 - p_1$$

or

$$\rho_1 v_1^2 + p_1 = \rho_0 v_0^2 + p_0 = P, \tag{7.80}$$

where P is called the *total flux of momentum* across the shock front.

The energy equation (7.78) becomes

$$\tfrac{1}{2}(v_1^2 - v_0^2) + e_1 - e_0 = p_0\tau_0 - p_1\tau_1 \qquad \text{where} \quad \tau_i = 1/\rho_i. \tag{7.81}$$

τ is called the *specific volume.* Recall that the specific entropy i is defined by $i = e + p\tau$. Using this expression in (7.81) gives Bernoulli's equation in the form

$$\tfrac{1}{2}v_1^2 + i_1 = \tfrac{1}{2}v_0^2 + i_0 = \tfrac{1}{2}\hat{q}^2. \tag{7.82}$$

Equation (7.82) is Bernoulli's equation given by (7.37) for a one-dimensional fluid. It was shown that the limit speed is the theoretical maximum value of the particle velocity (in this case v_i). We may write Bernoulli's equation in various forms by combining it with the other two conservation equations. For instance, using (7.76), (7.80) can be written as

$$m(v_1 - v_0)(v_1 + v_0) = m(v_1^2 - v_0^2) = m(p_0 - p_1)(v_1 + v_0),$$

so that

$$v_1^2 - v_0^2 = (p_0 - p_1)(\tau_0 + \tau_1). \tag{7.83}$$

Using (7.83), (7.82) becomes

$$\tfrac{1}{2}(p_1 - p_0)(\tau_0 + \tau_1) = \tfrac{1}{2}\Delta p(\tau_0 + \tau_1) = i_1 - i_0 = \Delta i. \tag{7.84}$$

Equation (7.84) tells us that the increase in the specific enthalpy Δi is equal to the work done by the pressure gradient Δp acting on the average specific volume $\tfrac{1}{2}(\tau_0 + \tau_1)$. Again, since $\Delta i = \Delta e + p_1\tau_1 - p_0\tau_0$, (7.84) may be written as

$$\tfrac{1}{2}(\tau_0 - \tau_1)(p_0 + p_1) = e_1 - e_0 = \Delta e. \tag{7.85}$$

This means that the increase of internal energy across the shock front equals the work done by the average pressure performing the compression $\tau_0 - \tau_1 = -\Delta\tau$. Equations (7.84) and (7.85) are very important relations that hold across the shock front. They are called the *Hugoniot relations.* They will be discussed below for an adiabatic gas.

Mechanical Shock Conditions. The mechanical shock conditions are defined as those conditions that arise from the first two conservation laws. Since they do not include the conservation of energy, they are valid for *any* continuous medium, regardless of the equation of state, as long as there is flow across the shock front (as long as $m \neq 0$). Moreover, the mechanical shock conditions do not depend on the thermodynamic nature of the medium. It is easily seen that the following relations summarize the mechanical shock conditions. They are given for reference.

$$
\begin{aligned}
&m=\rho_1 v_1=\rho_0 v_0, \qquad \rho_1 v_1^2+p_1=\rho_0 v_0^2+p_0=P,\\
&m\Delta v=-\Delta p, \qquad m^2=-\frac{\Delta p}{\Delta \tau}, \qquad v_0 v_1=\frac{\Delta p}{\Delta \rho},\\
&\tau_0\Delta p=-v_0\Delta v, \qquad \tau_1\Delta p=-v_1\Delta v,\\
&\Delta\tau\Delta p=-(\Delta v)^2, \qquad (\tau_0+\tau_1)\Delta p=-(v_1^2-v_0^2).
\end{aligned}
\tag{7.86}
$$

Hugoniot Relations for an Adiabatic Gas. In addition to the mechanical shock conditions, we need to consider the equation of state, or energy equation in its several forms, if we want to investigate a specific gas. We focus our attention on the Hugoniot relations given by (7.84) or (7.85). Consider (7.85). The disturbed state involves the quantities (p_1, ρ_1, e_1). Since these are variable in the disturbed state we suppress the subscript. We then define the Hugoniot function H which we set equal to zero, which is essentially (7.85). We obtain

$$H=H(p,\tau,e)=e-e_0+\tfrac{1}{2}(\tau-\tau_0)(p+p_0)=0. \tag{7.87}$$

Equation (7,87) characterizes all pairs of (p,τ) when values of (p_0,τ_0) are given in the undisturbed region for a given gas defined by e (for a known equation of state).

We now apply (7.87) to an adiabatic gas. The definition of specific heat at constant volume C_V is $e=C_V T$ where T is the temperature. Our adiabatic gas is a *perfect gas* which is defined by the condition $p\tau=RT$, where the gas constant $R=C_P-C_V$, (C_P is the specific heat at constant pressure). This allows us to get e as a function of p and τ, which is

$$e=\frac{p\tau}{\gamma-1}.$$

Inserting this expression of e into (7.87) yields

$$\frac{p\tau}{\gamma-1}-\frac{p_0\tau_0}{\gamma-1}+\tfrac{1}{2}(\tau-\tau_0)(p+p_0)=0,$$

which simplifies to

$$p(\tau-\mu^2\tau_0)-p_0(\tau_0-\mu\tau^2)=0. \tag{7.88}$$

Equation (7.88) is the *Hugoniot equation* for an adiabatic gas. For air $\gamma=1.4$, so that $\mu^2=(\gamma-1)/(\gamma+1)=\frac{1}{6}$ under standard conditions. Figure 7.14 shows a plot of p versus τ which is called a *Hugoniot curve.* We see that as τ decreases to its minimum value $\tau_m=\mu^2$, p increases asymptotically to infinity; and as τ increases to its maximum value $\tau_M=1/\mu^2$, p tends to zero.

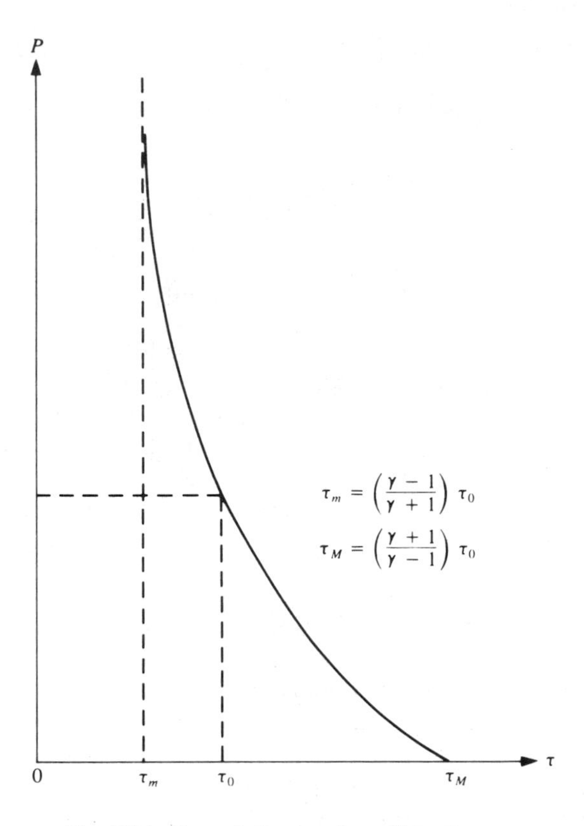

Fig. 7.14. Hugeniot curve for adiabatic gas.

Shock Conditions for an Adiabatic Gas

Prandtl's Relation. The relation between the velocities v_0 and v_1 (relative to the shock front), on either side of the shock front can be put in the following useful and elegant form due to Prandtl:

$$v_0 v_1 = c_*^2. \tag{7.89}$$

As seen, Prandtl's relation involves velocities only and therefore only relates kinematic quantities, not thermodynamic quantities such as pressure or density. We prove (7.89) by using Bernoulli's equation in the form given by (7.36), which can be applied on either side of the shock front as follows:

$$\mu^2 v_1^2 + (1-\mu^2)c_1^2 = c_*^2,$$

$$\mu^2 v_0^2 + (1-\mu^2)c_0^2 = c_*^2.$$

Note that c takes on different values in region (0) and (1). Multiplying the first equation by ρ_1, the second by ρ_0, and subtracting yields

$$\mu^2(\rho_1 v_1^2 - \rho_0 v_0^2) + (1-\mu^2)(\rho_1 c_1^2 - \rho_0 c_0^2) = c_*^2(\rho_1 - \rho_0).$$

For an adiabatic gas $c_i^2 = \gamma p_i/\rho_i$ $(i = 0, 1)$. Using this fact and the mechanical shock condition given by (7.80), we get

$$\mu^2(p_0 - p_1) + (1 - \mu^2)(p_1 - p_0) = c_*^2(\rho_1 - \rho_0),$$

which simplifies to

$$\frac{p_1 - p_0}{\rho_1 - \rho_0} = c_*^2.$$

Now one of the shock conditions in (7.86) is

$$\frac{p_1 - p_0}{\rho_1 - \rho_0} = v_0 v_1.$$

Relating this equation to the previous equation proves Prandtl's relation (7.89).

Shock Transition. In order to demonstrate the importance of Prandtl's relation, we now give without proof four basic properties of shock transition which means the relations governing pressure, density, and entropy across the shock front (we prove statement IV). Here, we follow [10, p. 141] rather closely where the authors define *shock strength* as any of the differences $\rho_1 - \rho_0$, $p_1 - p_0$, or $v_1 - v_0$. The proofs of the following statements are obtained by examining the Hugoniot relation coupled with the first law of thermodynamics. They are given in the above-mentioned reference.

I. The increase of entropy across a shock front is of third order in the shock strength.
II. The rise of pressure, density, and temperature across a shock front differs from reversible adiabatic changes of these quantities at most in terms of the third order in the shock strength. It is assumed here that the initial state and one quantity in the final state are the same for both processes.
III. Shocks are compressive. More precisely, density and pressure rise across the shock front.
IV. The flow velocity relative to the shock front is supersonic at the front side (undisturbed side) and subsonic at the back side. This statement clearly follows from the mechanical shock condition $(\tau_0 + \tau_1)(p_1 - p_0) = v_0^2 - v_1^2 = (v_0 - v_1)(v_0 + v_1)$ and statement III, thus showing that $v_0 > v_1$.

One of the important features of Prandtl's relation is that it allows us to go smoothly from a strong shock wave to a very weak shock wave or a sound wave. This is seen as follows: For a very weak shock wave we have $v_1 = v_0$, approximately, so that Prandtl's relation (7.89) shows that approximately we have $v_0^2 = v_1^2 = c_*^2 = c_0^2$, making use of (7.36). The shock condition $v_0 v_1 = (p_1 - p_0)/(\rho_1 - \rho_0)$ becomes approximately $\Delta p/\Delta\rho = v_0^2 = v_1^2 = c_0^2$. But $dp/d\rho = c_0^2$. This means that there exists an infinitely weak shock which propagates with the velocity of sound c_0. Using our definition of shock strength, namely $(p_1 - p_0)/p_0$, we see that the shock strength approaches $dp/p_0 = 0$, which is what we mean by an "infinitely weak" shock wave.

More Shock Relations for an Adiabatic Gas. We state for reference some additional relations for an adiabatic gas. The Hugoniot relation (7.88) can be put in the following forms:

$$\frac{p_1}{p_0} = \frac{\tau_0 - \mu^2 \tau_1}{\tau_1 - \mu^2 \tau_0} = \frac{\rho_1 - \mu^2 \rho_0}{\rho_0 - \mu^2 \rho_1}. \tag{7.90}$$

From equation (7.90) we obtain the following equations for the shock strength:

$$\begin{aligned} \frac{p_1 - p_0}{p_0} &= \frac{(1+\mu^2)(\tau_0 - \tau_1)}{\tau_1 - \mu^2 \tau_0} \\ &= \left[\frac{1+\mu^2}{1-\mu^2(\rho_1/\rho_0)}\right] \frac{\rho_1 - \rho_0}{\rho_0}. \end{aligned} \tag{7.91}$$

The second equation of (7.91) shows the relationship between the shock strength, as measured by the relative pressure ratio, to the relative density ratio. The following equation, which can also be obtained from the Hugoniot equation, gives the density ratio in terms of p_0 and p_1:

$$\frac{\rho_1}{\rho_0} = \frac{p_1 + \mu^2 p_0}{p_0 + \mu^2 p_1}. \tag{7.92}$$

Equation (7.92) shows that the density ratio satisfies the following inequality:

$$\mu^2 < \frac{\rho_1}{\rho_0} < \mu^{-2}.$$

Of course, the same inequality was given for the specific volume, by reversing the bounds. This was shown in Fig. 7.14 for the bounds of the Hugoniot curve for an adiabatic gas.

Another important relation is

$$m^2 = \frac{p_1 + \mu^2 p_0}{(1-\mu^2)\tau_0} = \frac{p_0 + \mu^2 p_1}{(1-\mu^2)\tau_1}. \tag{7.93}$$

Equation (7.93) can be derived by solving for the volume ratio τ_1/τ_0 in the Hugoniot equation (7.88), and inserting the result into the relation $\Delta p/\Delta\tau = -m^2$. Equation (7.93) is important because from it we obtain a simple relation between the pressure ratio p_1/p_0 and the Mach number $M_0 = |v_0|/c_0$. By using the relation $\rho_0 c_0^2 = \gamma p_0$ and the definition of M_0, (7.93) becomes

$$p_1 + \mu^2 p_0 = (1-\mu^2)\rho_0 v_0^2 = \gamma(1-\mu^2) p_0 M_0^2$$

or

$$\frac{p_1}{p_0} = \bar{p} = (1+\mu^2) M_0^2 - \mu^2. \tag{7.94}$$

Incidentally, (7.94) yields another proof that the pressure increases across the shock front or $\bar{p} > 1$, since $M_0 > 1$ for supersonic flow (which is the only type of flow giving shock waves). Another important shock relation, which involves u_1, c_1, and U is obtained from Prandtl's relation, Bernoulli's law, and the definition $v_i - u_i - U$. It is

$$\begin{aligned}(u_0 - U)(u_1 - U) &= \mu^2(u_0 - U)^2 + (1 - \mu^2)c_0^2,\\ &= \mu^2(u_1 - U)^2 + (1 - \mu^2)c_1^2,\end{aligned}$$

or

$$(1 - \mu^2)(U - u_0)^2 + (u_0 - u_1)(U - u_0) - (1 - \mu^2)c_0^2 = 0. \tag{7.95}$$

This quadratic in u_0 (if u_1 and c_0 are given) is equivalent to

$$\frac{u_1 - u_0}{c_0} = (1 - \mu^2)\left(\frac{v_0}{c_0} + \frac{c_0}{v_0}\right). \tag{7.96}$$

This gives the change in relative particle velocity across the shock front in terms of the relative velocity of the undisturbed state. Equation (7.96) gives another proof of the fact that v_0 is supersonic since we obtain the inequality

$$c_0^2 < (1 - \mu^2)v_0^2 < v_0^2, \qquad \mu^2 < 1.$$

A typical problem in one-dimensional shock wave propagation is: Given the data p_0, ρ_0, u_0, U for an adiabatic gas, we make the following calculations: First calculate $v_0 = u_0 - U$, then (since $c_0 = p_0/\rho_0$) calculate $M_0 = |v_0|/c_0$, then p_1 from (7.94); c_* from Bernoulli's law in the form $c_*^2 = \mu^2 v_0^2 + (1 - \mu^2)c_0^2$, then v_1 from Prandtl's relation $v_1 = c_*^2/v_0$, and finally, by again using Bernoulli's law, calculate c_1.

We now give an interesting application of the use of (7.95) by returning to our original model of a semi-infinite shock tube where the gas is initially quiescent and a shock wave is produced by suddenly moving the piston with a velocity u_P in the direction to compress the gas. Then the particle velocity in the disturbed region near the piston is $u_1 = u_P$ and $u_0 = 0$. We use (7.95) to solve for the shock wave velocity U in terms of u_P and obtain

$$U = \frac{u_P}{2(1 - \mu^2)} + \sqrt{\frac{c_0^2 + u_P^2}{4(1 - \mu^2)^2}}. \tag{7.97}$$

Obviously, (7.97) tells us that the shock wave velocity is supersonic ($U > c_0$). In fact, if $u_P \ll c_0$ then, from (7.97), U becomes approximately (by approximating the radicand)

$$U = c_0 + \frac{u_P}{2(1 - \mu^2)} + \frac{u_P^2}{8c_0(1 - \mu^2)^2}. \tag{7.98}$$

From (7.98) we see that as $u_P \to 0$, $U \to c_0$, which clearly means that the shock wave velocity tends (from above) to the velocity of sound as the piston velocity tends to zero.

Note the following silly and apparent contradiction: If we start with our shock tube in the quiescent state with the piston velocity zero, then nothing happens—there is no propagating shock wave. However, if we start things happening by giving the piston a finite velocity u_P we thereby generate a shock wave. If we then successively reduce u_P to zero, the shock wave still persists and tends to a sound wave. This is not a frivolous observation. In fact, situations like this come up again and again in fluid mechanics (and other branches of mathematical physics). For instance, in the flow of a viscous fluid around a body (the roughness of the surface producing friction), a boundary condition that must be satisfied is zero slip or no particle velocity relative to the body surface; no matter how small the viscosity coefficient is, we still have a viscous boundary layer in the fluid in the neighborhood of the surface. If we now consider a sequence of such flow problems for a smoother and smoother surface so that the viscosity coefficient tends to zero, then the same boundary condition persists although it becomes thinner. However, if we start a priori with an inviscid fluid (the viscosity coefficient is set equal to zero ab initio), then the boundary condition is a slip condition where there is no boundary layer so that the particle velocity at the surface is not zero. This type of problem opens up a whole host of problems in the field of asymptotic expansions.

Coming back to (7.98), we may cast this equation into a more convenient dimensionless form by using c_0 as a scale and replacing U and u_P by $\bar{U}$ and $\bar{u}_P$, where

$$\bar{u}_P = \frac{u_P}{c_0},$$

$$\bar{U} = \frac{\bar{u}_P}{2(1-\mu^2)} + \sqrt{1 + \left(\frac{\bar{u}_P}{2(1-\mu^2)}\right)^2}. \tag{7.99}$$

In this model u_P may be supersonic. If $\bar{u}_P$ is very large then (7.99) becomes asymptotically (again by approximating the radicand, this time for large $\bar{u}_P$)

$$\bar{U} = \frac{\bar{u}_P}{1-\mu^2} + \frac{1-\mu^2}{\bar{u}_P}. \tag{7.100}$$

This asymptotic equation tells us that as $\bar{u}_P \to \infty$, $\bar{U} \to \bar{u}_P/(1-\mu^2)$. This obviously means that the upper limit of U is $u_P/(1-\mu^2)$ for very large u_P.

Reflection of a Shock from a Rigid Wall, Finite Shock Tube

The problem of reflection of a shock wave from a rigid wall is one of great importance. Instead of directly impinging a shock wave on a wall, we shall take as our physical model a one-dimensional shock tube of length L with a piston initially at $x=0$ and closed at $x=L$ with a rigid partition (the wall). An *incident shock wave* is produced when the piston is suddenly moved to the right with a velocity $u_P < c_0$. This incident wave moves to the

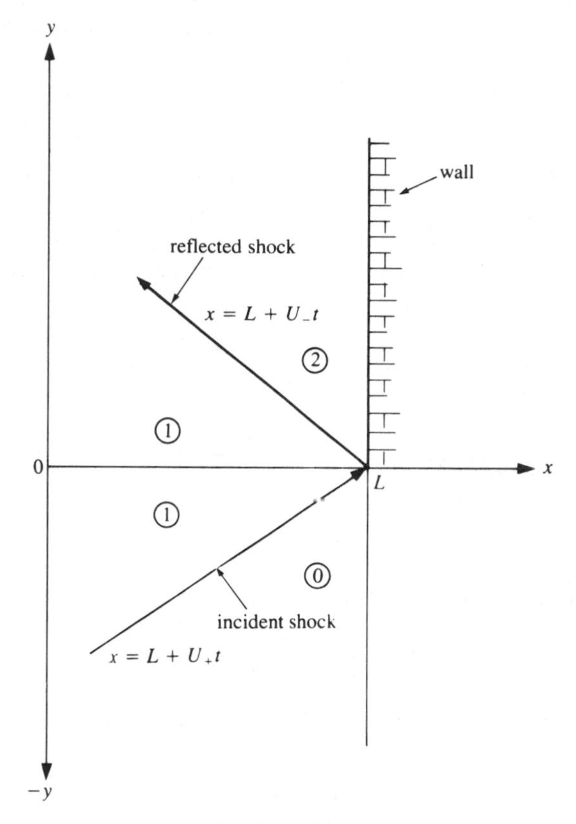

Fig. 7.15. Reflection of shock from wall.

right toward the wall with a velocity U_+ ahead of the piston. When the wave front hits the wall at $x = L$ it will be reflected toward the front end with a velocity U_- (the velocity of the *reflected shock wave*). The assumptions are that the wall is a perfectly reflecting barrier, the gas is quiescent in front of the incident shock front, and the particle velocity just behind the incident shock is u_P. To understand the physics of this phenomenon of reflection it is helpful to study Fig. 7.15 which shows the various regions in the (x, t) space in the neighborhood of the wall.

This figure shows the trajectories of the incident and reflected shock waves impinging on the rigid boundary at $x = L$ and $t = 0$. At some negative time the incident wave is generated from the moving piston; the wave travels with the trajectory $x = L + u_+ t$ where it is reflected from the wall at $t = 0$ and produces a reflected wave which travels with a trajectory $x = L + U_- t$. We stop the process before this reflected wave is reflected from the moving piston. The flow is separated into three regions:

1. An initial quiescent zone (0) ahead of the incident shock wave for $x > L + U_+ t$ and $t < 0$ where $p = p_0$, $\rho = \rho_0$, $u = u_0 = 0$.

2. Another quiescent zone (2) behind the reflected shock for $x > L + U_{-}t$ and $t > 0$ where $p = p_2$, $\rho = \rho_2$, $u_2 = 0$.
3. The region between the two shock waves, zone (1), where $u = u_1$, $p = p_1$, $\rho = \rho_1$.

The problem is to solve for the pressure ratios p_1/p_0, p_2/p_1, etc. In general, we want to find the flow and pressure field in region (2) in terms of data in region (0) and u_1. To this end, we use (7.95) by replacing u_0 with u_1 and c_0 with c_1. This equation then gives a quadratic for U where U_+ and U_- are the roots of this equation. We get

$$(U - u_1)^2 + \left(\frac{1}{1-\mu^2}\right) u_1(U - u_1) = c_1^2. \tag{7.101}$$

It is convenient to put (7.101) in dimensionless form by introducing the Mach numbers defined by

$$M_{\pm} = \frac{u_1 - U_{\pm}}{c_1}, \qquad \text{where} \quad M_+ < 0, \quad M_- > 0, \quad U_+ > 0, \quad U_- < 0. \tag{7.102}$$

Then $M_{\pm}$ are the roots of

$$M^2 - \frac{u_1 M}{c_1(1-\mu^2)} - 1 = 0. \tag{7.103}$$

It is clear that $M_+ M_- = -1$. By applying (7.94) to $M_{\pm}$, respectively, we obtain

$$\frac{p_0}{p_1} = (1+\mu^2)M_+^2 - \mu^2, \tag{7.104}$$

$$\frac{p_2}{p_1} = (1+\mu^2)M_-^2 - \mu^2, \qquad \text{where} \quad \mu^2 = \frac{\gamma - 1}{\gamma + 1}. \tag{7.105}$$

Recall that p_0 is the pressure in the quiescent region, p_1 is the pressure on the back side of the incident shock wave, and p_2 is the pressure on the back side of the reflected shock wave. From (7.104) and (7.105) we obtain the ratio of the excess pressure in the reflected to the incident shock wave, which is

$$\frac{p_2 - p_1}{p_1 - p_0} = 1 + \frac{1+\mu^2}{(p_0/p_1) + \mu^2}. \tag{7.106}$$

And finally, for a very strong incident shock $p_1/p_0 \gg 1$ so that (7.106) becomes approximately

$$\frac{p_2 - p_1}{p_1 - p_0} = 2 + \frac{1}{\mu^2} = 8 \qquad \text{for} \quad \mu = 1.4. \tag{7.107}$$

The approximation given by (7.107) is in marked contrast to the reflection of a linear wave where the reflected wave is merely doubled in intensity. This example shows the importance of studying the nonlinear properties of shock waves.

PART II. VISCOUS FLUIDS

Introduction

In Part I of this chapter we presented an exposition of one-dimensional compressible flow, two-dimensional steady flow, and shock wave phenomena. We emphasized the importance of characteristic theory in obtaining an insight into wave propagation in inviscid fluids. In the treatment of these subjects viscosity was neglected, except for the brief mention that we cannot neglect the thin viscous layer in the neighborhood of the shock front, in connection with the fact that a thermodynamically irreversible process must occur across the shock wave. The subject of wave propagation in fluids would be incomplete without a discussion of the properties of viscous fluid vis-à-vis wave propagation.

The general subject of the flow of a viscous fluid is quite extensive and obviously cannot be covered with any degree of completeness in this part of this chapter. Therefore, as in the first part of the chapter, we must be selective in our choice of material, omitting such topics as boundary layer theory and the theory of turbulence and viscous jets (which may be best left to works on aerodynamics), which are important to the understanding of propagating waves. Since a proper understanding of wave propagation in a viscous fluid presupposes a knowledge of the properties of the viscous medium, we therefore concentrate on some of the fundamental aspects of viscous flow emphasizing the dynamical aspects.

7.4. Viscosity: Elementary Considerations

We start the subject with some elementary ideas on the behavior of a viscous fluid. Every fluid offers some resistance, either to its flow or to the motion of an object in it. This resistance, which we loosely call the viscosity of the fluid (which will be made precise below), is large for substances like syrups and oil and small for water and air.

To get a feel for the concept of viscosity, let us take as a simple model a fluid contained between two parallel horizontal plates lying a distance h apart. The surfaces of the plates which are in contact with the fluid are rough, so that there is friction between those surfaces and the portion of the fluid in contact during the motion of the fluid. We consider planar flow in the (x, y) plane. Let the lower plate, which we represent by a horizontal line AB on the x axis, be fixed, and the upper plate CD be moving to the right with a velocity U. The plates are long enough so that we neglect end effects as long as we are not too near the ends. In order to move the plate CD, a tangential force P must be applied which, as a first approximation, is proportional to U, as shown by experiment. We express this simple relationship between P and U by

$$P = \frac{\mu A U}{h}, \tag{7.108}$$

where A is the area of each plate, and μ is a positive constant called the *dynamic viscosity coefficient.* Since the fluid performs a parallel gliding motion, the transverse particle velocity component v equals zero, and since we have planar flow, the velocity vector $\mathbf{v} = (u, 0, 0)$ where $u = u(x, y, t)$. In this section we shall consider only steady state flow so that we neglect the time-dependence. The fluid adheres to the upper and lower surfaces because of friction, as mentioned above, thus yielding the boundary conditions

$$u(x, 0) = 0, \qquad u(x, h) = 0. \tag{7.109}$$

To satisfy these boundary conditions for a fixed x, we must have

$$u = \frac{Uy}{h} \qquad \text{or} \qquad u_y = \frac{U}{h}. \tag{7.110}$$

This shows that (1) at each station the horizontal component of velocity u profile with respect to y is linear; (2) the cross-velocity gradient u_y is constant. Obviously, this statement is a consequence of (1). (Clearly these results hold for a given ratio U/h.) The xy component of the stress tensor τ_{xy} is due to the applied axial force P. This means

$$\tau_{xy} = \frac{P}{A}.$$

Using (7.108) and (7.110), this becomes

$$\tau_{xy} = \mu u_y. \tag{7.111}$$

Equation (7.111) tells us that the shear stress on the fluid due to the constant motion of the upper plate is proportional to the cross-velocity gradient u_y, the proportionality constant being the dynamic viscosity coefficient μ. This is an example of *simple shear flow.* We see from (7.111) that the dimensions of μ are stress x time or $ML^{-1}T^{-1}$ where (M, L, T) represent dimensions of mass, length, and time. A characteristic of simple shear flow is that the shear stress is constant at a given x since it is proportional to the constant velocity gradient.

The assumption, as mentioned above, is that U is small enough so that U is constant. The fluid is then called *Newtonian.* Also we assume that μ is independent of temperature (an isothermal process). If μ depends on u_y or temperature, then the fluid is said to be *non-Newtonian.* Sometimes it is convenient to use the *kinematic viscosity coefficient* ν, which is defined by

$$\nu = \frac{\mu}{\rho}. \tag{7.112}$$

The dimensions of ν are L^2T^{-1}.

7.5. Conservation Laws for a Viscous Fluid

We now turn to the more general treatment of the unsteady or time-dependent flow of a viscous fluid by re-examining the conservation laws in the setting of a viscous fluid. These conservation laws (mass, momentum, and energy) are the three fundamental laws of physics which, along with the laws of thermodynamics, lead to the field equations for any continuous medium: solid, liquid, or gas. From these field equations we derive the velocity, pressure, etc., fields for the medium considered, along with the definition of the medium which arises from the energy equation or equation of state. It is for this reason that we discussed the conservation laws in different settings. In Chapter 2 they were introduced in a broad physical setting and the distinction was made between the Lagrangian and Eulerian treatment. In Chapter 3 the conservation laws were treated from the point of view of weak solutions of PDEs which extended the domain of solutions to include jump discontinuities. This allows for global solutions, which laid the basis of the propagation of shock waves. In Chapter 5 they were treated from the point of view of an incompressible fluid (water), in the Eulerian representation, which is used in all problems involving fluid flow. In Chapter 6 and in this chapter they were examined again in a fluid dynamical setting. In Chapter 6 they were linearized in order to allow for the propagation sound waves. In this chapter the nonlinear conservation laws were used to derive the quasilinear PDEs which are the flow or field equations of one-dimensional compressible unsteady and two-dimensional steady flow. They were again investigated in this chapter in the setting of shock waves where the jump conditions were derived across the shock front. In the light of the above remarks, we re-introduce the conservation laws from the point of view of a viscous fluid.

The *continuity equation* or *conservation of mass* in the Eulerian representation (5.2) is repeated here in tensor form for the purpose of investigating viscous flow

$$(\rho v_j)_{,j} + \rho_t = 0, \tag{7.113}$$

where v_j is the jth component of the particle velocity. (Note that in the tensor notation the subscript ",j" represents differentiating with respect to x_j, the jth component of $\mathbf{x}$, and the repeated index "j" means summation over the three components of $\mathbf{v}$.)

We now manipulate *Euler's equation*, first for an inviscid fluid in order to prepare for the equations of motion for a viscous fluid. Euler's equation (5.6), for $\mathbf{F} = 0$, can be put in the tensor form

$$v_{i,j} = -v_j v_{i,j} - \left(\frac{1}{\rho}\right) p_{,i} \qquad \text{summed over } j. \tag{7.114}$$

This is the ith component of the equation of motion (Euler's equation) where the only external force is due to the pressure gradient. The *momentum flux* is a vector describing the rate of change of the momentum, whose ith component is $(\rho v_i)_t$. Using the continuity equation in the form (7.113), we can recast (7.114) into the form

$$\begin{aligned}(\rho v_i)_t &= -\rho v_j v_{i,j} - p_{,i} - v_i(\rho v_j)_{,j} \\ &= -p_{,i} - (\rho v_i v_j)_{,j}.\end{aligned} \tag{7.115}$$

In this form of Euler's equation the ith component of the momentum flux is equal to the ith component of the pressure gradient, plus the corresponding nonlinear terms arising from the convective part of the particle acceleration. We can combine the terms on the right-hand side of (7.115) by introducing the tensor $\mathbf{\Pi}$ whose ijth component is defined by

$$\Pi_{ij} = \delta_{ij} p + \rho v_i v_j. \tag{7.116}$$

Note that $\mathbf{\Pi}$ is a symmetric tensor. It has an important physical meaning which we describe below. Since $p_{,i} = \delta_{ij} p_{,j}$ by inserting (7.116) in (7.115) we obtain Euler's equation in the following form:

$$(\rho v_i)_t = -\Pi_{ij,j} \qquad \text{summed over } j. \tag{7.117}$$

We now obtain the meaning of Π_{ij} by integrating (7.117) over a given volume V_0. We get

$$\frac{\partial}{\partial t}\int_{V_0} \rho v_i \, dV = -\int_{V_0} \Pi_{ij,j} \, dV = \oint_{S_0} \Pi_{ij} \, dS_j, \tag{7.118}$$

where S_0 is the surface bounding V_0. The integrand $\Pi_{ij,j}$ in the right-hand volume integral of (7.118) is the tensor form of the divergence of Π_{ij}. We have thus applied the divergence theorem to transform this integral to the surface integral of the projection of Π_{ij} in the direction normal to the surface element whose jth component is dS_j. It is then clear that $\Pi_{ij} \, dS_j$ is the ith component of the momentum flowing across the surface element dS (normal to the surface element). Writing $dS_j = \nu_j \, dS$, where ν_j is the jth component of the outward-drawn normal $\boldsymbol{\nu}$ to the surface element dS, we write (7.117) as

$$\Pi_{ij}\nu_j = p\nu_j + \rho v_i v_j \nu_j = p\boldsymbol{\nu} + \mathbf{v}(\mathbf{v} \cdot \mathbf{v}).$$

This equation tells us that $\Pi_{ij}\nu_j$ is the jth component of the amount of momentum flowing across a unit surface area bounding V_0 per unit time. For this reason Π_{ij} is called the *momentum flux density tensor.* In particular, if the unit vector $\boldsymbol{\nu}$ is in the direction of the fluid flow, say the x direction, then only the longitudinal component of the momentum is transported in this direction, and the momentum flux density is $p + \rho u^2$. In a direction normal to the x component of velocity u, only the transverse component of the momentum is transported, and its flux density is just p.

Summarizing: The ith component of Euler's equation (7.117) tells us that the ith component of the rate of change of momentum (the inertial force) equals the corresponding component of the divergence of the momentum flux density. It will be seen below that this form of Euler's equation is useful in allowing us to generalize to a viscous fluid, and thus obtain the so-called "Navier–Stokes" equations which are fundamental in studying the properties of viscous flow.

Momentum Flux for a Viscous Fluid

The momentum flux density as given by (7.116) represents a completely reversible transfer of momentum. It is simply due to the pressure and the convective term $v_i v_j$ that expresses the mechanical transport of fluid from place to place. We therefore call this the *ideal momentum flux.* In order to generalize this momentum flux to take into account viscosity effects, we define a stress term $-\tau'_{ij}$ that describes the irreversible "viscous" transfer of momentum in the fluid. We can therefore write the momentum flux density tensor associated with a viscous fluid as

$$\Pi_{ij} = \delta_{ij} p + \rho v_i v_j - \tau'_{ij} = -\tau_{ij} + \rho v_i v_j, \tag{7.119}$$

where the tensor

$$\tau_{ij} = -\delta_{ij} p + \tau'_{ij}. \tag{7.120}$$

The stress tensor τ_{ij} gives the part of the momentum flux that is not due to the direct transfer of momentum associated with the mass of the moving fluid. This transfer of momentum associated with flow is given by the convective term $v_i v_j$. The tensor τ'_{ij} is called the *viscosity stress tensor* whose form we now determine.

We observe that a viscous fluid creates friction amongst the fluid particles which move with different velocities. This is called shear flow. Hence we make the hypothesis that τ'_{ij} depends only on the first derivatives of the velocity (the velocity gradient). The simplest assumption is that τ'_{ij} is a linear function of the component of the gradient $v_{i,j}$. There can be no terms in τ'_{ij} that are independent of $v_{i,j}$ since $\tau'_{ij} = 0$ for $\mathbf{v} = \text{const}$. Next, we observe that $\tau'_{ij} = 0$ for a pure rotation, since this type of motion causes no internal friction (within the linear theory), and hence no shear stress on a fluid particle. Suppose we have a pure rotational motion. Then $\mathbf{v} = \boldsymbol{\Omega} x \mathbf{r}$ where $\boldsymbol{\Omega}$ is the angular velocity. Written out, this becomes

$$v_1 = x_3 \Omega_2 - x_2 \Omega_3,$$

$$v_2 = x_1 \Omega_3 - x_3 \Omega_1,$$

$$v_3 = x_2 \Omega_1 - x_1 \Omega_2.$$

From these equations we observe that $v_{1,2} + v_{2,1} = 0, \ldots$, so that in general, $v_{i,j} + v_{j,i} = 0$ for rotational motion. Since, as mentioned above, τ'_{ij} involves no rotational motion, we assume that τ'_{ij} also depends linearly on $v_{i,j} + v_{j,i}$ as well as on $v_{i,j}$. The most general tensor of rank two (requiring a double index) satisfying these conditions may be put in the following convenient form:

$$\tau'_{ij} = \eta(v_{i,j} + v_{j,i} - \tfrac{2}{3}\delta_{ij}v_{k,k}) + \zeta\delta_{ij}v_{k,k} \qquad \text{summed over } k. \tag{7.121}$$

The constants η and ζ are called the *coefficients of viscosity*; they are both positive, as will be seen in the sections on energy dissipation and thermal conduction in a viscous fluid. Inserting (7.121) into (7.119) and (7.120), Euler's equation (7.117) becomes

$$\rho(v_{i,t} + v_j v_{i,j}) = -p_{,i} + [\eta(v_{i,j} + v_{j,i} - \tfrac{2}{3}\delta_{ij}v_{k,k})]_{,j} + (\zeta v_{k,k})_{,i}. \tag{7.122}$$

Equation (7.122) is the ith component of the equations of motion for a compressible viscous fluid in its most general form, since it allows for the dependence of the viscosity coefficients on the velocity gradient, which is also so for a non-Newtonian fluid. It is written in such a manner that the left-hand side represents the inertial force, where the ith component of the particle acceleration $\dot{v}_i$ is given by $v_{i,t}$ plus the convective term $v_j v_{i,j}$ which is nonlinear.

For a Newtonian viscous fluid the viscosity coefficients are constant. For this case, the divergence of the viscosity tensor becomes

$$\tau'_{ij,j} = \eta v_{i,jj} + (\zeta + \tfrac{1}{3}\eta)v_{k,ki} = \eta\nabla^2 v_i + (\zeta + \tfrac{1}{3}\eta)\,\mathrm{div}\,\mathbf{v}.$$

We may therefore write the equations of motion for a Newtonian viscous fluid in the vector form

$$\rho[v_t + (\mathbf{v}\cdot\mathbf{grad})\mathbf{v}] = -\mathbf{grad}\,p + \eta\nabla^2\mathbf{v} + (\zeta + \tfrac{1}{3}\eta)\,\mathbf{grad}\,\mathrm{div}\,\mathbf{v}. \tag{7.123}$$

For the case of an incompressible fluid, the flow is divergenceless so that $\mathrm{div}\,\mathbf{v} = 0$ and (7.123) becomes

$$\dot{\mathbf{v}} = \mathbf{v}_t + (\mathbf{v}\cdot\mathbf{grad})\mathbf{v} = -\left(\frac{1}{\rho}\right)\mathbf{grad}\,p + \nu\nabla^2\mathbf{v}, \tag{7.124}$$

where $\nu = \eta/\rho$ is the kinematic viscosity coefficient. Equation (7.124) is called the *Navier–Stokes* equation. This dynamical equation was first obtained by Navier using a sort of molecular approach involving the interaction of the molecules making up the viscous fluid. On the other hand, the method used by Stokes in deriving this equation does not appeal to any molecular hypothesis. Equation (7.124) tells us that the viscosity of an incompressible viscous fluid depends only on one coefficient ν. Table 7.2 gives the dynamic and kinematic viscosity coefficients η and ν for air and a few liquids at a temperature of 20°C in c.g.s. units. We remark that the dynamic viscosity coefficient of a gas at a given temperature is independent of the pressure, while the kinematic viscosity coefficient is inversely proportional to the pressure.

We now eliminate p from the Navier–Stokes equation by first putting (7.124) in the form

$$\mathbf{v}_t + \tfrac{1}{2}\,\mathbf{grad}\,v^2 - \mathbf{v}\mathbf{x}\,\mathbf{curl}\,\mathbf{v} = -\left(\frac{1}{\rho}\right)\mathbf{grad}\,p + \nu\nabla^2\mathbf{v}, \tag{7.125}$$

Table 7.2. Dynamic and kinematic viscosity coefficients at 20°C.

	η (g/cm sec)	ν (cm^2/sec)
air	0.00018	0.150
water	0.010	0.010
ethyl alcohol	0.018	0.022
glycerine	8.5	6.8
mercury	0.0156	0.0012

where we have used the well-known formula in vector analysis

$$(\mathbf{v} \cdot \mathbf{grad})\mathbf{v} = \tfrac{1}{2}\,\mathbf{grad}\; v^2 - \mathbf{v} \mathbf{x}\; \mathbf{curl}\; \mathbf{v},$$

which is derived in any standard work on vector analysis or applied mathematics such as [23].

Since $\mathbf{curl}(\mathbf{grad}\; p) = 0$, we take the curl of (7.125) to eliminate p. We obtain

$$(\mathbf{curl}\; \mathbf{v})_t = \mathbf{curl}(\mathbf{v} \mathbf{x}\; \mathbf{curl}\; \mathbf{v}) + \nu \nabla^2(\mathbf{curl}\; \mathbf{v}). \tag{7.126}$$

This vector form of the Navier–Stokes equation involves only the particle velocity and kinematic viscosity coefficient.

We now consider the *boundary conditions* on the Navier–Stokes equation, since these are necessary to solve any specific problem in viscous incompressible flow. Consider such a viscous fluid flowing over a stationary flat plate whose surface is rough. There are always forces of molecular attraction between the fluid and the surface of the plate. The result is that an infinitely thin or molecular layer of fluid adjacent to the surface adheres to it and is brought to rest. This is the boundary condition that the particle velocity of the fluid at the surface is zero. Furthermore, much experimental evidence on flow over airfoils shows that the viscous properties of the fluid are confined to a thin layer surrounding the surface of the plate. This layer is called the *boundary layer.* Outside this boundary layer the fluid is assumed to be inviscid. The boundary condition is

$$\mathbf{v} = 0 \qquad \text{on the surface.}$$

This means that both the normal and tangential components of $\mathbf{v}$ vanish on the boundary.

In the section on shock waves the decreasing to zero of the velocity of the piston (in causing the generation of a shock wave), brought out the subject of asymptotic expansions in connection with the changeover of flow from a viscous to an inviscid fluid over a plate. It was pointed out that an essentially different phenomenon occurs, when in one case the viscosity coefficient tends to zero, and in the other case where we set ab initio the viscosity coefficient equal to zero. We are now in a position to make those qualitative remarks quantitative, by an examination of the Navier–Stokes equation in connection with the boundary conditions. As mentioned previously, as the roughness of the plate gets less or the fluid gets less viscous, the kinematic viscosity coefficient gets smaller. But as long as this coefficient, however small, is not zero, the boundary condition is $\mathbf{v}=0$ (the condition of no slip on the surface). It turns out that the smaller this viscosity coefficient ν the thinner the boundary layer. However, as long as we have a boundary layer ($\nu>0$), no matter how thin, the boundary condition $\mathbf{v}=0$ still prevails. But if we, ab initio, set $\nu=0$, then the boundary layer does not exist and we therefore have inviscid flow over the plate, so that there is slip on the surface and only the normal component of the velocity is zero on the surface, in distinction to the case where we have a boundary layer where both the normal and the tangential components of $\mathbf{v}$ are zero on the surface. This gives the interesting situation that, as long as there is a finite viscosity in the fluid no matter how small, the fluid adheres to the surface; but if we have an inviscid fluid we suddenly switch over to the boundary condition of slip at the surface. Mathematically this means: However small ν is (as long as $\nu>0$), the Navier–Stokes equation (7.124) is valid. But if we set $\nu=0$, (7.124) is reduced to Euler's equation (5.6) or (7.115). Comparing Euler's equation with the Navier–Stokes equation, we see that the Navier–Stokes equation is a PDE which is second order in the spatial derivatives, while Euler's equation is first order in these derivatives. Since the number of boundary conditions depends on the order of the PDE, a boundary condition is lost in going from the Navier–Stokes equation (for viscous flow) to Euler's equation (for inviscid flow). What happens to that lost boundary condition and how is it lost? The answer is contained in a method called the method of *singular perturbations.* This method involves an *asymptotic expansion* in terms of ν. We now give a very simple example of a singular perturbation: Consider the damped harmonic oscillator with no forcing function. Its differential equation is $\ddot{x}+\gamma\dot{x}+\omega^2x=0$, where γ is the attenuation coefficient and ω is the frequency. x is a damped sinusoidal function of t for any positive values of γ and ω, and the solution depends on two ICs, for an IV problem, since the differential equation is second order. However, if we set $\omega=0$ then the differential equation becomes $\dot{x}+\gamma x=0$, which is a first-order differential equation, and hence the solution depends on only one IC. Moreover, this solution is no longer a damped sinusoidal function of t, but an exponentially damped function of t and hence nonoscillatory. Which IC is lost, and how does the oscillatory solution go over to the nonoscillatory solution or purely damped solution for zero frequency? These questions are answered by the method of singular perturbations. There is a vast literature on this subject; see, for example, [3].

Coming back to the boundary conditions for the flow of a viscous fluid over a surface, we now write down an expression for the force acting on a solid surface bounding the fluid. We consider an inertial coordinate system whose origin is on the surface, so that it is stationary. The force acting on an element of surface is equal to the momentum flux across the element. The momentum flux across the surface element $d\mathbf{S}$ (whose ith component is dS_i) is

$$\Pi_{ij}\, dS_j = (\rho v_i v_j - \tau_{ij})\, dS_j \qquad \text{summed over } j.$$

Writing $dS_i = \nu_i\, d\mathbf{S}$, where ν_i is the ith component of the outward-drawn normal $\boldsymbol{\nu}$ to the surface, and using the boundary condition $\mathbf{v}=0$ on the surface, the ith component of the force $\mathbf{P}$ acting on the element of surface becomes

$$P_i = -\tau_{ij}\nu_j = p\nu_i - \tau'_{ij}\nu_j. \tag{7.127}$$

The next class of boundary conditions occurs when there is a surface of separation between two immiscible viscous fluids. The conditions at the surface are that the velocities of the fluids must be equal (otherwise there would occur mixing of the fluids), and that the forces that each fluid exerts on the other at the surface must be equal and opposite. This condition can be written as

$$\nu_{1j}\tau_{1ij} = -\nu_{2j}\tau_{2ij},$$

where the indices 1 and 2 refer to the two fluids on either side of the surface. The normal vectors $\boldsymbol{\nu}_1$ and $\boldsymbol{\nu}_2$ are in the opposite directions, so that the boundary condition may be written as

$$\boldsymbol{\nu}_i\tau_{1ij} = \boldsymbol{\nu}_i\tau_{2ij}. \tag{7.128}$$

We defined a *free surface* as one on which the surface stresses or tractions are zero. Therefore, on a free surface the following condition holds:

$$\tau_{ij}\nu_j = \tau'_{ij}\nu_j - p\nu_i = 0. \tag{7.129}$$

For reference we write out in extended form the expressions for the components of the stress tensor τ_{ij} and the Navier–Stokes equations in cylindrical and spherical coordinates.

For the *cylindrical coordinate system* (r, θ, z) we have $\mathbf{v} = (u, v, w)$ in the (r, θ, z) directions, respectively. The components of τ_{ij} are

$$\begin{aligned}
\tau_{rr} &= -p + 2\eta\partial u/\partial r, & \tau_{\theta r} &= \eta[(1/r)\partial u/\partial\theta + \partial v/\partial r - v/r],\\
\tau_{\theta\theta} &= -p + 2\eta((1/r)\partial v/\partial\theta + u/r, & \tau_{\theta z} &= \eta[\partial v/\partial z + (1/r)\partial w/\partial\theta],\\
\tau_{zz} &= -p + 2\eta\partial w/\partial z, & \tau_{rz} &= \eta[\partial w/\partial r + \partial u/\partial z].
\end{aligned} \tag{7.130}$$

The components of the Navier–Stokes equations in cylindrical coordinates are

$$
\begin{aligned}
&\partial u/\partial t+u\partial u/\partial r+(v/r)\partial u/\partial\theta+w\partial u/\partial z-(1/r)v^2\\
&\qquad=-(1/\rho)\partial p/\partial r+\nu[\partial^2u/\partial r^2+(1/r^2)\partial^2u/\partial\theta^2\\
&\qquad\quad+\partial^2u/\partial z^2+(1/r)\partial u/\partial r-(2/r^2)\partial v/\partial\theta-(1/r^2)u],\\
&\partial v/\partial t+u\partial v/\partial r+(1/r)v\partial v/\partial\theta+w\partial v/\partial z+(1/r)uv\\
&\qquad=-(1/\rho r)\partial p/\partial\theta+\nu[\partial^2v/\partial r^2+(1/r^2)\partial^2v/\partial\theta^2+\partial^2v/\partial z^2\\
&\qquad\quad+(1/r)\partial v/\partial r+(2/r^2)\partial u/\partial\theta-(1/r^2)v],\\
&\partial w/\partial t+u\partial w/\partial r+(1/r)v\partial w/\partial\theta+w\partial w/\partial z\\
&\qquad=-(1/\rho)\partial p/\partial z+\nu[\partial^2w/\partial r^2+(1/r^2)\partial^2w/\partial\theta^2+\partial^2w/\partial z^2+(1/r)\partial w/\partial r].
\end{aligned}
\tag{7.131}
$$

The continuity equation in cylindrical coordinates is

$$\partial u/\partial r+(1/r)\partial v/\partial\theta+\partial w/\partial z+(1/r)u=0. \tag{7.132}$$

For a *spherical coordinate system* (r, θ, φ) the corresponding particle velocity components are (u, v, w). The components of the stress tensor become

$$
\begin{aligned}
\tau_{rr}&=-p+2\eta\partial u/\partial r,\\
\tau_{\varphi\varphi}&=-p+2\eta[(1/r\sin\theta)\partial w/\partial\varphi+u/r+(1/r)v\cot\theta],\\
\tau_{\theta\theta}&=-p+2\eta((1/r)\partial v/\partial\theta+(1/r)u),\\
\tau_{r\theta}&=\eta((1/r)\partial u/\partial\theta+\partial v/\partial r-(1/r)v),\\
\tau_{\theta\varphi}&=\eta[(1/r\sin\theta)\partial v/\partial\varphi+(1/r)\partial w/\partial\theta-(1/r)w\cot\theta],\\
\tau_{\varphi r}&=\eta[\partial w/\partial r+(1/r\sin\theta)\partial u/\partial\varphi-(1/r)w].
\end{aligned}
\tag{7.133}
$$

The equations of motion in spherical coordinates are

$$
\begin{aligned}
&\partial u/\partial t+u\partial u/\partial r+(1/r)v\partial u/\partial\theta+(1/r\sin\theta)w\partial u/\partial\varphi-(1/r)(v^2+w^2)\\
&\qquad=-(1/\rho)\partial p/\partial r+\nu[(1/r)\partial^2(ru)/\partial r^2+(1/r^2)\partial^2u/\partial\theta^2\\
&\qquad\quad+(1/r\sin\theta)^2\partial^2u/\partial\varphi^2+(1/r^2)\cot\theta\partial u/\partial\theta-(2/r^2)\partial v/\partial\theta\\
&\qquad\quad-(2/r^2\sin\theta)\partial w/\partial\varphi-(2/r^2)u-(2/r^2)v\cot\theta],\\
&\partial v/\partial t+u\partial v/\partial r+(1/r)v\partial v/\partial\theta+(1/r\sin\theta)w\partial v/\partial\varphi+(1/r)uv\\
&\qquad\quad-(1/r)w^2\cot\theta\\
&\qquad=-(1/\rho r)\partial p/\partial\theta+\nu[(1/r)\partial^2(rv)/\partial r^2+(1/r^2)\partial^2v/\partial\theta^2\\
&\qquad\quad+(1/r\sin\theta)^2\partial^2v/\partial\varphi^2+(1/r^2)\cot\theta\partial v/\partial\theta\\
&\qquad\quad-(2/(r\sin\theta)^2)\cos\theta\partial w/\partial\varphi+(2/r^2)\partial u/\partial\theta-(1/r\sin\theta)^2v],
\end{aligned}
$$

$$\begin{aligned}
&\partial w/\partial t + u\partial w/\partial r + (1/r)v\partial w/\partial\theta + (1/r\sin\theta)w\partial w/\partial\varphi + (1/r)uw \\
&\qquad + (1/r)vw\cos\theta \\
&\qquad = -(1/\rho r\sin\theta)\partial p/\partial\varphi + \nu[(1/r)\partial^2(rv)/\partial r^2 + (1/r^2)\partial^2 w/\partial\theta^2 \\
&\qquad + (1/r\sin\theta)^2\partial^2 w/\partial\varphi^2 + (1/r^2)\cot\theta\partial w/\partial\theta + (2/r^2\sin\theta)\partial u/\partial\varphi \\
&\qquad + (2/(r\sin\theta)^2)\cos\theta\,\partial v/\partial\varphi - (1/r\sin\theta)^2 w]. \qquad (7.134)
\end{aligned}$$

The continuity equation in spherical coordinates is

$$\partial u/\partial r + (1/r)\partial v/\partial\theta + (1/r\sin\theta)\partial w/\partial\varphi + (2/r)u + (1/r)v\cot\theta = 0. \qquad (7.135)$$

Conservation of Energy

We first determine the energy equation for an *ideal fluid* which means an inviscid fluid which obeys the perfect gas law. The energy equation arises from the law of conservation of energy. We consider a volume V_0 of fluid fixed in space (with reference to an inertial reference frame), and then determine how the energy equation of the fluid contained in that volume varies with time. The total energy per unit volume of the inviscid fluid is

$$\tfrac{1}{2}\rho\mathbf{v}\cdot\mathbf{v} + e = \tfrac{1}{2}\rho v^2 + e, \qquad (7.136)$$

where $\frac{1}{2}\rho v^2$ is the specific kinetic energy (energy per unit volume) and e is the specific internal energy. The rate of change of this kinetic energy is

$$\tfrac{1}{2}(\rho v^2)_t = \tfrac{1}{2}v^2\rho_t + \rho\mathbf{v}\cdot\mathbf{v}_t.$$

To calculate this quantity we now appeal to the continuity equation (5.2) and Euler's equation (5.6) (for $\mathbf{F}=0$) and obtain

$$\tfrac{1}{2}(\rho v^2)_t = -\tfrac{1}{2}v^2\,\mathrm{div}(\rho\mathbf{v}) - \mathbf{v}\cdot\mathbf{grad}\,p - \left(\frac{\rho}{2}\right)\mathbf{v}\cdot\mathbf{grad}\,v^2.$$

We now eliminate p from this equation by appealing to the definition of enthalpy i and the first law of thermodynamics which can be written as

$$di = T\,dS + \left(\frac{1}{\rho}\right)dp,$$

so that we may write $\mathbf{grad}\,p$ as

$$\mathbf{grad}\,p = \mathbf{grad}\,i - T\,\mathbf{grad}\,S.$$

In order to complete the calculation of the rate of change of the energy per unit volume of an ideal fluid given by (7.136), we also need to calculate $(\rho e)_t$ in terms of measurable quantities. To do this we again appeal to the first law and write it in the form $de = T\,dS - (p/\rho^2)\,d\rho$. Since $e + p = i$ we get $d(\rho e) = e d\rho + \rho de = i d\rho + T dS$, and finally obtain

$$(\rho e)_t = i\rho_t + \rho T S_t = -i \operatorname{div}(\rho \mathbf{v}) - \rho T \mathbf{v} \cdot \mathbf{grad}\, S.$$

Here we have used the adiabatic result defined by the condition of no heat exchange with the environment (for an ideal fluid), so that $\delta Q = T\,dS = 0$. This means that $S_t + \mathbf{v} \cdot \mathbf{grad}\, S = 0$. This follows from the fact that $dS/dt = S_t + uS_x + vS_y + wS_z$, where (S_x, S_y, S_z) are the components of **grad** S. Putting the above results together yields

$$[\tfrac{1}{2}\rho v^2 + \rho e]_t = -\operatorname{div}[\rho \mathbf{v}(\tfrac{1}{2}v^2 + i)]. \qquad (7.137)$$

Equation (7.137) represents the rate of change of the total energy of an ideal fluid. In order to grasp its meaning, it is useful to recast this equation in terms of the flux of energy across the surface S_0 bounding a given volume of fluid V_0. We obtain

$$\frac{\partial}{\partial t}\int_{V_0} (\tfrac{1}{2}\rho v^2 + \rho e)\, dV = -\int_{V_0} \operatorname{div}[\rho \mathbf{v}(\tfrac{1}{2}v^2 + i)]\, dV.$$

Applying Gauss's divergence theorem to the volume integral of the right-hand side, in the usual way, gives

$$\frac{\partial}{\partial t}\int_{V_0} (\tfrac{1}{2}v^2 + e)\, dV = -\oint_{S_0} \rho \mathbf{v}(\tfrac{1}{2}v^2 + i) \cdot d\mathbf{S}. \qquad (7.138)$$

The left-hand side of (7.138) represents the rate of change of the total energy of an ideal fluid contained in the volume V_0, which is equal to the decrease of this energy or the amount of energy flowing out normal to the surface bounding this volume (expressed by the right-hand side). The scalar product nature of the surface integral shows that this energy flux is also in the direction given by $\boldsymbol{v}$. It is for this reason that we call the expression

$$\rho \mathbf{v}(\tfrac{1}{2}v^2 + i) \qquad (7.139)$$

the *energy flux density vector.* It is clear that the expression (7.139) also shows that a unit mass of ideal fluid carries with it during its motion an amount of energy $\frac{1}{2}\rho v^2 + i$. The fact that the specific enthalpy i, and not the specific internal energy e, appears in the energy flux density is explained by the following simple argument: Using the definition $i = e + p/\rho$, we may write the flux of energy through the bounding surface as follows:

$$-\oint_{S_0} \rho \mathbf{v}(\tfrac{1}{2}v^2 + e) \cdot d\mathbf{S} - \oint_{S_0} \rho \mathbf{v} \cdot d\mathbf{S}.$$

The first term of this expression is the transport of the sum of the total energy per unit mass per unit time across the bounding surface by the fluid, while the second term is the work done by the pressure forces (the only external forces assumed to act on the perfect fluid) on the fluid within the bounding surface.

Energy Dissipation in a Viscous Fluid

We now extend the law of conservation of energy to an *incompressible viscous fluid.* To do this we must calculate the dissipation of energy due to viscosity. In order to simplify the calculations we make the assumption that the fluid is incompressible. We first calculate the rate of change of the kinetic energy and use the Navier–Stokes equation to bring in the viscosity stress tensor τ'_{ij}. The rate of change of the kinetic energy is

$$\tfrac{1}{2}(\rho v^2)_t = \rho \mathbf{v} \cdot \mathbf{v}_t,$$

recognizing that for an incompressible fluid $\rho_t = 0$. We now use the Navier–Stokes equation (7.124), coupled with the condition that, for an incompressible fluid, the ijth component of the divergence of the viscosity stress tensor reduces to

$$\Pi_{ij,j} = \nabla^2 v_i = v_{j,j}.$$

Upon solving for $\mathbf{v}_t$ in the Navier–Stokes equation and using this expression, we obtain

$$\mathbf{v}_t = -(\mathbf{v} \cdot \mathbf{grad})\mathbf{v} - \left(\frac{1}{\rho}\right) \mathbf{grad}\, p + \tau'_{ij,j}.$$

Using this equation, we obtain the following expression for the rate of change of the kinetic energy per unit volume (after a little manipulation):

$$(\tfrac{1}{2}\rho v^2)_t = -\rho(\mathbf{v} \cdot \mathbf{grad})\left(\tfrac{1}{2}v^2 + \frac{p}{\rho}\right) + \operatorname{div}(\mathbf{v} \cdot \tau') - \tau'_{ij} v_{i,j},$$

where $\mathbf{v} \cdot \tau'$ denotes the vector (the scalar product of a vector and a tensor) whose ith component is $v_j \tau'_{ji}$ (summed over j). Since the fluid is assumed incompressible, we have $\operatorname{div} \mathbf{v} = 0$, so that the above equation becomes

$$(\tfrac{1}{2}\rho v^2)_t = -\operatorname{div}\left[\rho \mathbf{v}\left(\tfrac{1}{2}v^2 + \frac{p}{\rho}\right) - \mathbf{v} \cdot \boldsymbol{\tau}'\right] - \tau'_{ij} v_{i,j}. \tag{7.140}$$

The expression in brackets in (7.140) is the total energy flux density of an incompressible viscous fluid. It consists of two terms:

1. $\rho \mathbf{v}(\frac{1}{2}v^2 + p/\rho)$, which we recognize as the energy flux in an ideal fluid. This is the part of the total energy flux due solely to the transfer of the mass of fluid; it neglects the dissipation of energy due to viscous damping.
2. $\mathbf{v} \cdot \boldsymbol{\tau}'$, which is the energy flux caused by the energy dissipation of the viscous fluid. Another way of looking at the situation is that the presence of viscous damping results in a transfer of momentum flux τ'_{ij}. This involves a transfer of energy that is equal to the scalar product of the momentum flux and the velocity.

We now integrate the kinetic energy change given by (7.140) over the volume V_0, and transform the volume integral of the divergence terms in the usual way by invoking the divergence theorem. We obtain

$$\frac{\partial}{\partial t}\int_{V_0} \tfrac{1}{2}\rho v^2\, dV = -\oint_{S_0}\left[\rho\mathbf{v}\left(\tfrac{1}{2}v^2+\frac{p}{\rho}\right)-\mathbf{v}\cdot\boldsymbol{\tau}'\right]\cdot d\mathbf{S}$$
$$-\int_{V_0} \tau'_{ij}v_{i,j}\, dV. \qquad (7.141)$$

The surface integral represents the rate of change of the kinetic energy of the fluid in V_0 due to the energy flux across the bounding surface S_0. The second term on the right-hand side is the volume integral and expresses the decrease in the kinetic energy per unit time due to viscous dissipation. We now let V_0 tend to infinity and use the boundary condition that the fluid is at rest at infinity. Then (7.141) becomes

$$\frac{\partial}{\partial t}\int_{V_0} \tfrac{1}{2}\rho v^2\, dV = T_t = -\int_{V_0} \tau'_{ij}v_{i,j}\, dV, \qquad (7.142)$$

where T stands for the kinetic energy of the fluid in V_0 (all the fluid in our system). We may further simplify (7.142) for an incompressible fluid by writing the *ij*th component of the viscosity stress tensor as

$$\tau'_{ij} = \eta(v_{i,j} + v_{j,i}). \qquad (7.143)$$

This is obtained by setting $\operatorname{div}\mathbf{v} = v_{k,k} = 0$ in (7.121). Using (7.143) we get

$$\tau'_{ij}v_{i,j} = \eta v_{i,j}(v_{i,j} + v_{j,i}) = \tfrac{1}{2}\eta(v_{i,j} + v_{j,i})^2. \qquad (7.144)$$

Inserting (7.144) into (7.142) gives

$$T_t = -\tfrac{1}{2}\int_{V_0} (v_{i,j} + v_{j,i})^2\, dV. \qquad (7.145)$$

Equation (7.145) expresses the rate of change of the kinetic energy of an incompressible viscous fluid where the fluid is at rest on the bounding surface (at infinity).

7.6. Flow in a Pipe, Poiseuille Flow

As an example of the motion of an incompressible viscous fluid, we consider the steady state flow of such a fluid in a pipe of circular cross section (although in the following analysis the cross section need not be circular). This is called *Poiseuille flow.*

Let the x axis be the axis of the cylinder. The flow is parallel to the x axis and the particle velocity $\mathbf{v} = (u, 0, 0)$ where the x component of the velocity $u = u(y, z)$. Obviously, the steady state condition means all the dependent variables are independent of time. Since $u_x = 0$, the continuity equation $\operatorname{div}\mathbf{v} = 0$ is satisfied identically. The y and z components of the

Navier-Stokes equation (7.124), when written out in extended form, yields

$$p_y = p_z = 0,$$

meaning that $p = p(x)$ so that the pressure is constant over a given cross section of the pipe. The x component of the Navier-Stokes equation becomes

$$\frac{dp}{dx} = \eta \nabla^2 u = \eta(u_{yy} + u_{zz}). \tag{7.146}$$

Since $u = u(y, z)$ and dp/dx is a function of x, (7.146) tells us that $dp/dx =$ const. We may therefore write the pressure gradient as

$$\mathbf{grad}\ p = \frac{\Delta p}{L}, \tag{7.147}$$

where Δp is the pressure difference between the ends of the pipe of length L.

The velocity distribution $u = u(y, z)$ is determined by solving the following boundary value problem:

$$u_{yy} + u_{zz} = \text{const.},$$

in the region $\mathscr{R}$ bounded by the closed curve Γ representing the circumference of a cross section of the pipe. The boundary condition is

$$u = 0 \quad \text{on } \Gamma.$$

We now consider a pipe of length L and circular cross section of radius R. We take the origin at the center of a circle at any station x, and express the Laplacian $u_{yy} + u_{zz}$ in polar coordinates. Upon using (7.147), the Navier-Stokes equation (7.124) becomes

$$\left(\frac{1}{r}\right)\frac{d}{dr}\left(r\frac{du}{dr}\right) = -\frac{\Delta p}{\eta L}. \tag{7.148}$$

Integrating (7.147) yields

$$u = -\left(\frac{\Delta p}{4\eta L}\right) r^2 + a \log r + b,$$

where a and b are constants to be determined. Since u must remain finite at the origin, we must have $a = 0$ ($\log r$ has a singularity at $r = 0$). b is determined by satisfying the boundary condition: $u = 0$ on $r = R$ (no slip). We then obtain

$$u = \left(\frac{\Delta p}{4\eta L}\right)(R^2 - r^2). \tag{7.149}$$

Equation (7.149) is the velocity distribution across the pipe in Poiseuille flow; it is a parabolic distribution in r, being a maximum at the origin and zero on the boundary.

We now determine the *discharge* Q defined as the mass of fluid flowing past a unit cross section per unit time. A mass of fluid equal to $2\pi\rho ru\,dr$ passes through an annular element $2\pi r\,dr$ of the cross section area per unit time. This definition of Q gives

$$Q = 2\pi\rho \int_0^R ru\,dr.$$

Upon using (7.149) and integrating we get

$$Q = \left(\frac{\pi \Delta p}{8\nu L}\right) R^4. \tag{7.150}$$

Equation (7.150) is called *Poiseuille's formula*. It says that the discharge is proportional to the fourth power of the radius of the pipe.

Poiseuille flow has important applications in the field of hemorheology, which is the study of blood flow—an important aspect of biophysics and bioengineering. However, since a blood vessel is a flexible tube rather than a rigid pipe, more complicated boundary conditions must be used that take into account the flexibility of the tube. Another complication is that the blood vessel is not straight, so that the axis of the tube is a nonconstant curve in space. Moreover, in arterial flow we no longer have steady flow, so that the time varying equations of motion must be satisfied for pulsatile flow.

7.7. Dimensional Considerations

In fluid mechanics many important results can be obtained by simple physical arguments. A consequence of this treatment is that certain dimensionless parameters such as the Reynolds number are introduced which offer a great simplification in allowing us to compare similar types of flows arising from the study of viscous flow over geometrically similar bodies.

In the following analysis we shall investigate the steady incompressible flow of a viscous fluid in which is immersed a solid body. The velocity at infinity U is called the *free-stream velocity*. What are the physical parameters that specify a given flow over the body? Clearly the flow is governed by solutions of the Navier–Stokes equation with the appropriate boundary conditions. The kinematic viscosity coefficient ν appears in the Navier–Stokes equation as well as the dependent variables $\mathbf{v}$, p, and ρ. The flow depends on the shape and dimensions of the immersed body, through the boundary conditions. The shape of the body is known, so that the body essentially depends on a linear dimension L (for example, the diameter of a sphere). This means that the flow is specified by the three parameters ν, U, and L. Recall that ν has the dimensions L^2/T and U has dimensions

L/T, where L is length and T is time. The only *dimensionless* combination of U, L, and ν is UL/ν or some function of it. This is the motivation for defining the *Reynolds number* R by

$$R=\frac{UL}{\nu}=\frac{\rho UL}{\eta}. \tag{7.151}$$

We shall see below that this definition of the Reynolds number has an important physical meaning. It is clear that, since in calculating the velocity field, the only dimensionless parameter is R or a function of R, the velocity distribution is obtained by solving the equations for steady incompressible flow and may be put in the following functional form:

$$\mathbf{v}=U\mathbf{f}\left(\frac{\mathbf{r}}{L}, R\right),$$

where $\mathbf{f}$ is a dimensionless vector function. From this function we obtain an important similarity principle which is illustrated by the following example: Suppose we have flows of two different velocities past two spheres of different radii. In addition, suppose these two flows have the same Reynolds number. Then since $\mathbf{v}/U=\mathbf{f}$, we see that the ratio of velocities $\mathbf{v}/U$ is the same function of the ratio $\mathbf{r}/L$ for each flow. Flows having the same value of $\mathbf{r}/L$ are called flows *of the same type*. Thus *flows of the same type with the same Reynolds number are similar*. This is called the *law of similarity* and was discovered by Osborne Reynolds in 1883. The flow patterns arc given by the streamlines. The law of similarity says that two flows having the same Reynolds number have geometrically similar streamlines for the case of streamline or nonturbulent flow.

And now for the physical interpretation of the Reynolds number R. We shall show that R can be given as the ratio of the inertial force to the viscous or frictional force acting on the fluid. In fact, for the example of the two spheres, the flow around the spheres is geometrically similar if the ratio of the inertial to the viscous forces is the same at every point in the flow, provided we neglect other forces. Looking at the Navier–Stokes equation (7.124), the left-hand side gives the inertial force $\rho\dot{\mathbf{v}}$ while the $\eta\nabla^2\mathbf{v}$ term on the right-hand side gives the viscous force. We therefore obtain

$$\frac{\text{inertial force}}{\text{viscous force}}=\frac{\rho\dot{\mathbf{v}}}{\eta\nabla^2\mathbf{v}}.$$

We recall that $\dot{\mathbf{v}}=\mathbf{v}_t+(\mathbf{v}\cdot\mathbf{grad})\mathbf{v}$, the second term of the right-hand side being the convective term which is nonlinear, a typical component being uu_x. This term can be expressed as a function of the following physical quantities: ρ, U, and L (the diameter of the sphere). Therefore the dimensions of a typical component of the convective term ρuu_x are $\rho U^2/L$. A typical term of the viscous force in the Navier–Stokes equation is ηu_{xx}. Using the same physical quantities and η, the dimensions of this term are

$\eta U/L^2$. Using the nonlinear term for the inertial force, we see that the ratio of the inertial to the viscous forces gives a measure of the effect of the nonlinear term. We also note that for steady flow the inertial term *is* the nonlinear term. Therefore this ratio becomes

$$\frac{\text{inertial force}}{\text{viscous force}} = \frac{\rho U^2/L}{\eta U/L^2} = R. \tag{7.152}$$

We note that for the asymptotic case of infinite Reynolds number the flow becomes inviscid, as we easily see by examining (7.152). On the other end of the spectrum, for the asymptotic case of zero Reynolds number, the flow is entirely viscous and the inertial force can be neglected.

In light of the current state of mathematical analysis and considering the difficult nature of the Navier–Stokes equation, there are no general analytical methods of solving this equation, except for special cases like Poiseuille flow and asymptotic cases involving large Reynolds numbers concerned with the technical problems of aerodynamics (high-speed aerodynamic flow for small viscosity), and the other end of the scale—small Reynolds numbers which involve low-speed flow of very viscous fluids, important in technical problems in the field of rheology. However, a whole host of numerical methods have been constructed which give detailed solutions for a large variety of cases. With the development of more powerful computers this may be the way to go in solving such nonlinear problems, if we are interested in solving specific problems in detail. In general, the classical analytical approaches that have been so successful in giving us an insight into linear problems are powerless when it comes to attacking problems involving such nonlinearities as those occurring in the Navier–Stokes equation. We must search for other techniques which, as mentioned, involve numerical methods. Investigators usually resort to numerical methods which solve special problems in great detail; but these methods are not very helpful in giving a deep physical insight into the more general aspects of viscous flow. Another approach, as alluded to, is to use the method of asymptotic expansions involving power series in the Reynolds number. Such an approach gives us a deep insight into the changeover of boundary conditions from zero slip for viscous flow to slip for inviscid flow, as mentioned in a previous section. Asymptotic expansions were treated in Chapter 5 in connection with the treatment of water waves using other expansion parameters.

Unsteady Flow

We now turn to unsteady or time-varying flow in connection with dimensional analysis considerations. Unsteady flows are characterized not only by the physical quantities ρ, U, and L, but by some time interval τ characteristic of the flow which determines how the flow changes with time.

For instance, for oscillatory flows we may take τ as the period of oscillation. From these four parameters, in addition to the Reynolds number, we may construct another dimensionless number S defined by

$$S = \frac{U\tau}{L}. \tag{7.153}$$

S is called the *Strouhal number.* In comparing two unsteady fluids, similar motion occurs when the numbers R and S have the same values.

7.8. Stokes's Flow

Sir George Stokes (one of the developers of the Navier–Stokes equation) was interested in linearizing this equation for the steady state condition in order to determine what types of flow problems could be solved. As pointed out above, the quadratic or nonlinear term $(\mathbf{v} \cdot \mathbf{grad})\mathbf{v}$ is the inertial term for steady state flow. Therefore, for viscous flow, neglecting the quadratic term is equivalent to setting the Reynolds number equal to zero. We know that a very small Reynolds number means the slow motion of a very viscous fluid. Therefore Stokes set himself the problem of attempting to obtain an approximate solution to the slow motion of a highly viscous fluid, or an exact solution to the limiting case of zero Reynolds number. To fix his ideas, he investigated such a flow in which is embedded a sphere with a rough surface. This problem has important applications in rheology and the materials sciences in general, for example, the problem of settling of suspensions, the action of colloids, etc. If the Reynolds number is set equal to zero the quadratic term is then set equal to zero in the steady state Navier–Stokes equation, which then becomes

$$\mathbf{grad}\, p = \eta \nabla^2 \mathbf{v}, \tag{7.154}$$

subject to the incompressibility condition

$$\operatorname{div} \mathbf{v} = 0. \tag{7.155}$$

Equations (7.154) and (7.155) completely define the flow field by allowing us to solve for the velocity and pressure fields. To these equations we must, of course, append the boundary condition of zero slip on the surface of the sphere and the boundary condition of uniform fluid motion at infinity (with respect to the sphere).

We now investigate the slow motion of a sphere moving with a uniform velocity $\mathbf{U}$ in a still incompressible viscous fluid. It is convenient to transform this problem to one where the coordinates are fixed at the center of the sphere. This is easily done by subtracting $\mathbf{U}$ from the original problem. This allows us to consider the problem of finding the flow field and the stress on a stationary sphere in the fluid whose free-stream velocity is $\mathbf{U}$ (boundary condition at infinity). To solve the original problem we merely subtract $\mathbf{U}$

from the solution for the case of the fixed sphere. Therefore $\mathbf{v}$ is interpreted as the perturbed fluid velocity due to flow around the fixed sphere in the flow field whose free-stream velocity is $\mathbf{U}$.

To formulate this problem we use a spherical coordinate system (r, θ, φ) where θ is the polar angle and φ is the azimuth angle. Taking the curl of (7.154) and recognizing that **curl grad** $p=0$, we obtain

$$\nabla \cdot \nabla \nabla \mathbf{x} \mathbf{v}=0,$$

since the operators $\nabla \cdot \nabla$ and $\nabla \mathbf{x} \equiv \mathbf{curl}$ commute. But **curl** $\mathbf{v}$ is twice the mean angular velocity $\boldsymbol{\omega}$, so that we get

$$\nabla^2 \boldsymbol{\omega}=0. \tag{7.156}$$

Equation (7.156) tells us that the components of the angular velocity are harmonic functions since they satisfy Laplace's equation. It is evident from symmetry that the vortex lines (described below) are circles in planes normal to the x axis with centers on the axis.

At this point we describe the meaning of *vortex lines*. A vortex line is defined as a curve whose tangent has everywhere the direction of the axis of rotation of the fluid elements. If $(\omega_x, \omega_y, \omega_z)$ are the rectangular components of $\boldsymbol{\omega}$, then the differential equations whose solutions yield the vortex lines are

$$\frac{dx}{\omega_x}=\frac{dy}{\omega_y}=\frac{dz}{\omega_z}.$$

If vortex lines are drawn through every point on a small closed curve, the fluid contained in the tube thus formed is said to constitute a *vortex filament*. The tube is taken to be so thin that the angular velocity is practically the same for all points on any cross section. If A is the cross-sectional area, the quantity ωA is defined as the *vortex strength*. There are three fundamental laws of vortex motion in an incompressible fluid in which the external force is derivable from a potential function (in our case we only consider pressure), which we shall merely state. For a derivation of these laws the reader is referred to standard works on hydrodynamics such as [25, p. 202]. These laws, which are due to Helmholtz and Lord Kelvin, are:

1. The same particles of fluid constitute a vortex filament at all times.
2. The strength of a vortex filament is constant with respect to time.
3. The strength of a vortex filament is constant throughout the filament.

Continuing with Stokes's flow, we write the angular velocity in spherical coordinates. In the usual notation, let $\mathbf{i}$, $\mathbf{j}$, $\mathbf{k}$ be unit vectors in the x, y, z directions, respectively. Now $\boldsymbol{\omega}$ is in the $(\mathbf{j}, \mathbf{k})$ plane since its axis is along the x axis. In terms of spherical coordinates we have

$$\boldsymbol{\omega}=-\mathbf{j} f(r, \theta) \sin \varphi+\mathbf{k} f(r, \theta) \cos \varphi, \tag{7.157}$$

where f is a function of r, θ to be determined. Laplace's equation reduces to

$$\frac{\partial}{\partial r}\left(\frac{r^2 \partial f}{\partial r}\right)+\left(\frac{1}{\sin\theta}\right)\frac{\partial}{\partial\theta}\left(\sin\theta\frac{\partial f}{\partial\theta}\right)-\frac{f}{\sin^2\theta}=0. \tag{7.158}$$

f may be separated into the product of a function of r and a function of θ. The strength of a vortex filament is zero for $\theta=0$ and π, and must increase continuously as we pass from either of these limits to the plane $\theta=\pi/2$. Hence f is of the form

$$f(r,\theta)=g(r)\sin\theta.$$

Inserting this expression into Laplace's equation (7.158) yields the following ODE for f:

$$\frac{d}{dr}\left(\frac{r^2\,dg}{dr}\right)-2g=0,$$

which is satisfied by either r or $1/r^2$, as can easily be seen by assuming a solution of the form $g=r^n$ and inserting this expression into the differential equation; this immediately gives $n=1, -2$. Since $\boldsymbol{\omega}$ must vanish for $r=\infty$, this boundary condition demands that g be of the form $1/r^2$. Then (7.157) becomes

$$\begin{aligned}\boldsymbol{\omega}&=A\left(-\mathbf{j}\left(\frac{1}{r^2}\right)\sin\theta\sin\varphi+\mathbf{k}\left(\frac{1}{r^2}\right)\sin\theta\cos\varphi\right)\\&=A\left(-\frac{\mathbf{j}z}{r^3}+\frac{\mathbf{k}y}{r^3}\right)\\&=A\,\mathbf{curl}\,\frac{\mathbf{i}}{\mathbf{r}},\end{aligned} \tag{7.159}$$

where A is an arbitrary constant and we have used the transformation $y=r\sin\theta\sin\varphi$, $z=r\sin\theta\cos\varphi$. Since $\boldsymbol{\omega}=\frac{1}{2}\,\mathbf{curl}\,\mathbf{v}$, we obtain $\mathbf{v}$ from (7.159), which is

$$\mathbf{v}=\frac{\mathbf{i}2A}{r}+\mathbf{grad}\,\phi, \tag{7.160}$$

where ϕ is any scalar function of the coordinates. We may add this last term since, as we recall, **curl grad** $\phi=0$. $\mathbf{v}$ as given by (7.160) satisfies the equation of motion (7.154) and the symmetry conditions imposed on $\boldsymbol{\omega}$. We must now force (7.160) to satisfy the continuity equation (7.155) which we write in spherical coordinates as

$$\operatorname{div}\mathbf{v}=-\frac{2A\cos\theta}{r}+\nabla^2\phi=0 \qquad \text{or} \qquad \nabla^2(A\cos\theta+\phi)=0, \tag{7.161}$$

since $\partial/\partial x(1/r) = -x/r^3$ and $-(2/r^2)\cos\theta = \nabla^2 \cos\theta$. Equation (7.161) tells us that

$$\phi = -A\cos\theta + \Phi,$$

where Φ is a solution of Laplace's equation. We observe that ϕ must contain the factor $\cos\theta$ in order to satisfy the boundary condition at the surface of the sphere for all values of θ, which means that ϕ must be a linear combination of $r\cos\theta$ and $(1/r^2)\cos\theta$. Therefore ϕ becomes

$$\phi = \left(-A + Br + \frac{C}{r^2}\right)\cos\theta,$$

so that (7.160) becomes

$$\mathbf{v} = \mathbf{i}\left(\frac{2A}{r}\right) - A\,\mathbf{grad}\cos\theta + B\,\mathbf{grad}(r\cos\theta) + C\,\mathbf{grad}\left(\frac{1}{r^2}\right)\cos\theta. \tag{7.162}$$

We put this equation in component form by recognizing that (u, v) are the components of $\mathbf{v}$ in the directions of increasing (r, θ). We get

$$u = \left(B + \frac{2A}{r} - \frac{2C}{r^3}\right)\cos\theta,$$

$$v = -\left(B + \frac{A}{r} + \frac{C}{r^3}\right)\sin\theta,$$

where the constants A, B, C can be determined as follows: First, we satisfy the boundary condition $r = \infty$, which is

$$u(\infty, \theta) = U, \qquad v(\infty, 0) = 0. \tag{7.163}$$

This yields $B = U$. Let R be the radius of the sphere. The boundary condition on the surface of the sphere is zero slip, which means

$$u(R, \theta) = 0, \qquad v(R, \theta) = 0. \tag{7.164}$$

From these boundary conditions we get

$$A = -\tfrac{3}{4}RU, \qquad B = U, \qquad C = -\tfrac{1}{4}R^3U. \tag{7.165}$$

The velocity components then become

$$\begin{aligned} u &= \left(1 - \frac{3}{2}\frac{R}{r} + \frac{1}{2}\frac{R^3}{r^3}\right)U\cos\theta, \\ v &= -\left(1 + \frac{3}{4}\frac{R}{r} - \frac{1}{4}\frac{R^3}{r^3}\right)\sin\theta. \end{aligned} \tag{7.166}$$

The magnitude of $\boldsymbol{\omega}$ becomes

$$\omega = -\frac{3}{4}\left(\frac{RU}{r^2}\right)\sin\theta. \tag{7.167}$$

Since we have satisfied all the boundary conditions and the solution of Laplace's equation is unique, we have completely solved the kinematic portion of the problem. This means that we have solved for the velocity field for the incompressible flow of a viscous fluid whose free-stream velocity is $\mathbf{U}$, around a sphere of radius R.

The next step is to consider the dynamic part of the problem by calculating the stress on the surface of the sphere. The mean pressure is obtained at once from the steady state Navier–Stokes equation by setting $\mathbf{v}=0$ in (7.124) and using (7.162) and (7.165). We get

$$\mathbf{grad}\; p = \tfrac{3}{4}\eta\nabla^2\mathbf{v} = -\eta\nabla^2\,\mathbf{grad}\cos\theta = \tfrac{3}{4}\eta RU\nabla^2\,\mathbf{grad}\cos\theta,$$

so that we obtain

$$p = \tfrac{3}{4}\eta RU\nabla^2\cos\theta + p_0 = -\tfrac{3}{2}\eta RU\left(\frac{1}{r^2}\right)\cos\theta + p_0, \tag{7.168}$$

where p_0 is the pressure at infinity (atmospheric pressure). Equation (7.168) gives the pressure in the fluid surrounding the sphere. By setting $r=R$ we get the pressure on the surface. Clearly the boundary condition on the pressure at infinity is satisfied by setting $r=\infty$.

To calculate the stresses on the surface of the sphere, we need to determine the strain field—the components of the strain tensor. It will simplify the analysis if we choose a right-handed rectangular coordinate system such that the x axis is directed along the radius vector $\mathbf{r}$, and y is normal to it in the direction of increasing θ. Let $\boldsymbol{\tau}$ be the stress tensor. $\boldsymbol{\tau}$ may be written as the following two-dimensional symmetric matrix:

$$\tau = \begin{pmatrix} \tau_{xx} & \tau_{xy} \\ \tau_{yx} & \tau_{yy} \end{pmatrix},$$

where the diagonal elements represent the normal components of the stress tensor and the off-diagonal elements, the shear components. The reason that we have a two-dimensional stress matrix is due to our coordinate system and the symmetric nature of the sphere. The stress $\boldsymbol{\tau}_r$ on a surface normal to r is

$$\boldsymbol{\tau}_r = \boldsymbol{\tau}_i = \mathbf{i}\cdot\boldsymbol{\tau} = \mathbf{i}\tau_{xx} + \mathbf{j}\tau_{xy}. \tag{7.169}$$

The strain tensor is represented by the symmetric matrix

$$\boldsymbol{\varepsilon} = \begin{pmatrix} \varepsilon_{xx} & \varepsilon_{xy} \\ \varepsilon_{yx} & \varepsilon_{yy} \end{pmatrix}.$$

In fluid dynamics, instead of using this strain tensor directly, we deal with the *strain rate tensor* which is defined by $\boldsymbol{\varepsilon}_t$. To see how this works, we define the displacement vector $\mathbf{d}=(\xi, \eta, \zeta)$, where ξ, η, and ζ represent the components of the displacement of a fluid particle from its equilibrium position in the x, y, and z directions, respectively, and are functions of x,

y, z, t. For a particle in equilibrium $\mathbf{d}=0$. We note that the components of the strain matrix can be obtained from the corresponding components of the gradient of the displacement vector. The elements of the strain matrix are

$$\varepsilon_{xx}=\frac{\partial \xi}{\partial x}, \qquad \varepsilon_{xy}=\frac{\partial \xi}{\partial y}+\frac{\partial \eta}{\partial x},$$

$$\varepsilon_{yx}=\varepsilon_{xy}, \qquad \varepsilon_{yy}=\frac{\partial \eta}{\partial y}.$$

The particle velocity $\mathbf{v}$ is related to the displacement vector by $\mathbf{d}_t=\mathbf{v}$ (reverting to subscript notation for partial derivatives) so that the components of the strain rate tensor become

$$\varepsilon_{xx,t}=u_x, \qquad \varepsilon_{xy,t}=u_y+v_x, \qquad \varepsilon_{yy,t}=v_y.$$

Equation (7.169) tells us that the only components of the stress tensor we need are τ_{xx} and τ_{xy}. These are related to the corresponding components of the strain rate tensor by the following equations:

$$\tau_{xx}=-p+2\eta\varepsilon_{xx,t}, \qquad \tau_{xy}=\eta\varepsilon_{xy,t}. \tag{7.170}$$

Using (7.166) and recalling the definition of our coordinate system, the components of the strain rate tensor become

$$\begin{aligned}\varepsilon_{xx,t}&=u_x=\frac{3}{2}\left(\frac{R}{r^2}\right)\left[1-\left(\frac{R}{r}\right)^2\right]U\cos\theta,\\ \varepsilon_{xy,t}&=u_y+v_x=-\frac{3}{2}\left(\frac{R^3}{r^4}\right)U\sin\theta.\end{aligned} \tag{7.171}$$

On the spherical surface $r=R$ so that (7.171) becomes

$$\varepsilon_{xx,t}=0, \qquad \varepsilon_{xy,t}=-\frac{3}{2}\left(\frac{U}{R}\right)\sin\theta \qquad \text{for} \quad r=R. \tag{7.172}$$

Inserting (7.168) and (7.172) into (7.170) yields

$$\begin{aligned}\tau_{xx}&=\frac{3}{2}\left(\frac{\eta U}{R}\right)\cos\theta-p_0,\\ \tau_{xy}&=-\frac{3}{2}\left(\frac{\eta U}{R}\right)\sin\theta.\end{aligned} \tag{7.173}$$

It is clear that τ_{xx} is the normal stress in the plane containing $\mathbf{r}$ in the direction normal to the surface; if positive, then it is in the direction away from the surface and is tensile, if negative, then it is directed inward and is compressive. For example, if $\theta=\pi/2$ then $\tau_{xx}=-p_0$ and is thus a pure compression due to the atmospheric pressure p_0. τ_{xy} is the shear stress in the plane normal to $\mathbf{r}$ in the direction tangent to the surface.

Let **F** be the resultant force on the spherical surface. It is clear from symmetry that **F** is in the direction of the fluid motion at infinity given by the free stream velocity **U**. We calculate **F** by integrating with respect to θ over the surface and obtain the following expression for the magnitude of **F**:

$$F=\int_0^{\pi} \tau_{xx}\cos\theta\, 2\pi R^2 \sin\theta\, d\theta - \int_0^{\pi} \tau_{xy}\sin\theta\, 2\pi R^2\sin\theta\, d\theta$$

$$=2\pi\eta RU+4\pi\eta RU=6\pi\eta RU. \tag{7.174}$$

Equation (7.174) is known as *Stokes's law.* Since the *drag* of the viscous fluid on the sphere (which is the force due to the viscous resistance of the flow over the sphere) can depend only on the motion of the fluid relative to the sphere, the resistance offered by a quiet fluid to the sphere moving with a velocity **U** is also given by (7.174).

The solution we have just obtained for the dynamic part of the Stokesian flow (Stokes's law) does not satisfy the boundary condition of $\mathbf{v}=0$ at infinity, even if the Reynolds number is very small. In order to see this we perform an order of magnitude analysis. This is done by first estimating the magnitude of the nonlinear term $(\mathbf{v}\cdot\mathbf{grad})\mathbf{v}$ in the Navier-Stokes equation (7.124). Recall that if this term is neglected the steady state Navier–Stokes equation becomes $\mathbf{grad}\, p=\eta\nabla^2\mathbf{v}$. At large distances from the sphere the derivatives of the velocity are of the order UR/r^2. From this we see that the nonlinear term $(\mathbf{v}\cdot\mathbf{grad})\mathbf{v}$ is of the order U^2R/r^2. The linear terms retained in the Navier–Stokes equation such as $(1/\rho)\,\mathbf{grad}\, p$ are of the order $\eta RU/\rho r^3$. The condition that the linear terms are much greater than the nonlinear terms is given by the following inequality:

$$\frac{\eta RU}{\rho r^3}\gg\frac{U^2R}{r^2}.$$

Clearly this inequality holds *only* at distances r such that

$$r\ll\frac{\nu}{U},\qquad \nu=\frac{\eta}{\rho}. \tag{7.175}$$

As we see from (7.175) the ratio of the kinematic viscosity coefficient to the free stream velocity is the upper limit on the distance from the center of the sphere such that Stokes's law holds. Clearly, in the limiting case of zero Reynolds number, this upper limit is infinite.

To obtain the velocity distribution at large distances from the sphere we must take into account the term $(\mathbf{v}\cdot\mathbf{grad})\mathbf{v}$ which was omitted in deriving Stokes's law. At large distances $\mathbf{v}\sim\mathbf{U}$, so that we can approximate the operator $\mathbf{v}\cdot\mathbf{grad}$ by $\mathbf{U}\cdot\mathbf{grad}$, and thus obtain the approximation

$$(\mathbf{U}\cdot\mathbf{grad})\mathbf{v}=-\left(\frac{1}{\rho}\right)\mathbf{grad}\, p+\nu\nabla^2\mathbf{v}. \tag{7.176}$$

Thus the nonlinear Navier–Stokes equation is reduced to the linear equation (7.176) by the above approximation, which is called the *Oseen approximation*, due to C.W. Oseen (1910). We shall not investigate the solution corresponding to the Oseen approximation, which is discussed in some detail in [25, p. 608] including a discussion of flow around a cylinder. We merely mention that the velocity distribution and drag thus obtained for the Stokesian flow can be used as first-order approximations in expansions in powers of the Reynolds number. The next order approximation to the drag is contained in the more exact equation

$$F = 6\pi\eta UR\left[1 + \left(\frac{3UR}{8\nu}\right)\right]. \tag{7.177}$$

An example of Stokesian flow is the case of a sphere of radius R and density σ falling under the action of a gravitational force at a very small Reynolds number in a still viscous fluid of density ρ. There are two external forces acting on the sphere:

(1) the force of gravity given by

$$\tfrac{4}{3}R^3\sigma g;$$

(2) the buoyancy force given by

$$\tfrac{4}{3}\rho R^3 g.$$

When a steady state condition is reached, the difference between these forces must equal the drag given by Stokes's law (7.174), yielding

$$\tfrac{4}{3}R^3 g(\sigma - \rho) = 6\pi\eta RU.$$

From this equation we solve for the steady state velocity acquired by the sphere, which is

$$U = \frac{2(\sigma - \rho)R^2 g}{9\eta}, \tag{7.178}$$

where g is the acceleration of gravity.

7.9. Oscillatory Motion

Next, we investigate the oscillatory motion of a viscous fluid due to the oscillations of a body immersed in the fluid. Since the vibrating body produces waves in the fluid, it is clear that we no longer can use the steady state approximation to the Navier–Stokes equation but must also consider the $\rho\mathbf{v}_t$ term as part of the inertial force.

In order to study wave propagation in a viscous fluid due to the oscillating body, it is of interest to start with a simple but useful example. To this end, we consider an incompressible fluid bounded by an infinite planar surface in the form of a plate which acts as a simple harmonic oscillator, oscillating in its own plane with a frequency ω. The problem is to determine the resulting motion of the fluid and thereby determine the properties of the traveling waves. To fix our ideas, we let the plate lie in the (y, z) or $x=0$ plane and have it oscillate along the y axis. The fluid region $\mathcal{R}$ is $x>0$. The velocity V of the oscillating plate is clearly in the y direction and, of course, is time harmonic. It may be put in the form

$$V = \mathrm{Re}[V_0 \exp(-i\omega t)],$$

where Re[ ] means the real part of [ ], and the constant $V_0 = Ae^{-i\delta}$, where A is the amplitude and δ is the phase angle. For simplicity, we proceed as if V were complex, since only linear operations are involved, and then take the real part of the final result. (This technique is used whenever we deal with complex quantities and we want results in terms of real quantities.) Therefore we write

$$V(t) = V_0 \exp(-i\omega t). \tag{7.179}$$

The velocity of the fluid $\mathbf{v} = (u, v, w)$ must satisfy the boundary condition

$$\mathbf{v}(0, t) = V(t) \quad \text{or} \quad u(0, t) = 0, \quad v(0, t) = V(t), \quad w(0, t) = 0 \quad \text{at} \quad x = 0. \tag{7.180}$$

From symmetry $\mathbf{v}$ is spatially dependent only on x. Since the continuity equation for an incompressible fluid is $\operatorname{div} \mathbf{v} = 0$, we have $u_x = 0$. Invoking the boundary condition (7.180) gives $u = 0$ in $\mathcal{R}$. The nonlinear term $(\mathbf{v} \cdot \mathbf{grad})\mathbf{v}$ of the Navier–Stokes equation vanishes since $u = 0$ and v is independent of y and z. This means that the Navier–Stokes equation is linearized, thus becoming

$$\mathbf{v}_t = -\left(\frac{1}{\rho}\right) \mathbf{grad}\, p + \nu \nabla^2 \mathbf{v}. \tag{7.181}$$

Since $\mathbf{v}$ is a function of (x, t) and $u(x, t) = 0$, the x component of (7.181) yields $p_x = 0$ or $p = \text{const.}$ so that we obtain a constant pressure distribution throughout the fluid. The y component of (7.181) becomes

$$v_t = \nu v_{xx}. \tag{7.182}$$

Equation (7.182) is a parabolic second-order linear PDE and would be the one-dimensional Fourier unsteady heat equation if v were the temperature. Since the parabolic equation represents a diffusion process, (7.182) can be interpreted as representing the diffusion of the flux of fluid from the boundary $x = 0$. Therefore, for this model of oscillatory flow the y component of the fluid velocity has a diffusionlike property.

The z component of (7.181) yields the same PDE for w as (7.182). However, the boundary conditions at $x = 0$ are

$$w(0, t) = 0, \qquad w_x(0, t) = 0.$$

Since initially the fluid is at rest and we need two boundary conditions and one initial condition for the solution of the parabolic PDE, the homogeneous initial and boundary conditions for w tell us that $w(x, t) = 0$ throughout the fluid. The particle velocity then becomes $\mathbf{v} = (0, v(x, t), 0)$. We can also see on a physically intuitive basis that the fluid velocity must be in the y direction, since the cause of the fluid motion is the plate which oscillates in the y direction.

We now assume a time-harmonic solution for $v(x, t)$ of the form

$$v = v_0 \exp[i(\omega t - kx)], \tag{7.183}$$

where v_0 is the amplitude of v and k is the *wave number* defined by $k = 2\pi/\lambda$ where λ is the wavelength. Inserting (7.183) into (7.182) allows us to solve for k which becomes

$$k_\pm = (1 \pm i)\sqrt{\frac{\omega}{2\nu}}. \tag{7.184}$$

Upon using the boundary condition (7.179), (7.183) becomes

$$v(x, t) = V_0 \exp\left[-\sqrt{\frac{\omega}{2\nu}}x\right] \exp\left[i\left(\omega t - \sqrt{\frac{\omega}{2\nu}}x\right)\right], \qquad x > 0, \quad t > 0, \tag{7.185}$$

where we have taken k to have a negative imaginary part in order to ensure stability; otherwise, the velocity would increase without limit in the fluid, which is an unstable situation. The solution for $v(x, t)$, given by (7.185), tells us that the oscillating boundary (the plate) produces a transverse progressing wave, since the fluid velocity v is normal to the direction of wave propagation (the x direction). The most important property of the wave is that it is rapidly damped as we go into the fluid away from the oscillating plate. In fact, over a distance of one wavelength, the amplitude diminishes by a factor of $e^{2\pi} \sim 540$. The *depth of penetration* of the wave δ is defined as the distance x from the plate where the amplitude fall off to e^{-1} of V_0. We see from (7.185) that

$$\delta = \sqrt{\frac{2\nu}{\omega}}, \tag{7.186}$$

which tells us that the depth of penetration depends on the ratio of the kinematic viscosity to the frequency of the vibrating plate. The higher the ratio the less the damping or the stronger the diffusion process.

The phase velocity or wave velocity is obtained by first setting the exponent $i(\omega t - \sqrt{\omega/2\nu}\, x)$ equal to a constant, and then forming $dx/dt = c$. This gives

$$c = \sqrt{2\nu\omega}. \tag{7.187}$$

For a given frequency of vibration of the plate the more viscous the fluid the more the traveling transverse wave is damped, but the greater is its velocity.

We now calculate the shear stress acting on the vibrating plate due to the viscosity of the fluid. This stress clearly lies along the surface of the plate in the y direction, and is therefore τ_{xy}. Recalling that $\tau_{xy} = \eta v_x$ and using the solution for v given by (7.185) we get, upon setting $x = 0$,

$$\tau_{xy}(0, t) = -\sqrt{\frac{\omega\eta\rho}{2}}(i+1)V(t), \tag{7.188}$$

which shows that there is a constant phase angle between the simple harmonic motion of the plate and the shear stress on the plate. For example, if $V = V_0 \cos \omega t$, then

$$\tau_{xy} = -\sqrt{\omega\eta\rho}\, V_0 \cos\left(\omega t - \frac{\pi}{4}\right),$$

giving a phase difference between the frictional force and the plate velocity.

Finally, we easily calculate the time average of the energy dissipated due to viscous damping. The energy dissipated per unit area of the plate per unit time (the power lost per unit area), is equal to the mean value of the product of the shear stress τ_{xy} and the velocity v. The value of this average energy loss is

$$\tfrac{1}{2} V_0^2 \sqrt{\frac{\omega\eta\rho}{2}}.$$

We now consider the general case of an oscillating body of arbitrary shape. For an arbitrary body the nonlinear term $(\mathbf{v} \cdot \mathbf{grad})\mathbf{v}$ in the Navier–Stokes equation is not necessarily zero as it was for the above case of the vibrating plate. The following question therefore arises: Under what conditions can we neglect the nonlinear term? To answer this question we go back to the linearized Navier–Stokes equation obtained from (7.124) by setting this nonlinear term equal to zero. We take the **curl** of this equation, recognizing that **curl grad** $p = 0$, and obtain

$$(\mathbf{curl\, v})_t = \nu \nabla^2\, \mathbf{curl\, v}. \tag{7.189}$$

Equation (7.189) tells us that the vorticity given by **curl v** satisfies the three-dimensional unsteady Fourier heat conduction or diffusion equation. This clearly means that the vorticity exponentially decays in the fluid as we go away from the oscillating body. In other words, the motion of the fluid caused by the oscillating body is rotational in the boundary layer which is in the neighborhood of the surface. The fluid motion is irrotational at large distances from the body (outside the boundary layer) thus giving potential flow in this region. Since, as mentioned, the decay of vorticity is exponential, the boundary between rotational and potential flow in the fluid is arbitrary, and must be defined according to a prescribed criterion such as the distance away from the body surface where the vorticity decays to e^{-1} of its maximum value. This is the depth of penetration δ of the rotational flow which is of the order $\sqrt{\nu/\omega}$.

Two important limiting cases now arise: The quantity δ may be either small or large compared to a characteristic length L associated with the body. We now investigate each case.

Case 1. $\delta \gg L$. This implies that $L^2\omega \ll \nu$ from (7.186). Suppose also that the Reynolds number is very small so that we can neglect the nonlinear part of the inertial term $(\mathbf{v} \cdot \mathbf{grad})\mathbf{v}$. If A is the amplitude of oscillations of the body, the vibrational velocity of the body is of the order A. The Reynolds number then becomes

$$R = \frac{\omega AL}{\nu}.$$

The combination of very small Reynolds number and very large depth of penetration yields the following inequalities:

$$L^2\omega \ll \nu, \qquad \frac{\omega AL}{\nu} \ll 1. \tag{7.190}$$

This means that we have the condition where the body oscillates with a very low frequency thus giving essentially a steady state, so that we can neglect the $\rho\mathbf{v}_t$ part of the inertial force in the Navier–Stokes equation.

Case 2. $\delta \ll L$. We now find the condition for which the nonlinear or convective term $(\mathbf{v} \cdot \mathbf{grad})\mathbf{v}$ can be neglected compared to the linear acceleration term $\mathbf{v}_t$, subject to this inequality. In this case we do not need the restriction that the Reynolds number must be small. To estimate the magnitude of the nonlinear term we observe that the operator $\mathbf{v} \cdot \mathbf{grad}$ denotes differentiation in the direction of $\mathbf{v}$. Near the surface of the body $\mathbf{v}$ is nearly tangent to the surface. In this direction the velocity changes appreciably only over distances of the order of the characteristic length L of the body. We therefore get the following expression for the order of magnitude of the nonlinear term

$$(\mathbf{v} \cdot \mathbf{grad})\mathbf{v} \sim \frac{v^2}{L} \sim \frac{A^2\omega^2}{L},$$

since the velocity is of the order $A\omega$. On the other hand, $\mathbf{v}_t$ is of the order $A\omega^2$. We now attempt to satisfy the inequality

$$(\mathbf{v} \cdot \mathbf{grad})\mathbf{v} \ll \mathbf{v}_t,$$

in order to be able to neglect the nonlinear term. Comparing the order of magnitude of the nonlinear term with the linear term, we see that this inequality is satisfied if $A \ll L$. This means that for the case of small depth of penetration compared to the characteristic length of the body, our condition for neglecting the nonlinear term is that the amplitude of the oscillating body must be very small compared to the characteristic body length. Therefore this case yields the following conditions for neglecting the nonlinear term:

$$L^2\omega \gg \nu, \qquad A \ll L. \tag{7.191}$$

We are now in a position to discuss the flow field around an oscillating body of arbitrary shape for the case of small depth of penetration which leads to condition (7.191). The flow is rotational only in a thin layer near the surface. Outside this layer the flow is irrotational as well as incompressible, meaning that **curl** $\mathbf{v}=0$ and **div** $\mathbf{v}=0$, so that

$$\nabla^2\mathbf{v}=0. \tag{7.192}$$

This means potential flow exists so that the fluid is ideal or inviscid. It is clear that a potential function $\varphi(x, t)$ exists such that $\mathbf{v}=\mathbf{grad}\ \varphi$. It follows that the Navier–Stokes equation reduces to the linearized Euler equation

$$\rho\mathbf{v}_t=-\mathbf{grad}\ p, \tag{7.193}$$

where ρ is constant. Since the layer of rotational flow is thin, in solving (7.192) to get the flow outside this layer we need only take as boundary condition $\mathbf{v}=0$ on the surface of the body, where $\mathbf{v}$ is the fluid velocity relative to the body. Clearly this is the boundary condition of no-slip flow on the surface. However, we have a problem: Laplace's equation for $\mathbf{v}$ (7.192) and Euler's equation (7.193) cannot be solved for this boundary condition. The best we can do is require that the normal component of the fluid velocity be zero on the boundary. If we were to solve for $\mathbf{v}(\mathbf{x}, t)$ outside this layer by attempting to satisfy the boundary condition $\mathbf{v}\cdot\boldsymbol{\nu}=0$, where $\boldsymbol{\nu}$ is the normal to the surface, we would find that the tangential component of $\mathbf{v}$ must change rapidly as we go in the normal direction toward the edge of the layer. We now determine the nature of the variation of this component as we go through the layer from the surface to the edge. The reason for doing this is to determine the tangential component of $\mathbf{v}$ at the edge of the layer as a boundary condition for the potential flow outside the layer. (The thickness of the layer will be defined below.) We start by approximating a small portion or element of the surface of the body by a planar surface, so that we can use the results of the previous investigations of the flow of a viscous fluid over an oscillating plate. We construct a rectangular coordinate system whose origin is at the center of the planar element with the x axis directed normal to the surface into the fluid, and the y axis directed parallel to the tangential component of the velocity of the body. The fluid velocity relative to the body is $\mathbf{v}=(0, v, 0)$ where $v(x, t)$ is the tangential component of $\mathbf{v}$. The boundary condition at the surface is $v(0, t)=0$. In the layer $v(x, t)$ behaves as

$$v(x, t)=v_0\exp(-i\omega t)\left[1-\exp\left(-(1+i)x\sqrt{\frac{\omega}{2\nu}}\right)\right], \tag{7.194}$$

where (7.185) was adapted to satisfy the boundary condition. We now define the edge of the layer to be located at $x=\delta$, which means that the depth of penetration is interpreted as the thickness of the layer. Then from (7.194) we get $v(\delta, t)$ as the tangential component of $\mathbf{v}$ at the edge of the layer. The required boundary condition is obtained by setting $v(\delta, t)=V(t)$, where $V(t)$ is a prescribed function of t given by (7.179). We then solve for v_0 in (7.194),

$$v_0=\frac{V_0}{[1-\exp(-(1+i)\delta\sqrt{\omega/2\nu})]}. \tag{7.195}$$

7.10. Potential Flow†

We mentioned above that the flow in the region outside the rotational layer is irrotational incompressible so that the fluid velocity is equal to the gradient of a potential function and also satisfies Laplace's equation. This important type of flow is worth investigating for its own sake.

We consider two-dimensional flow in the (x, y) plane so that the powerful method of complex variables can be applied. Let the fluid velocity $\mathbf{v}(x, y, t)=(u, v)$. The potential function φ then satisfies the following:

$$u=\varphi_x, \qquad v=\varphi_y. \tag{7.196}$$

Since the fluid is incompressible, the continuity equation $\mathbf{div}\,\mathbf{v}=0$ becomes

$$u_x+v_y=0. \tag{7.197}$$

Equations (7.196) and (7.197) imply that φ satisfy the two-dimensional Laplace equation

$$\varphi_{xx}+\varphi_{yy}=0. \tag{7.198}$$

Since the flow is irrotational, $\mathbf{curl}\,\mathbf{v}=0$, which becomes

$$v_x-u_y=0. \tag{7.199}$$

The continuity equation (7.197), combined with the condition of irrotationality (7.199), yields

$$u_{xx}+u_{yy}=0, \qquad v_{xx}+v_{yy}=0, \tag{7.200}$$

meaning that each component of $\mathbf{v}$ satisfies Laplace's equation in two dimensions (u and v are harmonic functions). Euler's equations then become

$$\rho u_t=-p_x, \qquad \rho v_t=-p_y. \tag{7.201}$$

† Also refer to Chapter 5, Section 5.2 on potential flow.

Differentiating the first equation with respect to x and the second with respect to y, adding, and using the continuity equation, yields $p_{xx}+p_{yy}=0$; the pressure also satisfies Laplace's equation.

The Stream Function

The continuity equation (7.197) implies that a function $\psi(x, y, t)$ exists called the *stream function* which has the following properties:

$$u=\psi_y, \qquad v=-\psi_x. \tag{7.202}$$

From equation (7.202) we see that the continuity equation is satisfied identically. We also see that in order for (7.202) to satisfy the irrotationality condition (7.199) ψ must also satisfy the two-dimensional Laplace equation or $\psi_{xx}+\psi_{yy}=0$.

To give an interpretation for ψ and to show that ψ indeed deserves the name of stream function, we consider the case of two-dimensional steady state flow. Suppose we have solved for the potential and have thereby determined $\varphi=\varphi(x, y)$. Consider a family of curves in the (x, y) plane for which $\varphi=\varphi_i$, where along each curve φ_i is a given constant (in general, different for each curve). Each curve is called an *equipotential curve.* In three dimensions such a family of curves is generalized to a family of *equipotential surfaces.* Now consider a family of curves in two dimensions such that $\psi(x, y)=\psi_j$, where ψ_j is a given constant along each curve. It follows that each point in the (x, y) plane has associated with it a given value of (φ_i, ψ_j) $(i, j=1, 2, \ldots)$. Now $d\varphi=0$ on the curve $\varphi=\varphi_i$ and $d\psi=0$ on the curve $\psi=\psi_j$. Expanding these differentials gives

$$\begin{aligned} d\varphi &= \varphi_x\,dx+\varphi_y\,dy=u\,dx+v\,dy=0 \quad \text{on} \quad \varphi=\varphi_i, \\ d\psi &= \psi_x\,dx+\psi_y\,dy=-v\,dx+u\,dy=0 \quad \text{on} \quad \psi=\psi_j, \end{aligned} \tag{7.203}$$

upon using (7.196) and (7.202). The second equation of (7.203) yields

$$\frac{dy}{dx}=\frac{v}{u} \qquad \text{for} \quad \psi=\psi_j=\text{const.} \tag{7.204}$$

Now consider a fluid particle trajectory in the (x, y) plane. The tangent to the trajectory at each point on the curve is given by $dy/dx=v/u$, since $dy=v\,dt$ and $dx=u\,dt$. Therefore (7.204) tells us that a curve of constant ψ has everywhere a slope dy/dx equal to the tangent of $\mathbf{v}$. It is for this reason that ψ is called the stream function. That curve along which the stream function is constant is called a *streamline* because the curve gives the direction of $\mathbf{v}$. For example, in *laminar flow* of a viscous fluid the family of streamlines or family of curves $\psi=\psi_j$ $(j=1, 2, \ldots)$ are nonintersecting or parallel, while in *turbulent flow* the streamlines are intersecting. Equation (7.204) is the differential equation for the streamlines. Coming back to

(7.203), the first equation yields

$$\frac{dy}{dx} = -\frac{u}{v} \quad \text{for} \quad \varphi = \varphi_i = \text{const.} \tag{7.205}$$

Equation (7.205) is the differential equation for the family of equipotential curves. Comparing (7.204) and (7.205) reveals the fact that the equipotential curves and the streamlines form a system of *orthogonal trajectories* (a curve of constant φ cuts a curve of constant ψ orthogonally).

If we draw a curve between two points A and B in the (x, y) plane, we assert that the mass flux of fluid Q flowing across this curve is given by the difference in values of the stream function at these points, regardless of the shape of the curve (Q only depends on the endpoints). This assertion is easily proved as follows: Let $\boldsymbol{\nu}$ be the unit normal to the curve at any point on the curve. Then the normal component of $\mathbf{v}$ is $\mathbf{v} \cdot \boldsymbol{\nu}$, and we have

$$Q = \rho \int_A^B \mathbf{v} \cdot \boldsymbol{\nu}\, d\mathbf{S} = \rho \int_A^B (-v\,dx + u\,dy) = \rho \int_A^B d\psi = \rho(\psi_B - \psi_A). \tag{7.206}$$

We may easily derive the fluid particle velocity $\mathbf{v}$ from the stream function ψ. Let $AB = dS$ be a differential arc dS of a curve. We resolve $\mathbf{v}$ into components along and perpendicular to dS. The component along dS contributes nothing to the flux Q across the curve dS. But the component normal to dS $v_n = Q/dS = \psi_B - \psi_A = d\psi/dS$. In rectangular coordinates, by considering infinitesimal increments dx, dy, we see that u, v satisfy (7.202). In polar coordinates (r, θ) we have

$$u = -\left(\frac{1}{r}\right)\psi_\theta, \qquad v = \psi_r, \tag{7.207}$$

where the components u, v are the components of $\mathbf{v}$ in the direction of increasing r and increasing θ, respectively. Instead of giving some examples of two-dimensional potential flow, we reserve this for the following section on complex variables.

Complex Variables

We now digress to give a brief description of those elements of complex variable theory (called function theory) needed for the investigation of two-dimensional potential flow. There are many excellent works on function theory. We recommend [7] to the interested reader who wants to get a further background in complex variable theory.

We introduce the complex z plane where $z = x + iy$. Let $w = f(z)$ be an *analytic function* of z. By this we mean that the derivative df/dz exists in the region $\mathcal{R}$ of analyticity, and has a unique value at each point in $\mathcal{R}$ independent of the direction of differentiation. This means that we shall have the same value for df/dz if we differentiate, for example, with respect to x (keeping y constant) or differentiate with respect to y (keeping x constant). In defining z we see that x is the real part and y is the imaginary part of z. Therefore, if we write $w = U + iV$, where U and V are differentiable functions of x and y, U is the real part and V is the imaginary part of w. As an example: If $f(z) = 1/z$ then $df/dz = -1/z^2$ except for the *singularity* at $z = 0$ where f is infinite, meaning that f is analytic everywhere except at the origin $z = 0$. The formula $w = f(z)$ means that we have a mapping from the z to the w plane. Examples abound: If $f = \sin z$, then $U = \sin x \cosh y$, $V = \cos x \sinh y$. If $f = \cos z$, then $U = \cos x \cosh y$, $V = -\sin x \sinh y$. If $f = z^n$, then $U = r^n \cos n\theta$, $V = r^n \sin n\theta$ (where $z = re^{i\theta}$). If $f = \log z$, then $U = \log z$, $V = \theta$. If $f = 1/z$, then $U = r^{-1} \cos \theta$, $V = -r^{-1} \sin \theta$.

We now capitalize on the statement that df/dz is independent of the direction of differentiation by writing

$$\frac{dw}{dz} = \frac{df}{dz} = U_x + iV_x = -i(U_y + iV_y),$$

since $d/dz = \partial/\partial x = -i\partial/\partial y$. Equating real and imaginary parts of the above equation yields

$$U_x = V_y, \qquad U_y = -V_x. \tag{7.208}$$

The system (7.208) is called the *Cauchy-Riemann equations*. They are fundamental equations in function theory; they represent the conditions for $f(z)$ to be an analytic function of z. From (7.208) we can easily derive the fact that both U and V are harmonic functions of x, y, meaning that they both satisfy Laplace's equation (in this case, obviously in two dimensions).

We now come to the physical interpretation of $f(z)$ by identifying U with the potential $\varphi(x, y)$ and V with the stream function $\psi(x, y)$. This gives

$$w = f(z) = \varphi + i\psi. \tag{7.209}$$

With this definition of w we see that the Cauchy-Riemann equations (7.208) are compatible with the definitions of φ and ψ given by (7.196) and (7.202). We may now go further and get the velocity components u, v in terms of df/dz. From (7.209), the Cauchy-Riemann equations, and (7.196) and (7.202), we have

$$\frac{dw}{dz} = \frac{df}{dz} = \varphi_x + i\psi_x = u - iv = |\mathbf{v}| \exp(-i\theta), \qquad \mathbf{v} \cdot \mathbf{v} = u^2 + v^2, \qquad \tan \theta = \frac{v}{u}.$$

If we define the complex conjugate of dw/dz by $\overline{dw/dz}$ (replace i by $-i$) we obtain

$$\overline{\frac{dw}{dz}} = u + iv = |\mathbf{v}| \exp(i\theta), \qquad \tan \theta = \frac{v}{u}. \tag{7.210}$$

w is called the *complex potential* and $\overline{dw/dz}$ is called the *complex velocity*. Equation (7.210) tells us that the modulus of the complex velocity is equal to the magnitude of $\mathbf{v}$ and the argument gives the angle between $\mathbf{v}$ and the x axis.

For flow past a solid in two dimensions, the flow must be tangent to the contour of the surface (we have inviscid flow), so that the contour must be a streamline which we may take as $\psi(x, y) = \psi_j = 0$. Then the flow past a given contour reduces to that of determining an analytic function $f(z)$ that has real values on the contour.

EXAMPLE. Let $f(z) = z^2$. Then $U = x^2 - y^2$, $V = 2xy$, and we obtain

$$w = \varphi + i\psi = x^2 - y^2 + 2ixy \qquad \text{or} \qquad \varphi = x^2 - y^2, \qquad \psi = 2xy,$$

from which we get

$$u = 2x, \qquad v = -2y.$$

The streamlines are rectangular hyperbolas $xy = \psi_j$ having the x, y axis as asymptotes, and the equipotential curves are $x^2 - y^2 = \varphi_i$ are also hyperbolas which are orthogonal to the streamlines. Figure 7.16 shows the streamlines and equipotential lines in the first quadrant of the (x, y) plane. This figure shows that the selection of z^2 for the complex potential has the physical significance of giving the two-dimensional potential flow field around a rectangular corner whose vertex is the origin. If we look at a rectangular drop of fluid contained in $ABCD$ in the figure, we see that u is equal to one constant on AD and another constant on BC. Similarly, v is a constant on AB and a constant on DC. This is seen from the solution for $\mathbf{v}$. Hence

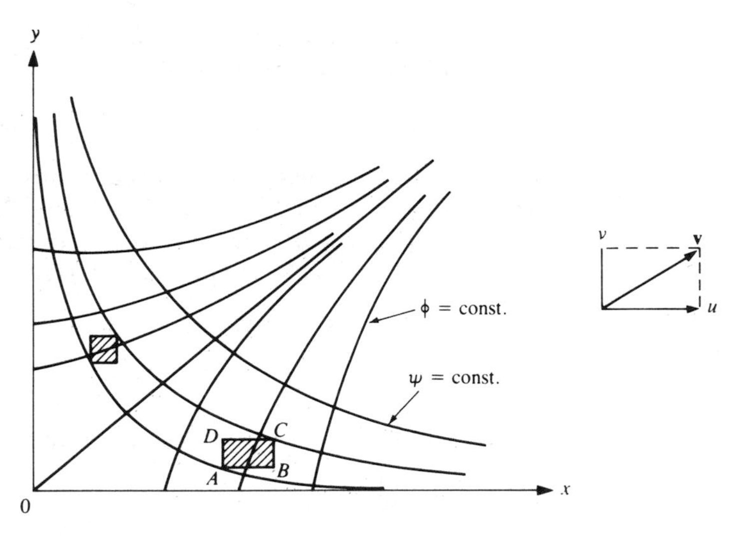

Fig. 7.16. Streamlines and equipotential curves as orthogonal hyperbolas.

the rectangle $ABCD$ remains a rectangle as we go from one to another set of streamlines. Also, the area of the rectangle remains constant (according to the continuity equation), although the shape changes.

We do not intend to go into problems of inviscid flow in any further detail. Lamb [25] gives a detailed discussion of such problems and also problems in three-dimensional flow where complex variable theory cannot be used. Davis [13] gives a discussion of the elements of complex variable theory in the setting of continuum mechanics.

CHAPTER 8

Wave Propagation in Elastic Media

Introduction

In previous chapters we investigated the properties of waves propagating in water and in inviscid and viscous fluids. It was pointed out that the conservation law that distinguishes one medium from another is the energy equation. It is the conservation law that contains the appropriate equation of state or constitutive equation which defines the medium. For example, an adiabatic equation of state was used to define a fluid such as air and a different equation of state for water. The adiabatic condition cannot be used across a shock wave since we must allow for a jump in entropy, etc. It was further pointed out in previous chapters that the approach used in mathematically describing the conservation laws was the Euler representation. This representation is more useful in dealing with large particle motions of the fluid. It was also stated that the conservation laws contain the fundamental physics of a given situation in the sense that from these conservation equations, which are called the field equations, we can obtain the velocity, pressure fields, etc., which give the wave properties. Since the field equations were couched in the Eulerian coordinates, the various fields that were derived from their solutions were also expressed as functions of these Eulerian coordinates. In principle, we can map back into the Lagrangian coordinates and thereby obtain the particle trajectories. In this chapter we shall continue our investigations of wave propagation by considering an elastic solid. It is more useful to express the conservation laws in the Lagrange representation because of the relative small particle motion and somewhat rigid boundaries. Since the phenomenon of wave propagation in solids has a long history which expresses an important aspect of the development of scientific progress, we shall open this chapter with a brief historical survey of this field.

Historical Introduction to Wave Propagation

The history of wave propagation in elastic solids is a fascinating chapter in the history of scientific thought. We can only give a brief outline here. Our information is based essentially on the historical introduction given by A.E.H. Love [28] with some added comments on some modern developments. The first mathematical physicist to study the nature of the resistance of solids was Galileo. Since he was not in possession of any law connecting deformation with the forces that produced them, he treated solids as inelastic bodies. Nevertheless, his investigations set the direction for further investigators. In 1660 Robert Hooke provided the ncessary foundation for the theory of his fundamental law relating stress to strain for small amplitude deformations. This was one of the landmarks in the history of elasticity. Hooke in England as well as Mariotte in France investigated experimentally what we now call the "stress-strain relations". Coulomb, in studying the torsional rigidity of fibers was the first to investigate shear strain which he considered in connection with rupture. Thomas Young defined the modulus of elasticity and was the first to consider shear as an elastic strain. He defined the "modulus of elasticity" as "a column of the same substance capable of producing a pressure on its base which is the weight causing a certain degree of compression, as the length of the substance is to the diminution of its length." This quote was taken from "The Miscellaneous Works of Dr. Young" by Kelland (1845), as mentioned by Love. However, the theory of propagation of elastic waves essentially starts with the investigations of the vibrations of solid bodies. Euler and Daniel Bernoulli obtained the differential equations of the lateral vibrations of bars. They determined the normal modes of vibration and the period or secular equations corresponding to different sets of boundary conditions. E.F.F. Chladni in Germany investigated these vibrational modes experimentally, as well as the torsional and longitudinal vibrations of bars. Navier was the first to investigate the equations of equilibrium and the vibrations of elastic solids in a systematic way. This was documented in the Memoirs of the Paris Academy of Sciences in 1827. He started from Newton's concept of material bodies as being made up of "molecules" possessing finite size and shape acting upon each other by central forces. Navier regarded these Newtonian "molecules" as material points, and assumed that the elastic reactions arise from variations in the intermolecular forces resulting from the change in molecular configuration caused by a deformation. His method consisted of deriving the equations of motion for a given molecule under the action of the forces causing displacement. The material is assumed to be isotropic and the equations of equilibrium and vibration contain a single material parameter of the same nature as Young's modulus. He then formed an expression for the work done in a small relative displacement by all the forces acting upon the molecule by all the other molecules. By an application of the calculus of variations he deduced these differential equations and also the boundary conditions that hold at the surface of the body. This memoir is very important for giving us a first generalization of its kind. However, it was criticized for not deriving the forces acting on a molecule in a proper way. His expressions involve triple summations which he replaced by triple integrations; and the validity of this procedure was suspected by some other investigators.

A new direction of activity took place in connection with wave propagation phenomena vis-à-vis the propagation of light waves. Fresnel in France announced certain conclusions regarding interference patterns of polarized light. He showed that this interference or diffraction phenomenon could be explained only by the hypothesis that light involves transverse vibrations. He demonstrated how a medium consisting of "molecules" connected by central forces (à la Newton) could execute such transverse vibrations, and thereby transmit transverse waves which would thus exhibit interference or diffraction patterns. Young in England also exhibited and explained these diffraction patterns in terms of the wave nature of light. Before the time of Young and Fresnel, neither the supporters nor the opponents of this undulatory theory of light appear to have conceived of light waves being transverse and thus they could not account for diffraction phenomena. Note that this concept of light involving molecular motion, which started with Newton and then was used by Fresnel to show light as a transverse wave, is superseded by developments in modern physics in which the concept of the *photon* was introduced. The photon has zero mass and travels with the speed of light (the greatest speed in the universe in a vacuum). The physics of Young, Fresnel, and the other investigators in this pre-quantum mechanical era are not to be taken seriously in light of modern physics, which introduced the quantum mechanical intepretation of light as exhibiting both wavelike and particlelike properties. This *duality nature of light* could not be conceived by these nineteenth century workers. However, what lasts is not their physics but their profound mathematical insight into the properties of waves. Their heritage was in giving us an extensive mathematical arsenal of methods which are of inestimable value to the modern investigators in quantum mechanics.

Now two great French mathematical physicists come on the scene; Cauchy and Poisson. Cauchy's pure mathematics was of the highest order. Among other things he developed the field of complex variables or function theory, some of which we used previously in inviscid two-dimensional flow. But, in addition, his excursions into the realm of mathematical physics led him to discover most of the elements of the theory of elasticity by 1822. He introduced the notion of stress at a point which is determined by the tractions per unit cross-section area across all plane elements through the point. He showed that the stress is expressible by means of six components (which we now call a symmetric tensor), and he introduced the "principle planes of stress". He also expressed the state of strain near a point in terms of the six components of the strain (the symmetric strain tensor). He then determined the equations of motion by which the stress components are connected with the forces distributed through the material. By means of the stress–strain relations (generalization of Hooke's law) he then eliminated the stress components from the equations of motion and thereby derived the equations of motion in terms of the displacements. He assumed the material to be isotropic and linear in the sense that the deformations are small enough to allow for the hypothesis of a linear relation between the stress and the strain tensors. His equations of motion are now used by people investigating the field of wave propagation in an isotropic linearly elastic medium (the so-called *infinitesimal theory of elasticity*). At a later date, he extended his theory to include crystal structure, wherein he made use of the hypothesis of material points between which there are certain attractive or repulsive forces. Cauchy applied his theory of elasticity to the propagation of light waves in crystals as well as isotropic elastic solids.

In Poisson's investigations, he first obtained the equations of equilibrium in terms of stress components. He then estimated the traction across any plane resulting from the "intermolecular forces" that he introduced, thus obtaining expressions for the stresses in terms of the strains. These involved summations over all "molecules" within a certain sphere of influence about a given molecule. The equations of motion that Poisson thereby obtained were identical to those of Navier. He then used his theory of elasticity to investigate the propagation of waves through an isotropic elastic medium and found two types of waves which, at great distances from the source, were practically "longitudinal" and "transverse".

Green sought a new foundation for the theory of elasticity developed by Cauchy and Poisson, as he was dissatisfied with the hypotheses they used. He appealed to Lagrange's *Mecanique Analytique* wherein he used Lagrange's method of the calculus of variations coupled with the principle of conservation of energy to minimize the potential energy for a system in equilibrium. It is worthwhile quoting him: "In whatever way the elements of any material system may act upon each other, if all the internal forces exerted be multiplied by the elements of their respective directions, the total sum for any assigned portion of the mass will always be the exact differential of some function. (It appears that he is writing about the virtual work done on a body; the exact differential being the work done by the potential function for the conservative system.) But this function is known, we can immediately apply the general method given in the *Mecanique Analytique* (the variational method of Lagrange), and which appears to be more especially applicable to problems that relate to the motions of systems composed of an immense number of particles mutually acting on each other. One of the advantages of this method, of great importance, is that we are necessarily led by the mere process of the calculation, and with little care on our part, to all the equations and conditions which are requisite and sufficient for the complete solution of any problem to which it may be applied (an extravagant statement!)" Green expanded this potential, which we now know as the *strain-energy function*, in terms of the components of the strain tensor and deduced the equations of elasticity, which contained 21 constants in the general case. In the case of isotropy it reduced to two constants, and the equations are the same as those of Cauchy. Lord Kelvin based the argument for the existence of Green's strain energy function on the first and second laws of thermodynamics.

At this stage in the historical development of wave propagation in solids (the latter half of the nineteenth century) it appears that the great Irish mathematical physicist Sir W.R. Hamilton should step into the picture. However, Love makes no reference to Hamilton in his historical introduction although in Chapter VII he does deduce the equations of motion from the *Hamilton principle.* The foundations for Hamilton's principle lay in the principle of virtual work of D'Alembert and the equations of motion of Lagrange (both of whom were also not mentioned by Love in his historical introduction). Hamilton's principle was based on the calculus of variations. He determined the necessary condition for the minimization of the functional which is the time integral of the difference between the kinetic and potential energies. The equations of motion were obtained from this principle. Instead of using the Lagrange equations of motion, which is a second-order partial differential equation (PDE) involving the generalized coordinates and velocities, Hamilton derived a set of two first-order PDEs involving the generalized coordinates and momentum. The necessary constraints on a dynamical system were easily incorporated in Hamilton's equations of motion by the introduction of cyclic coordinates by use of canonical transformations. Hamilton and Jacobi of Germany developed a theory of PDEs which involve Hamilton's principle function. The Hamilton-Jacobi theory is fundamental in a study of the relationship between geometric optics and the propagation of light waves.

The theory of free vibrations of elastic bodies requires the integration of the equations of vibratory motion taking into account the prescribed boundary conditions of stress or displacement. Poisson gave the solution of the problem of free radial vibrations of a solid sphere, and Clebsch founded the general theory on the model of Poisson's solution. This theory applied the notion of principal coordinates and normal functions to systems with an infinite number of degrees of freedom. The problem of determining the properties of propagating waves in an elastic solid requires a different type of investigation from those concerned with normal modes of vibration. In the case of an isotropic medium Poisson adapted methods which involve a synthesis of solutions of simple harmonic motion, and he obtained a solution expressing the displacement at any time in terms of the initial distribution of displacement and velocity. Later on Stokes, using a different method, showed that Poisson's two waves are waves of irrotational dilatation and waves of equivoluminal distortion (rotational waves). Cauchy and Green discussed the propagation of plane waves through a crystalline medium, and obtained equations for the wave speed in terms of the direction of the normal to the wave front. In general, the wave surface has three sheets; when the medium is isotropic all the sheets are spheres, and two of them are coincident. Christoffel discussed the advance through the medium of a surface of discontinuity. At any instant, the surface separates two portions of the medium in which the displacements are expressed by different formulas. He showed that the surface moves normal to itself with a velocity which is determined, at any point, by the direction of the normal to the surface, according to the same law that holds for plane waves propagated in that direction. Besides the irrotational and rotational waves which can be propagated through an isotropic elastic solid, Lord Rayleigh investigated a third type of wave which can be propagated over the surface. The velocity of this surface wave is less than either of the other two.

The modern theory of wave propagation in solids, which is not contained in Love's historical introduction, is intimately connected with the theory of hyperbolic PDEs. Essentially, modern theory of wave propagation relies quite heavily on the classical analysis of Riemann where the introduced the theory of characteristics to the solution of wave propagation problems in fluids in his famous treatise on shock waves. The theory of characteristics is the cornerstone in the nonlinear theory of hyperbolic PDEs. Cauchy also investigated the so-called *Cauchy problem*, which is the problem of obtaining solutions of the wave equation given the initial displacement and particle velocity (the *initial value problem*). The Cauchy problem was put in a modern setting by the great French mathematical physicist Hadamard. His book entitled *Lectures on Cauchy's Problem*, Dover, 1952, is recommended to the interested reader of sufficient mathematical background, since it uses modern mathematical methods which are rather abstract. It has a wealth of material on the researches of Riemann, Cauchy, Volterra, Kirchhoff, etc., put in a modern setting. It does not contribute directly to the field of wave propagation in elastic media.

During World War II and in the late 1940s and 1950s there was much activity on the subject of stress wave propagation in solids in connection with impact loading problems. This subject had its impetus in investigations into the field of armor penetration with respect to designing tougher and more durable armor and in developing armor-piercing weapons. Investigators of the highest caliber, such as J. von Neumann, Kistiakowsky, K.O. Friedrichs of New York University, T. von Kármán, and a host of others, developed the theory of shock wave propagation and detonation waves (especially von Neumann, Kistiakowsky, and Friedrichs), and stress wave propagation in solids due to impact loading which was spearheaded by von Kármán in the United States and independently by G.I. Taylor in England. The reader is referred to the book entitled *Stress Waves in Solids* by H. Kolsky, Dover, 1963. In this monograph Kolsky presents both a theoretical and experimental account of wave propagation in solids in elastic and dissipative media. He summarizes the experimental work carried on up to 1952. There was also much development in the field of plasticity and viscoelastic wave propagation in connection with the need for engineers to design plastics that could withstand high loading rates and other tough environmental conditions. Davis [13] gives an exposition of viscoelasticity and wave propagation in viscoelastic media.

A modern development in the field of PDEs of the hyperbolic type, which has been directly applied to wave propagation in elastic media, is the method of *spherical means* developed by Fritz John. This powerful method of solving the fourth-order PDE in three-dimensional space is described in *Methods of Mathematical Physics*, vol. II, p. 706.

During World War II and after there was an explosion in the field of numerical analysis, primarily in connection with shock and detonation wave propagation in solids where nonlinear effects must be taken into account. The method of characteristics played a key role in numerical methods which was spearheaded by the seminal work of von Neumann where, among other things, he laid down his famous criterion for the stability of shock waves. This was done with respect to gases but the methodology applies to shock wave propagation in solids. The book entitled *Supersonic Flow and Shock Waves* by Courant and Friedrichs, Interscience, 1948, is a classic in the field of shock wave propagation in gases with a chapter on nonlinear wave propagation in solids. There is not much in this work on numerical methods. This discipline is treated in detail in works such as *Finite Difference Methods for Partial Differential Equations* by G.E. Forsythe and W.R. Wasow, Wiley, 1960 and *Difference Methods for Initial Value Problems* by R.D. Richtmyer, Interscience, 1957. Davis [12] also discusses finite difference methods from the point of view of continuum dynamics. In the 1940s analogue computers were in their heyday. But with the development of digital computers, the analogue computers became passé because of the tremendous speed and versatility of the digital computers. von Neumann laid the foundations for these high-speed computers with his pioneering mathematical work on numerical methods and the theory of automata. One of the first digital computers was the ENIAC at the Ballistics Research Laboratory, Aberdeen Proving Ground, Maryland. The Computing Facility at New York University (now called the Courant Institute of Mathematical Sciences), in the course of their work for the Atomic Energy Commission, did much work in the area of numerical methods of solution of the PDEs occurring both in gases and solids. Many other institutions and investigations were involved in this area, but this field takes on a life of its own and cannot be further developed here.

8.1. Fundamental Concepts of Elasticity

In this section we shall review the fundamental ideas of elasticity which are necessary for an understanding of wave propagation in an elastic medium. In the chapter on fluid mechanics we dealt with wave propagation phenomena in gases. Although we investigated nonlinear wave propagation by the use of characteristic theory so that we could not use the classical methods so fruitful in solving linear problems, still the propagating waves had not much variety especially in an inviscid fluid governed by an adiabatic equation of state. The interesting concept of jump discontinuities across shock waves extended the nature of solutions of the field equations governing shock waves to weak solutions. However, in general, the physics of wave propagation in gases did not give us such a variety of waves. Now when we come to the study of wave propagation in elastic solids we shall find a richer variety of waves such as irrotational, rotational surface waves, etc. To prepare the way for the study of this type of wave phenomena, we shall present, in a summary way, such concepts as deformation, strain, stress, and the stress–strain relations for an elastic body. The reader is referred to [28] for a more detailed account of this subject using an extended notation, to [35] for an approach using tensor notation, and to [13] for an account of this subject from the point of view of a combined matrix and tensor notation. We shall follow this latter method of approach here.

Deformation and Strain

An elastic solid is defined as a deformable solid continuum which suffers no energy loss when it goes back to its undeformed or equilibrium state. We now discuss the properties of a deformable continuum in order to develop the concepts of deformation and strain. The nature of deformation and strain is a purely geometric property of the medium and is independent of whether it is a solid liquid or gas. All we demand is that we have a *continuous medium.* We define such a medium (which we call a *continuum*) as one which contains a continuous distribution of matter in the sense that the molecular and crystalline structure of the material is neglected. This means that mathematically we can define a volume element of the medium as $dV = dx\,dy\,dz$ where the material contained in dV is continuous even though dV is a differential volume element. This concept of a continuum is based, in a sense, on an averaging process where we take advantage of the large number of molecules in a differential volume element to smear out the effects of individual molecules.

We describe a deformable continuum. To this end, we consider two states or configurations of the body (body and medium are used interchangeably):

(1) its initial configuration characterized by being in equilibrium or in its undeformed state;
(2) its deformed configuration or state.

By a *deformed state* of a body we mean the following: Given any two neighboring points P_1 and P_2 in the body in its initial or equilibrium state. Under a transformation to the deformed state $P_1 \to P_1'$ and $P_2 \to P_2'$ and the distance between the two undeformed points changes in the deformed state.

If $P_1'P_2' < P_1P_2$ then that part of the body undergoes a compression. If $P_1'P_2' > P_1P_2$ then it is in tension. We use the Lagrange representation whereby a point in the undeformed state is given by the coordinate system $\mathbf{a} = (a_1, a_2, a_3)$ which is the radius vector representing the Lagrange coordinates. In the deformed state that point goes to the position given by $\mathbf{x} = (x_1, x_2, x_3)$ which is the radius vector—the Eulerian coordinates. The coordinates of the above points are $P_1 : \mathbf{a}_1$, $P_2 : \mathbf{a}_2$ (undeformed state), $P_1' : \mathbf{x}_1$, $P_2' : \mathbf{x}_2$ (in the deformed state). The *displacement vector* $\mathbf{u} = (u_1, u_2, u_3)$ is defined by $\mathbf{u} = \mathbf{x} - \mathbf{a}$ so that $P_i' - P_i = \mathbf{u}_i$ $(i = 1, 2)$. Since we use the Lagrange representation $\mathbf{x} = \mathbf{x}(\mathbf{a}, t)$ and $\mathbf{u} = \mathbf{u}(\mathbf{a}, t)$. Now the undeformed line element $P_1P_2 = \mathbf{a}_2 - \mathbf{a}_1$ and the corresponding deformed line element $P_1'P_2' = \mathbf{x}_2 - \mathbf{x}_1 = \mathbf{u}_2 - \mathbf{u}_1 - (\mathbf{a}_2 - \mathbf{a}_1)$. Clearly if $\mathbf{u}_2 - \mathbf{u}_1 = 0$ then there is no deformation since $P_2'P_1' = P_2P_1$. If $\mathbf{u}_2 - \mathbf{u}_1 > 0$ we have tension, if $\mathbf{u}_2 - \mathbf{u}_1 < 0$ we have compression. We have thus defined deformation in terms of the behavior of the displacement vector $\mathbf{u}$.

One-Dimensional Case

Before we define the full three-dimensional strain matrix we first study the one-dimensional case. We take as our physical model a long thin bar or solid cylinder of deformable material of unit cross-sectional area. In its deformed state the bar is either in axial compression or tension, and the cross-sectional area is assumed not to change—this is in keeping with our assumption of a long thin bar which is the one-dimensional case. The bar is represented by a continuous filament of matter. The mathematical model is: The undeformed bar is represented by the a axis. The deformed state is represented by the x axis using a common origin. The a and x axes are coincident, both representing the axis of the bar. We define a particle by a continuous distribution of matter (filament) in the interval $[a, a + da]$ in the equilibrium state. Let dm_a be the mass of this particle. Let dV_a be the volume occupied by the particle. Then $dV_a = da$, where da is the length of the interval. Note that we have a unit cross-sectional area allowing for the dimensions of volume da to be L^3. The axial dimension of the bar is represented by the transformation $a \to x(a, t)$, where a is the Lagrange coordinate and t is time, since we are dealing with time-dependent deformations. The particle occupying the interval $[a, a + da]$ in the undeformed state then occupies the interval $[x, x + dx]$ in the deformed state in which the particle has an element of mass dm_x. Now by the conservation of mass $dm_x = dm_a$. The volume occupied by this particle in the deformed state is $dV_x = dx$ (no change of cross-sectional area). The displacement vector $\mathbf{u} = (u, 0, 0)$, where u is the component of displacement in the axial direction (which is the a or x axis since they are coincident). In the Lagrangian representation, u becomes

$$u(a, t) = x(a, t) - a. \tag{8.1}$$

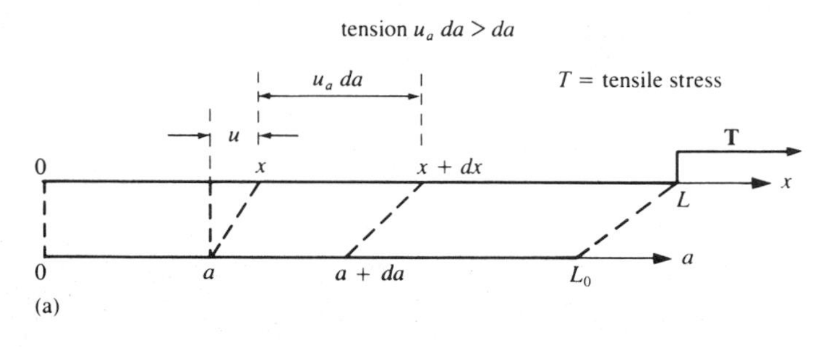

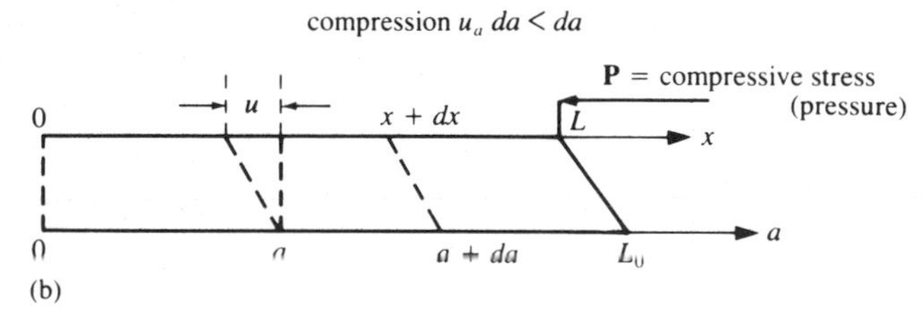

Fig. 8.1. One-dimensional strain: (a) tensile, (b) compressive.

Figure 8.1 shows how the interval along the a axis in the undeformed state is transformed to the corresponding interval along the x axis in the deformed state for tension and compression. The figure shows that the endpoints transform as follows [using (8.1)]:

$$a \to a + u, \qquad a + da \to a + da + u + du, \qquad du = u_a\,da.$$

We define the strain ε for the one-dimensional case as

$$\varepsilon = \frac{dV_x - dV_a}{dV_a} = \frac{dx - da}{da} = x_a - 1 = u_a. \tag{8.2}$$

We call $\varepsilon = u_a$ the *one-dimensional linear strain.* In obtaining (8.2) we differentiated with respect to a keeping t constant. Since the cross-sectional area is constant, (8.2) tells us that the one-dimensional strain is defined as the ratio of the difference in arc-length elements to the undeformed arc-length element. Clearly $\varepsilon = \varepsilon(a, t)$.

Let ρ_a be the density of the particle occupying the volume da and let ρ_x be the density of the *same particle* in the volume dx. The conservation of mass, as mentioned above, is $dm_a = dm_x$. Since $\rho = dm/dV$ for each state and since $dx = (1+\varepsilon)\,da$, we can represent the conservation of mass by the equation

$$\frac{\rho_x}{\rho_a} = R = \frac{1}{1+\varepsilon}. \tag{8.3}$$

R is called the *compression ratio.* It is described by Fig. 8.2 which shows a plot of the compression ratio versus strain. Note that ε asymptotically

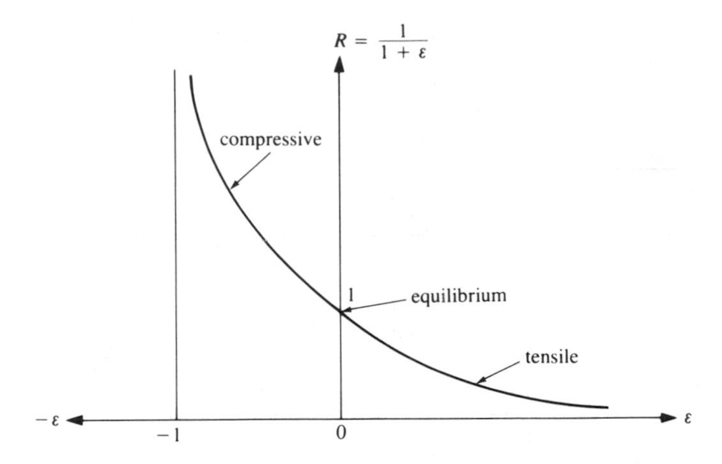

Fig. 8.2. Compressive rate versus strain.

approaches -1 as the strained volume approaches zero. This corresponds to infinite density of the compressed particle. In the tensile region there is an upper limit to the tensile strain beyond which the bar fractures. The following regions exist:

$$\varepsilon \begin{cases} >0 \text{ tensile strain} & dV_x > dV_a, \\ =0 \text{ equilibrium} & dV_x = dV_a, \\ <0 \text{ compressive strain} & dV_x < dV_a. \end{cases}$$

A more general definition of one-dimensional strain (which involves a nonlinear term) is given by the difference of the squares of the elements of the arc length divided by the square of the undeformed arc-length element. Specifically, we define by

$$\varepsilon = \frac{1}{2} \frac{(dx)^2 - (da)^2}{(da)^2}. \tag{8.4}$$

The reason for the factor $\frac{1}{2}$ will be clear when we work out the linear approximation. Using the definition of u given by (8.1), (8.4) becomes

$$\varepsilon = u_a + \tfrac{1}{2}u_a^2 = \varepsilon_L + \varepsilon_N, \tag{8.5}$$

where $\varepsilon_L = u_a$ is the linear part of the strain and $\varepsilon_N = \frac{1}{2}u_a^2$ is the nonlinear part. Turning to (8.4) the linear approximation becomes

$$(dx)^2 - (da)^2 = (dx - da)(dx + da) \sim 2da(dx - da).$$

This approximation is true if $(dx - da)/da \ll 1$. This is the justification for the expression given by (8.2) where ε is interpreted as ε_L.

EXAMPLE. Consider a bar of length L fixed at $a=x=0$ and pulled by an amount $\Delta L>0$ at the end $a=L$. We keep L constant. We have the following intervals:

$$0<a<L \quad \text{(equilibrium state)}, \qquad 0<x<L+\Delta L \quad \text{(deformed state)}.$$

Let the transformation $a\to x$ be

$$x=\left(1+\frac{\Delta L}{L}\right)a+\frac{ba^2}{2},$$

where b is a small positive constant such that $b\ll L$. Then the displacement is $u=a(\Delta L/L)+ba^2/2$ so that $u_a=\Delta L/L+ba$. This tells us that $\varepsilon_L=\Delta L/L$, which is the relative extension of the bar, since we neglect the term ba which is the product of two small terms (second order). Calculating u_a^2 gives the nonlinear part of the strain $\varepsilon_N=\frac{1}{2}(\Delta L/L)^2+ba$, where we have neglected the third-order term $ba(\Delta L/L)$. This tells us that ε_N has a term equal to $\frac{1}{2}\varepsilon_L^2$ plus ba which is also a second-order term.

Three-Dimensional Strain

For the equilibrium state let $\mathbf{a}=(a, b, c)$ and for the deformed state let $\mathbf{x}=(x, y, z)$. Consider a particle of the medium occupying a volume element $dV_a=da\,db\,dc$ in the initial state. Clearly the particle is composed of a continuous distribution of matter. In the deformed state that same particle occupies the volume element $dV_x=dx\,dy\,dz$. For a fixed t let P: (a, b, c) be the coordinates of a point in dV_a and let P': (x, y, z) be the coordinates of a corresponding point in dV_x which is the position of P under a deformation given by the transformation $\mathbf{a}\to\mathbf{x}$ so that $x=x(a, b, c)$, $y=y(a, b, c)$, $z=z(a, b, c)$, where the components of $\mathbf{x}$ are differentiable functions of the components of $\mathbf{a}$ and similarly for the inverse mapping. The relation between dV_a and dV_x is given by

$$dV_x=\det(\mathbf{J})\,dV_a, \tag{8.6}$$

where $\det(\mathbf{J})$ is the determinant of $\mathbf{J}$ which is the Jacobian of the transformation $\mathbf{a}\to\mathbf{x}$ and is given by the following matrix:

$$\mathbf{J}=\frac{(x, y, z)}{(a, b, c)}=\begin{pmatrix} x_a & x_b & x_c \\ y_a & y_b & y_c \\ z_a & z_b & z_c \end{pmatrix} \tag{8.7}$$

(see [13, Chap. 1] for details). The principle of conservation of mass has the same form as the one-dimensional case: $\rho_x\,dV_x=\rho_a\,dV_a$, so that the compression ratio in three dimensions becomes

$$R=\frac{\rho_x}{\rho_a}=\frac{1}{\det(\mathbf{J})}. \tag{8.8}$$

To develop the three-dimensional strain matrix we consider a curve C_a embedded in the body in the initial state. Under the mapping $\mathbf{a} \rightarrow \mathbf{x}$, C_a maps into the curve C_x in the deformed state, where C_x is composed of the same particles as C_a. Using matrix notation, let the column matrix $d\mathbf{a}$ have components da, db, dc. Let $d\mathbf{a}^*$ be the transpose of the matrix $d\mathbf{a}$. We can therefore write the column matrix $d\mathbf{a}$ as $d\mathbf{a} = (da\, db\, dc)^*$. Similarly, the column matrix $d\mathbf{x} = (dx\, dy\, dz)^*$ which is the transpose of the row matrix $d\mathbf{x}^*$. Let $d\mathbf{s}_a$ be the element of arc length of C_a and let $d\mathbf{s}_x$ be the corresponding element of arc length of C_x under the transformation $d\mathbf{a}\; d\mathbf{x}$. This transformation is given by

$$d\mathbf{x} = \mathbf{J}\, d\mathbf{a}, \tag{8.9}$$

where $\mathbf{J}$ is given by (8.7). The magnitude of $d\mathbf{s}_a$ is the positive square root of

$$d\mathbf{s}_a \cdot d\mathbf{s}_a = (d\mathbf{s}_a)^2 = (d\mathbf{a})^*(d\mathbf{a}). \tag{8.10a}$$

Similarly, the magnitude of $d\mathbf{s}_x$ is the root of

$$d\mathbf{s}_x \cdot d\mathbf{s}_x = (d\mathbf{s}_x)^2 = (d\mathbf{x})^*(d\mathbf{x}). \tag{8.10b}$$

The transpose of (8.9) is $(d\mathbf{x})^* = (d\mathbf{a})^*\mathbf{J}^*$. Using this expression and (8.9) yields

$$(d\mathbf{s}_x)^2 = (d\mathbf{x})^*(d\mathbf{x}) = (d\mathbf{a})^*\mathbf{J}^*\mathbf{J}(d\mathbf{a}). \tag{8.11}$$

Suppose the transformation $\mathbf{a} \rightarrow \mathbf{x}$ has the property that for every curve C_a all arc lengths are unchanged in going to a corresponding C_x. This means $(d\mathbf{a})^*(d\mathbf{a}) = (d\mathbf{x})^*(d\mathbf{x})$. From this we deduce $\mathbf{J}^*\mathbf{J} = \mathbf{E}$, the 3×3 identity matrix. It follows that $\mathbf{J}$ is the rotation matrix, yielding a rigid body rotation (no deformation).

This special case suggests that we use the expression $\mathbf{J}^*\mathbf{J} - \mathbf{E}$ as a measure of strain. In fact, it is easily seen that the three-dimensional generalization of (8.4) is obtained from

$$\begin{aligned}(d\mathbf{x})^*(d\mathbf{x}) - (d\mathbf{a})^*(d\mathbf{a}) &= (d\mathbf{a})^*\mathbf{J}^*\mathbf{J}(d\mathbf{a}) - (d\mathbf{a})^*(d\mathbf{a}) \\ &= (d\mathbf{a})^*(\mathbf{J}^*\mathbf{J} - \mathbf{E})(d\mathbf{a}),\end{aligned}$$

from which we get the three-dimensional symmetric strain matrix

$$\varepsilon = \tfrac{1}{2}(\mathbf{J}^*\mathbf{J} - \mathbf{E}). \tag{8.12}$$

An easy extension of the one-dimensional case tells us that the linear approximation is given by

$$\frac{ds_x - ds_a}{ds_a} = \left(\frac{d\mathbf{a}}{ds_a}\right)^* \varepsilon_L \left(\frac{d\mathbf{a}}{ds_a}\right).$$

Strain Matrix as a Function of the Displacement Vector

The displacement vector $\mathbf{u} = (u, v, w)$ where u, v, w are the components of $\mathbf{u}$ in the a, b, c directions, respectively. We have $\mathbf{u} = \mathbf{x} - \mathbf{a}$ or $u = x - a$, $v = y - b$, $w = z - c$. Let $\mathbf{K} = (u, v, w)/(a, b, c)$ be the Jacobian of $\mathbf{u}$ with respect to $\mathbf{a}$. This means

$$\mathbf{K} = \begin{pmatrix} u_a & u_b & u_c \\ v_a & v_b & v_c \\ w_a & w_b & w_c \end{pmatrix}. \tag{8.13}$$

Since $x_a = u_a + 1$, $x_b = u_b$, etc., we have $\mathbf{J} = \mathbf{E} + \mathbf{K}$ so that $\mathbf{J}^*\mathbf{J} - \mathbf{E} = (\mathbf{E} + \mathbf{K}^*)(\mathbf{E} + \mathbf{K})$ and we obtain the following expression for the symmetric strain matrix

$$2\varepsilon = \mathbf{K} + \mathbf{K}^* + \mathbf{K}\mathbf{K}^* = 2\varepsilon_L + 2\varepsilon_N, \tag{8.14}$$

so that $\varepsilon_L = \frac{1}{2}(\mathbf{K} + \mathbf{K}^*)$ and $\varepsilon_N = \frac{1}{2}\mathbf{K}\mathbf{K}^*$, where the elements of ε_L are

$$\begin{aligned} &\varepsilon_{Laa} = u_a, && \varepsilon_{Lbb} = v_b, && \varepsilon_{Lcc} = w_c, \\ &\varepsilon_{Lab} = \tfrac{1}{2}(v_a + u_b), && \varepsilon_{Lbc} = \tfrac{1}{2}(w_b + v_c), && \varepsilon_{Lac} = \tfrac{1}{2}(u_c + w_a), \end{aligned} \tag{8.15a}$$

and the elements of ε_N are

$$\begin{aligned} \varepsilon_{Naa} &= \tfrac{1}{2}[(u_a)^2 + (v_a)^2 + (w_a)^2], \\ \varepsilon_{Nbb} &= \tfrac{1}{2}[(u_b)^2 + (v_b)^2 + (w_b)^2], \\ \varepsilon_{Ncc} &= \tfrac{1}{2}[(u_c)^2 + (v_c)^2 + (w_c)^2], \\ \varepsilon_{Nab} &= \tfrac{1}{2}[u_a u_b + v_a v_b + w_a w_b], \\ \varepsilon_{Nbc} &= \tfrac{1}{2}[u_b u_c + v_b v_c + w_b w_c], \\ \varepsilon_{Nac} &= \tfrac{1}{2}[u_a u_c + v_a v_c + w_a w_c]. \end{aligned} \tag{8.15b}$$

The diagonal elements of both ε_L and ε_N are the normal components of the strain while the off-diagonal elements are the components of the shear strain. [13] has a more complete discussion of strain including principal axes transformations, generalized and curvilinear coordinates, and many worked-out examples.

Stress Tensor

To grasp the physical meaning of the stress tensor we consider a body occupying a volume V enclosed by a surface S embedded in the medium. We define the vector $\mathbf{T}$ as the *stress vector.* $\mathbf{T}$ represents the surface force per unit area acting at a point P on the surface. Let dS be an element of surface area on S. The force $\mathbf{T}\,dS$ is defined as representing the action of the portion of the material outside V (on the positive or outside of S) on the negative side (on the material in V). Hence if a unit normal $\boldsymbol{\nu}$ is drawn

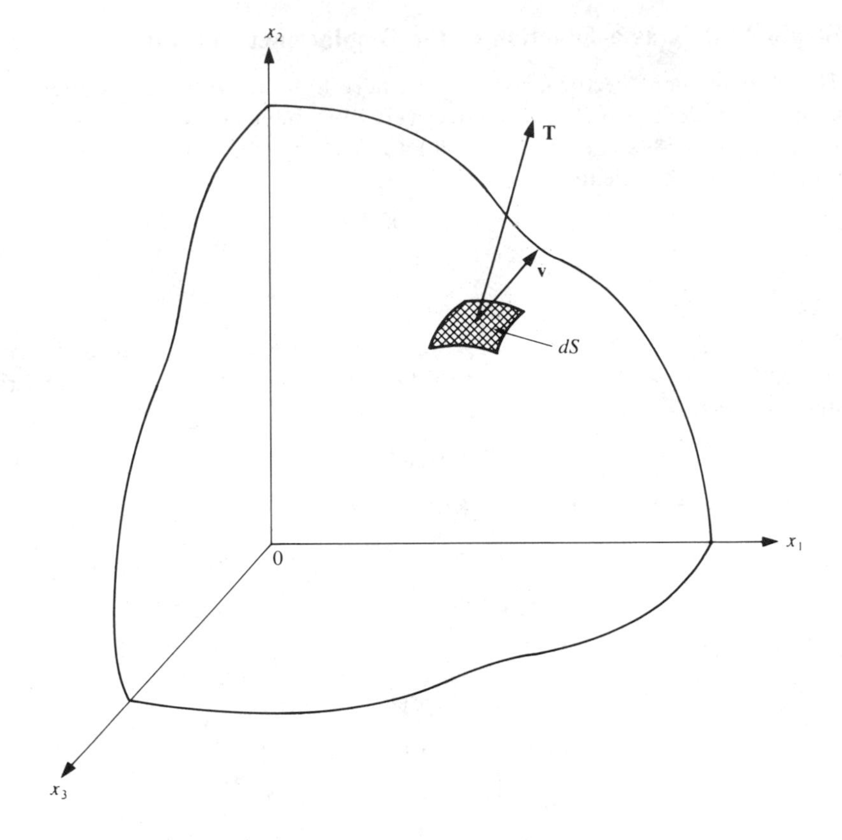

Fig. 8.3. Action of stress vector **T** on element of surface *ds*.

to the surface element dS at P, as shown in Fig. 8.3, then the action of the material lying on the negative side of the normal upon that on the positive side is $-\mathbf{T}\,dS$. It is clear that the surface forces in a solid are more complicated than those in an ideal fluid (where these forces are normal to the surface). In a solid the surface forces depend on two parameters: the orientation of the surface given by $\boldsymbol{\nu}$ and the direction of the surface stress vector. In order to display these two parameters we define the stress tensor acting on the surface by $\mathbf{T}^{\boldsymbol{\nu}}$, where the direction of the stress vector is given by $\mathbf{T}$ and the superscript $\boldsymbol{\nu}$ identifies the orientation of the surface element in terms of its normal.

Cauchy showed that the state of stress at any point in the body is *completely characterized* by the specification of the stress tensor or matrix. To demonstrate this stress principle of Cauchy we construct the tetrahedron PABC, as shown in Fig. 8.4, as follows: Let P be any point in the medium. We take a coordinate system defined by $\mathbf{x} = (x_1, x_2, x_3)$ with P as the origin. From P we draw a small vector of length h and direction $\boldsymbol{\nu}$. We now

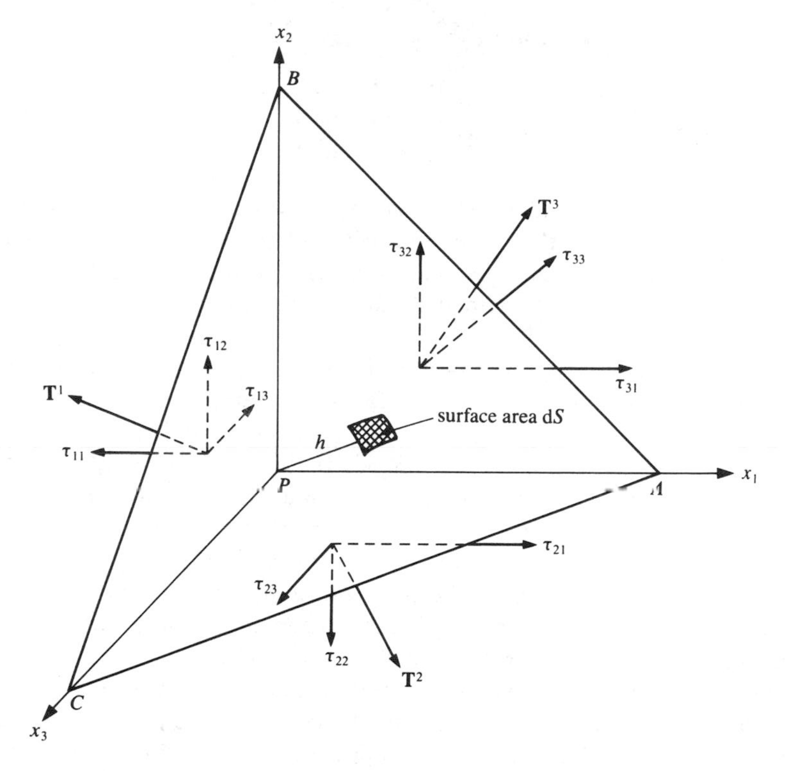

Fig. 8.4. Components of stress tensor τ_{ij} and stress vectors $\mathbf{T}^i$ acting on a tetrahedron.

construct the planar surface ABC such that this vector intersects the surface ABC at the point Q at which we erect a surface element dS. ABC is then a distance h away from P and $\boldsymbol{\nu}$ is normal to ABC. We have thus constructed the tetrahedron $PABC$. Let $\mathbf{T}^1$ be the stress vector acting on the planar face PBC (in the direction of $-x_1$). Let $\mathbf{T}^2$ be the stress vector acting on the face PAC (direction of $-x_2$), and let $\mathbf{T}^3$ be the stress vector acting on the face PAB (direction of $-x_3$). Clearly this means that $\mathbf{T}^i$ is the stress vector acting on the planar surface normal to the x_i axis. This is shown in the figure. Let $\mathbf{e}_i$ be the unit vector in the x_i direction. We then resolve $\mathbf{T}^i$ into its components along the coordinate axes.

$$\mathbf{T}^i = \mathbf{e}_j \tau_{ij} \quad \text{summed over } j, \qquad i, j = 1, 2, 3. \tag{8.16}$$

τ_{ij} is the ijth component of the stress tensor or the ijth element of the 3×3 matrix representing the stress tensor. Note that the subscript i gives the orientation of the surface element dS which is given by $d\mathbf{S} \cdot \mathbf{e}_i$. The subscript j identifies the direction of the ijth component of the stress vector. For example, τ_{23} is the component of the stress tensor in the x_3 direction acting on the element of surface area whose outward-drawn normal is in the x_2 direction.

The stress matrix is given by

$$\boldsymbol{\tau} = \begin{pmatrix} \tau_{11} & \tau_{12} & \tau_{13} \\ \tau_{21} & \tau_{22} & \tau_{23} \\ \tau_{31} & \tau_{32} & \tau_{33} \end{pmatrix}. \tag{8.17}$$

Like the strain matrix, the stress matrix is symmetric so that $\tau_{21} = \tau_{12}$, $\tau_{23} = \tau_{32}$, $\tau_{13} = \tau_{31}$, and there are six distinct elements of the stress matrix or stress tensor. The diagonal elements are the normal components of the stress and the off-diagonal elements represent the components of the shear stress.

To give a geometric interpretation of the elements of the stress matrix we construct a parallelepiped whose faces are parallel to the coordinate axes. Let $\mathbf{T}^i$ be the stress vector acting on a face of the parallelepiped normal to the x_i axis. This is shown in Fig. 8.5, where the components τ_{ij} are shown as well as the corresponding stress vectors. We use the following *standard convention* in regard to the signs of the scalars τ_{ij}: Draw an exterior normal to a given face of the parallelepiped. Then the components τ_{ij} of the stress tensor are reckoned positive if the corresponding components of force act in the direction of increasing x_1, x_2, x_3, when the normal has the same direction as the positive direction of the axis to which the face is normal. On the other hand, if the exterior normal to a given face is in the opposite direction to that of the positive coordinate axis, then the positive values of τ_{ij} are associated with forces directed opposite to the positive directions of the axes.

We now turn our attention to demonstrating Cauchy's principle which tells us that the stress can be calculated on any element of surface passing through the point P. This means we want a connection between the elements of the stress matrix and the stress vector $\mathbf{T}^\nu$. To this end, we again turn to the tetrahedron in Fig. 8.4. Let the magnitude of the area of face ABC be

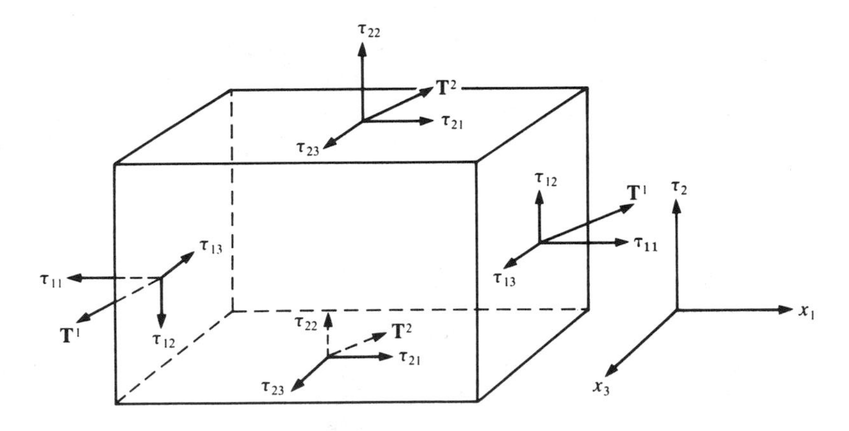

Fig. 8.5. Components of stress tensor and stress vectors on parallelepiped.

S so that $\mathbf{S} = \boldsymbol{\nu} S$. Then the face normal to the x_i axis has the area $S_i = l_i S$, where l_i is the direction cosine of x_i with respect to $\boldsymbol{\nu}$. Since we assume that no external forces or moments act on the tetrahedron, it is in equilibrium, so that the resultant force and moment on the material within the tetrahedron are zero. The vanishing of the moments leads to the symmetry of the stress matrix, meaning $\tau_{ij} = \tau_{ji}$. Clearly this means that the component of the stress tensor in the x_j direction (acting on the face whose normal is in the x_i direction), is equal to the component of the stress tensor in the x_i direction (acting on the face whose normal is in the x_j direction). The x_i component of the force due to the stress vector acting on the face ABC is $(T_i^{\nu} + \varepsilon_i)S$, where ε_i is the ith component of any stress vector of order h, meaning: $\lim_{h \to 0} \varepsilon_i = 0$. The corresponding component of force acting on the faces of areas S_j $(j = 1, 2, 3)$ is given by $(-\tau_{ji} + \varepsilon_i)S_j$. Let F_i be the x_i component of the specific body force. Its contribution is $(F_i + e_i)\rho(hS/3)$, where ρ is the density of the material, $hS/3$ is the volume of the tetrahedron, and e_i is the ith component of any specific body force of order h. For the tetrahedron to be in equilibrium we must set the sum of these forces equal to zero. We obtain

$$(T_i^{\nu} + \varepsilon_i)S + (-\tau_{ij} + \varepsilon_i)l_j S + (\rho h S/3)(F_i + e_i) = 0.$$

Dividing by S and letting $h \to 0$ yields $T_i = \tau_{ij} l_j$ (summed over j). This is *Cauchy's principle of stress* which tells that the stress can be calculated on any element of surface passing through P whose normal is $\boldsymbol{\nu}$ by specifying the stress matrix $\boldsymbol{\tau}$.

Stress–Strain Relations

In this section we formulate the constitutive equations that define an elastic solid undergoing small amplitude strain (this is called *infinitesimal elasticity*). For infinitesimal elasticity the fundamental assumption, which is based essentially on experimental evidence, is that the stress tensor is a linear function of the strain tensor. This is the *generalized form of Hooke's law.*

To develop the generalized Hooke's law we first recall that, according to Cauchy's principle, the stress field in a continuous medium is completely determined by the stress matrix whose ijth element is τ_{ij}. We assume that for an elastic material there is a one-one analytic relation between the stress tensor (or matrix) and the strain tensor. This means

$$\tau_{ij} = T_{ij}(\varepsilon_{11}, \varepsilon_{12}, \ldots, \varepsilon_{33}). \tag{8.18}$$

We further impose the restriction that in the initial or equilibrium configuration $\tau_{ij} = \varepsilon_{ij} = 0$. Hooke's law is obtained by expanding the right-hand side of (8.18) in a power series in the ε_{ij}'s and approximating this expansion by retaining only the linear terms. This means that each τ_{ij} is a linear combination of all the ε_{ij}'s. We now take advantage of the symmetry of the stress matrix $\boldsymbol{\tau}$ and the strain matrix $\boldsymbol{\varepsilon}$ by introducing the following simplified notation:

$$\varepsilon_{11} = \varepsilon_1, \qquad \varepsilon_{22} = \varepsilon_2, \qquad \varepsilon_{33} = \varepsilon_3, \qquad \varepsilon_{12} = \varepsilon_4, \qquad \varepsilon_{23} = \varepsilon_5, \qquad \varepsilon_{13} = \varepsilon_6, \tag{8.19}$$

and a similar notation for the ε_{ij}'s. The linear expansion then becomes

$$\tau_i = c_{ij}\varepsilon_j \quad \text{(summed over } j\text{)}, \qquad i, j = 1, 2, 3. \tag{8.20}$$

There are 36 c_{ij}'s which are the elastic or material constants. It will be shown from certain symmetry relations that the number of these constants can be drastically reduced. In fact, they reduce to the two Lamé constants for an isotropic elastic solid. The system (8.20) is a set of six equations where each τ_i is a linear combination of the six ε_i's. It is the tensor representation of the generalized Hooke's law, and can be put in the following matrix form:

$$\mathbf{t} = \mathbf{Ce}, \tag{8.21}$$

where $\mathbf{t}$ is a 6×1 column matrix whose ith component is τ_i, $\mathbf{C}$ is a 6×6 nonsingular matrix whose ijth element is c_{ij}, and $\mathbf{e}$ is a 6×1 column matrix whose ith element is ε_i. Inverting (8.21) allows us to solve for $\mathbf{e}$, obtaining

$$\mathbf{e} = \mathbf{Kt}, \tag{8.22}$$

where $\mathbf{K} = \mathbf{C}^{-1}$. The elements of $\mathbf{K}$ are called the *elastic moduli* of the material.

We now consider the *strain-energy density function* W which is a scalar potential introduced by Green in 1839 in the *Transactions of the Cambridge Philosophical Society*. It is a consequence of the first law of thermodynamics and the law of conservation of energy for an elastic or nondissipative medium. The existence and properties of W were given by Love in 1944. The essential result is that W can be represented as a quadratic function of the strain matrix. In matrix notation this becomes

$$W = \tfrac{1}{2}\mathbf{e}^*\mathbf{Ce}. \tag{8.23}$$

Writing out this quadratic expression gives

$$W = \tfrac{1}{2}c_{ij}\varepsilon_i\varepsilon_j. \tag{8.24}$$

Since W is a scalar we have $W = W^*$ (W is symmetric in i, j). Applying this fact to (8.23) yields the result $\mathbf{C} = \mathbf{C}^*$, thus reducing the number of elements in $\mathbf{C}$ to 21.† This is easily seen: Since $\mathbf{C}$ is a symmetric 6×6 matrix the number of elements in the upper triangular part of $\mathbf{C}$ (including the diagonal) is $6+5+4+3+2+1=21$. Since the τ_i's are given by (8.20) we

† A note on the material or elastic constants given by $\mathbf{C}$: In elastostatic problems, where a body is in equilibrium, the isothermal process is valid since it corresponds to the case of slow loading and unloading. However, for wave propagation problems an adiabatic process is the best approximation to the physical situation. The elastic constants given by $\mathbf{C}$ cannot be the same for these two cases.

get the following fundamental property of W:

$$\frac{\partial W}{\partial \varepsilon_i} = \tau_i. \tag{8.25}$$

Equation (8.25) can also be derived from thermodynamic considerations.

Hooke's Law for an Isotropic Material

An isotropic elastic material is defined as one whose elastic constants c_{ij} are invariant with respect to any rotation of the reference frame. Let $\mathbf{R}$ be the 3×3 rotation matrix which rotates $\mathbf{x}$ into $\mathbf{x}'$ so that $\mathbf{x}'=\mathbf{R}\mathbf{x}$, also $\mathbf{x}=\mathbf{R}^*\mathbf{x}'$. The strain and stress matrices in the rotated coordinate system $\mathbf{x}'$ are given by the *similarity transformations*

$$\boldsymbol{\varepsilon}' = \mathbf{R}\boldsymbol{\varepsilon}\mathbf{R}^*, \qquad \boldsymbol{\tau} = \mathbf{R}\boldsymbol{\tau}\mathbf{R}^*. \tag{8.26}$$

See [13, Chap. 1] for details on the similarity transformation. We first rotate $\mathbf{x}$ through a right angle about the x_1 axis. We have $\mathbf{x}'=\mathbf{R}\mathbf{x}$ which becomes

$$\begin{pmatrix} x_1' \\ x_2' \\ x_3' \end{pmatrix} = \begin{pmatrix} 1 & 0 & 0 \\ 0 & 0 & -1 \\ 0 & 1 & 0 \end{pmatrix} \begin{pmatrix} x_1 \\ x_2 \\ x_3 \end{pmatrix}.$$

Let f_i $(i=1,2,\ldots,6)$ stand for the ith component of $\boldsymbol{\varepsilon}$ or $\boldsymbol{\tau}$ in the unrotated or $\mathbf{x}$ reference frame and f_i' stand for the ith component of $\boldsymbol{\varepsilon}'$ or $\boldsymbol{\tau}'$ in the rotated or $\mathbf{x}'$ reference frame. The similarity transformation given by (8.26) becomes

$$\begin{pmatrix} f_1' & f_4' & f_6' \\ f_4' & f_2' & f_5' \\ f_6' & f_5' & f_3' \end{pmatrix} = \begin{pmatrix} 1 & 0 & 0 \\ 0 & 0 & -1 \\ 0 & 1 & 0 \end{pmatrix} \begin{pmatrix} f_1 & f_4 & f_6 \\ f_4 & f_2 & f_5 \\ f_6 & f_5 & f_3 \end{pmatrix} \begin{pmatrix} 1 & 0 & 0 \\ 0 & 0 & 1 \\ 0 & -1 & 0 \end{pmatrix}$$

$$= \begin{pmatrix} f_1 & -f_6 & f_4 \\ -f_6 & f_3 & -f_5 \\ f_4 & -f_5 & f_2 \end{pmatrix}.$$

Similarly a rotation about the x_3 axis through 90° gives the following transformation for $\boldsymbol{\varepsilon}'$ and $\boldsymbol{\tau}'$:

$$f_1'=f_2, \qquad f_2'=f_1, \qquad f_3'=f_3, \qquad f_4'=-f_4, \qquad f_5'=f_6, \qquad f_6'=-f_5.$$

Hooke's law applied to the rotated system is obtained from (8.20) by replacing ε_i by ε_i' and τ_j by τ_j', the c_{ij}'s remain invariant because the medium is isotropic. Applying the above relations between the f_i''s and f_i's for the two rotations yields the following relations amongst the elastic constants:

$$c_{12}=c_{13}=c_{23}, \qquad c_{11}=c_{22}=c_{33}, \qquad c_{44}=c_{55}=c_{66}.$$

The other c_{ij}'s are zero. This reduces the number of distinct c_{ij}'s to 3.

We now rotate the $\mathbf{x}$ coordinate system through an angle of 45° about the x_3 axis according to the transformation $x_1' = \frac{1}{2}(x_1 + x_2)$, $x_2' = \frac{1}{2}(-x_1 + x_2)$, $x_3' = x_3$. Inserting this rotation into the similarity transformations (8.26) yields

$$\tau_4' = \tfrac{1}{2}(-\tau_1 + \tau_2), \qquad \varepsilon_4' = \tfrac{1}{2}(-\varepsilon_1 + \varepsilon_2).$$

From (8.20) and the above relations for the c_{ij}'s we get

$$\tau_4 = c_{44}\varepsilon_4, \qquad \tau_4' = c_{44}\varepsilon_4'.$$

This gives

$$\tfrac{1}{2}(-\tau_1 + \tau_2) = c_{44}(-\varepsilon_1 + \varepsilon_2).$$

Using the relation

$$\tau_i = c_{i1}\varepsilon_1 + c_{i2}\varepsilon_2 + c_{i3}\varepsilon_3$$

for $i = 1, 2$ and the relations $c_{22} = c_{11}$, $c_{12} = c_{23} = c_{13}$, we get

$$\tfrac{1}{2}(-\tau_1 + \tau_2) = \tfrac{1}{2}(c_{11} - c_{22})(-\varepsilon_1 + \varepsilon_2) = c_{44}(-\varepsilon_1 + \varepsilon_2).$$

Writing μ for c_{44} and using the same analysis for the other components of the shear stress and strain, we get

$$\tau_{ij} = 2\mu\varepsilon_{ij}, \qquad i \neq j, \qquad i, j = 1, 2, 3. \tag{8.27}$$

We may write the first equation of (8.20) as

$$\tau_{11} = c_{12}(\varepsilon_{11} + \varepsilon_{22} + \varepsilon_{33}) + (c_{11} - c_{12})\varepsilon_{11} = \lambda\Theta + 2\mu\varepsilon_{11},$$

where $\lambda = c_{12}$ and we have set

$$\Theta = \varepsilon_{ii} = \varepsilon_{11} + \varepsilon_{22} + \varepsilon_{33}. \tag{8.28}$$

Θ is the *dilatation.* It is the trace (sum of the diagonal elements) of the strain matrix and is a measure of the relative change in volume due to a compression or dilatation. The elastic constants μ and λ are the *Lamé constants.*

Hooke's law for an isotropic elastic medium can now be put in the tensor form

$$\tau_{ij} = \lambda\delta_{ij}\Theta + 2\mu\varepsilon_{ij}. \tag{8.29}$$

where δ_{ij} is the Kronecker delta. Let Φ be the trace of the stress matrix so that

$$\Phi = \tau_{ii} = \tau_{11} + \tau_{22} + \tau_{33}. \tag{8.30}$$

Setting $i = j$ in (8.29), summing over i, and using the fact that $\delta_{ii} = 3$, gives the following relation between Φ and Θ:

$$\Phi = (3\lambda + 2\mu)\Theta. \tag{8.31}$$

The inverse of (8.29) gives the components of the strain tensor as a function of the components of the stress tensor. We get

$$\varepsilon_{ij} = -\frac{\lambda\delta_{ij}}{2\mu(3\lambda+2\mu)}\Phi + \frac{\tau_{ij}}{2\mu}. \tag{8.32}$$

It is useful to write out Hooke's law given by (8.29) in extended form using the (x, y, z) coordinate system. We get

$$\begin{aligned}
\tau_{xx} &= (\lambda+2\mu)\varepsilon_{xx} + \lambda(\varepsilon_{yy}+\varepsilon_{zz}),\\
\tau_{yy} &= (\lambda+2\mu)\varepsilon_{yy} + \lambda(\varepsilon_{xx}+\varepsilon_{zz}),\\
\tau_{zz} &= (\lambda+2\mu)\varepsilon_{zz} + \lambda(\varepsilon_{xx}+\varepsilon_{yy}),\\
\tau_{xy} &= 2\mu\varepsilon_{xy},\\
\tau_{yz} &= 2\mu\varepsilon_{yz},\\
\tau_{xz} &= 2\mu\varepsilon_{xz}.
\end{aligned} \tag{8.33}$$

The first three equations of (8.33) tell us that each normal component of the stress tensor is a linear combination of all the normal components of the strain tensor. The last three equations tell us that each component of the shear stress is proportional to the corresponding component of the shear strain.

It is also instructive to invert (8.33) and thus write out the strain components as functions of the stress components in extended form. In order to get a representation of this form of Hooke's law that has physical significance, we first introduce the positive constants E and σ defined by

$$E = \frac{\mu(3\lambda+2\mu)}{\lambda+\mu}, \qquad \sigma = \frac{\frac{1}{2}\lambda}{\lambda+\mu}. \tag{8.34}$$

E is the *Young's modulus* and σ is the *Poisson's ratio* of the isotropic elastic material. The inversion of (8.33) can then be given as

$$\begin{aligned}
\varepsilon_{xx} &= \frac{1}{E}[\tau_{xx} - \sigma(\tau_{yy}+\tau_{zz})],\\
\varepsilon_{yy} &= \frac{1}{E}[\tau_{yy} - \sigma(\tau_{zz}+\tau_{xx})],\\
\varepsilon_{zz} &= \frac{1}{E}[\tau_{zz} - \sigma(\tau_{xx}+\tau_{yy})],\\
\varepsilon_{xy} &= \frac{1}{E}(1+\sigma)\tau_{xy},\\
\varepsilon_{yz} &= \frac{1}{E}(1+\sigma)\tau_{yz},\\
\varepsilon_{xz} &= \frac{1}{E}(1+\sigma)\tau_{xz}.
\end{aligned} \tag{8.35}$$

We have reviewed the concepts of deformation, strain, stress, and the relation between stress and strain for an infinitesimal elastic solid given by the generalized form of Hooke's law and its specialization to an isotropic solid. These ideas embody the field of elastostatics. Note that we left out the equilibrium equations which is a special case of the equations of motion to be treated below. In order to discuss wave propagation in an elastic medium we need to extend these concepts to a dynamical or time-varying situation. To this end, we must develop the equations of motion for an elastic solid.

8.2. Equations of Motion for the Stress Components

Consider a continuous distribution of matter occupying a volume V_0 enclosed by a surface S_0. In the interior of V_0 a particle occupying a volume element dV enclosed by a surface element of area dS is acted upon by external forces, the resultant of which induces an inertial force due to the acceleration of the particle. This inertial force is $\rho \mathbf{a}\, dm$, where ρ is the particle density and $\mathbf{a}=(u_{1,tt}, u_{2,tt}, u_{3,tt})$ is the particle acceleration, where $\mathbf{u}=(u_1, u_2, u_3)$ is the displacement vector so that $\mathbf{v}=(u_{1,t}, u_{2,t}, u_{3,t})$ is the particle velocity. Let $\rho\mathbf{F}\, dV$ be the resultant of the body forces in dV, where $\mathbf{F}$ is the specific body force (force per unit mass), and let $\mathbf{T}\, dS$ be the resultant of the surface forces on dS. Then the vector sum of these forces must equal the inertial force on the particle occupying dV, thus yielding

$$\mathbf{F}\rho\, dV+\mathbf{T}\, dS=\rho\mathbf{a}\, dV,$$

as the vector equation of motion for the particle in dV. Integrating over the volume V_0 and using (8.16), we may write the equation of motion as the three scalar equations

$$\int_{V_0} \rho F_i\, dV+\oint_{S_0} \tau_{ij} l_j\, dS=\int_{V_0} \rho u_{i,tt}\, dV. \tag{8.36}$$

The surface integral in (8.36) is now transformed to a volume integral by using the divergence theorem giving

$$\oint_{S_0} \tau_{ij} l_j\, dS=\int_{V_0} \tau_{ij,j}\, dV.$$

Using this equation (8.36) becomes

$$\int_{V_0} (\rho F_i+\tau_{ij,j})\, dV=\int_{V_0} \rho u_{i,tt}\, dV. \tag{8.37}$$

The system (8.37) consists of three equations for $i=1, 2, 3$. Since the material is continuous, if the region of integrations shrinks uniformly to an interior point, we obtain the differential form of the equations of motion, which we put in the tensor representation

$$\tau_{ij,j} = \rho(-F_i + u_{i,tt}) \quad \text{summed over } j, \qquad i, j = 1, 2, 3, \tag{8.38}$$

where F_i is the ith component of the specific body force. It is instructive to write (8.38) in extended form using an (x, y, z) coordinate system and letting $\mathbf{F} = (X, Y, Z)$, $\mathbf{u} = (u, v, w)$, etc. We get

$$\begin{aligned}
\frac{\partial \tau_{xx}}{\partial x} + \frac{\partial \tau_{xy}}{\partial y} + \frac{\partial \tau_{xz}}{\partial z} &= -X + \frac{\partial^2 u}{\partial t^2}, \\
\frac{\partial \tau_{yx}}{\partial x} + \frac{\partial \tau_{yy}}{\partial y} + \frac{\partial \tau_{yz}}{\partial z} &= -Y + \frac{\partial^2 v}{\partial t^2}, \\
\frac{\partial \tau_{zx}}{\partial x} + \frac{\partial \tau_{zy}}{\partial y} + \frac{\partial \tau_{zz}}{\partial z} &= -Z + \frac{\partial^2 w}{\partial t^2}.
\end{aligned} \tag{8.39}$$

The systems (8.38) or (8.39) represent the conservation of linear momentum and are therefore Newton's law of motion for a continuum. Moreover, they are the *linearized* equations of motion, since the nonlinear terms for the particle acceleration which represent the convective terms are neglected. They are valid for any continuum—solid, liquid or gas—as long as we maintain the assumption of a continuous distribution of mass. As mentioned, these equations are linear. This means that the Lagrangian representation that was used has the same form as the Euler representation; the difference between these two representations only appears in the nonlinear terms.

We have three equations involving the six components of the stress tensor and the three components of the displacement vector. Therefore we have an insufficient number of equations to solve for the stress field in the volume V_0. This is not surprising since we have an incomplete description of the situation in the sense that we have no equations defining the elastic material. The requisite equations are the *constitutive equations* which, as we showed above, are the linear relationship between the stress and strain tensors given by Hooke's law.

8.3. Equations of Motion for the Displacement, Navier Equations

In this section we derive the linear equations of motion for an elastic solid in terms of the components of the displacement vector. These are sometimes called the *Navier equations.* We shall show that the Navier equations can be uncoupled into two vector PDEs:

(1) An equation of motion for the dilatation which represents an irrotational displacement field yielding irrotational waves that propagate with a characteristic velocity that depends on the elastic constants and the density. Since the field is irrotational, a potential can be introduced whose gradient gives the displacement vector.

(2) An equation of motion for the rotation vector which represents a rotational displacement field yielding rotational on equivoluminal waves that propagate with a wave velocity that is less than that for a dilatation wave.

We now derive the Navier equations. The relationship between the stress and strain for an infinitesimal isotropic elastic body is given by Hooke's law (8.29). However, in many problems in elasticity it is more convenient to obtain the equations of motion for the displacement (the Navier equations), and then derive the stress field from the definition of strain and Hooke's law. To this end, we use the definition of infinitesimal strain whose components were obtained in the form given by (8.15a). For a consistent notation we continue to use $\mathbf{x} = (x_1, x_2, x_3)$ as independent spatial variables instead of the Lagrangian coordinates $\mathbf{a}$; since for the linearized theory there is no difference in the expressions for the stress, strain, and equations of motion in the Lagrangian and Eulerian representations. We therefore use as the definition of infinitesimal strain:

$$\varepsilon_{ij} = \tfrac{1}{2}(u_{i,j} + u_{j,i}). \tag{8.40}$$

The dilatation Θ given by (8.28) can be written as

$$\Theta = \varepsilon_{ii} = u_{i,i} = \operatorname{div} \mathbf{u}. \tag{8.41}$$

Equations (8.38) and (8.29) yield

$$\mu \nabla^2 u_i + \frac{(\lambda + \mu)\partial\Theta}{\partial x_i} = \rho(-F_i + u_{i,tt}), \qquad i = 1, 2, 3, \tag{8.42}$$

where (in tensor notation) $\nabla^2 u_i = u_{i,jj}$. The system of three equations given by (8.42) is called the Navier equations. It is a set of three coupled linear second-order PDEs for the components of the displacement vector $\mathbf{u}$. It is of interest to write out this tensor equation in extended form.

$$\begin{aligned} \mu \nabla^2 u + (\lambda + \mu)\Theta_x &= \rho(-X + u_{tt}), \\ \mu \nabla^2 v + (\lambda + \mu)\Theta_y &= \rho(-Y + v_{tt}), \\ \mu \nabla^2 w + (\lambda + \mu)\Theta_z &= \rho(-Z + w_{tt}), \end{aligned} \tag{8.43}$$

where the dilatation can be written as

$$\Theta = \operatorname{div} \mathbf{u} = u_x + v_y + w_z.$$

We now write the Navier equation given by (8.42) or (8.43) as a single vector equation. To this end, let $(\mathbf{i}, \mathbf{j}, \mathbf{k})$ be unit vectors in the (x, y, z) directions. Multiply the first of (8.43) by $\mathbf{i}$, the second by $\mathbf{j}$, the third by $\mathbf{k}$, and add. The result is

$$\mu \nabla^2 \mathbf{u} + (\lambda + \mu)\, \mathbf{grad}\, \Theta = \rho(-\mathbf{F} + \mathbf{u}_{tt}). \tag{8.44}$$

A well-known result in vector analysis is

$$\nabla^2 \mathbf{u} = \mathbf{grad} \operatorname{div} \mathbf{u} - \mathbf{curl}\, \boldsymbol{\psi}, \tag{8.45}$$

where $\boldsymbol{\psi} = \mathbf{curl}\, \mathbf{u}$. See, for example, [20] for the derivation of (8.44). The rotation vector $\boldsymbol{\psi}$, when written out, becomes

$$\boldsymbol{\psi} = \mathbf{i}(w_y - v_z) + \mathbf{j}(u_z - w_x) + \mathbf{k}(v_x - u_y).$$

Inserting (8.45) into (8.44) yields another form for the vector Navier equation, namely,

$$-\mu\, \mathbf{curl}\, \boldsymbol{\psi} + (\lambda + \mu)\, \mathbf{grad}\, \Theta = \rho(-\mathbf{F} + \mathbf{u}_{tt}). \tag{8.46}$$

In (8.46) $\nabla^2 \mathbf{u}$ is replaced by $\mathbf{curl}\, \boldsymbol{\psi}$. This form of the Navier equation has the advantage of allowing us to either eliminate $\boldsymbol{\psi}$ or Θ and thus obtain dilatational or rotational waves.

Equation of Motion for Θ

We take the divergence of each term of (8.46) and use the fact that $\mathbf{div\, curl}\, \boldsymbol{\psi} = 0$, obtaining the following second-order PDE for the dilatation:

$$(\lambda + 2\mu)\nabla^2 \Theta = \rho[-\operatorname{div} \mathbf{F} + \Theta_{tt}]. \tag{8.47}$$

We see that (8.47) is the three-dimensional nonhomogeneous wave equation for Θ where the wave speed is given by $\sqrt{(\lambda + 2)}/\rho$. Since the rotation vector is not contained in (8.47), this is the equation of motion for the dilatation whose solution yields the irrotational displacement field.

Equation of Motion for $\boldsymbol{\psi}$

Next we take the **curl** of each side of (8.46) and use the fact that $\mathbf{curl\, grad}\, \Theta = 0$. We get

$$\mu \nabla^2 \boldsymbol{\psi} = \rho(-\mathbf{curl}\, \mathbf{F} + \boldsymbol{\psi}_{tt}). \tag{8.48}$$

Equation (8.48) is the three-dimensional nonhomogeneous wave equation for the rotation vector (note the dilatation is missing) whose solution gives the rotational displacement field yielding rotational or equivoluminal waves (since the dilatation does not appear in this form of the equations of motion, it is equivalent to a constant volume situation).

8.4. Propagation of a Plane Elastic Wave

In this section we shall investigate the propagation of a plane wave in an infinite elastic medium. A *plane wave* is defined as one whose wave front is a planar surface perpendicular to the direction of the propagating wave.

We first consider the case where the direction of wave propagation is along the x axis. The displacement vector becomes $\mathbf{u} = \mathbf{u}(x, t)$, and the wave front is parallel to the (y, z) plane. We shall demonstrate the following two types of waves:

(1) A *longitudinal wave*, where $\mathbf{u} = (u, 0, 0)$ is in the direction of wave propagation.
(2) A *transverse wave*, where $\mathbf{u} = (0, v, w)$ so that $\mathbf{u}$ is in a plane perpendicular to the direction of wave propagation.

Since $\Theta = u_x$, for the case $\mathbf{F} = 0$, the extended form of the Navier equations (8.43) become

$$c_L^2 u_{xx} = u_{tt},$$
$$c_T^2 v_{xx} = v_{tt}, \tag{8.49}$$
$$c_T^2 w_{xx} = w_{tt},$$

where

$$c_L = \sqrt{\lambda + \frac{2\mu}{\rho}}, \qquad c_T = \sqrt{\frac{\mu}{\rho}}. \tag{8.50}$$

c_L is the *longitudinal wave speed* and c_T is the *transverse wave speed.* The first equation of (8.49) is the one-dimensional wave equation for u, showing that this component is propagated longitudinally in the x direction with a wave speed c_L. The second and third equations show that the lateral components v and w are propagated laterally or transversely with a wave speed c_T. Since the dilatation $= \operatorname{div} \mathbf{u} = u_x$, it is clear that the transverse wave propagation represented by the second and third equations of (8.49) is divergenceless, meaning that there is no volume change. This is a special case of (8.44) where $\Theta = \mathbf{F} = 0$ and $\mathbf{u} = \mathbf{u}(x, t)$. This case of the propagation of a plane elastic wave is an example of a more general method of decomposing waves into longitudinal and transverse waves, which we give below.

More General Decomposition

To give a more general method of decomposing elastic waves into longitudinal and transverse wave forms, we start by writing the vector form of the Navier equation (8.44) as

$$c_T^2 \nabla^2 \mathbf{u} + (c_L^2 - c_T^2)\, \mathbf{grad}(\operatorname{div} \mathbf{u}) = \mathbf{u}_{tt}, \tag{8.51}$$

where we used (8.50) and set $\mathbf{F} = 0$ for simplicity. We now decompose (8.51) or split this equation into two vector equations by decomposing $\mathbf{u}$. To do this we define the two displacement vectors $\mathbf{u}_T$ and $\mathbf{u}_L$ as follows: set

$$\mathbf{u} = \mathbf{u}_T + \mathbf{u}_L, \tag{8.52}$$

where $\mathbf{u}_T$ satisfies the equation

$$\operatorname{div} \mathbf{u}_T = 0 \tag{8.53}$$

and $\mathbf{u}_L$ satisfies the equation

$$\mathbf{curl}\ \mathbf{u}_L = 0. \tag{8.54}$$

Since $\mathbf{u}_T$ is defined by (8.53) as being divergenceless, we know from vector analysis that there exists a rotation vector (as given previously) such that

$$\mathbf{u}_T = \mathbf{curl}\ \boldsymbol{\psi}, \tag{8.55}$$

$\boldsymbol{\psi}$ is sometimes called the *vector potential.* On the other hand, $\mathbf{u}_L$ is irrotational as shown by (8.54) so that there exists a scalar function $\phi(x, y, z, t)$ called the *scalar potential* such that

$$\mathbf{u}_L = \mathbf{grad}\ \phi. \tag{8.56}$$

Using (8.55) and (8.56) we may write (8.52) as

$$\mathbf{u} = \mathbf{curl}\ \boldsymbol{\psi} + \mathbf{grad}\ \phi. \tag{8.57}$$

We see that (8.57) represents the fact that the displacement vector is decomposed into a divergenceless vector and irrotational vector.

We now insert (8.52) into (8.51), take the divergence of each term of the resulting equation, and use (8.53). We obtain, after a little manipulation,

$$\operatorname{div}(c_L^2 \nabla^2 \mathbf{u}_L - \mathbf{u}_{L,tt}) = 0. \tag{8.58}$$

Now the curl of the expression in parentheses of (8.58) is also zero. It is a fact that any vector whose divergence and curl both vanish is identically a zero vector. We therefore obtain

$$c_L^2 \nabla^2 \mathbf{u}_L = \mathbf{u}_{L,tt}. \tag{8.59}$$

Equation (8.59) tells us that $\mathbf{u}_L$ satisfies the vector wave equation with the wave speed c_L. Since $\mathbf{u}_L = \mathbf{grad}\ \phi$, it is clear that the scalar potential ϕ also satisfies the wave equation with the same wave speed. All solutions of (8.59) yield longitudinal waves that are irrotational (since $\boldsymbol{\psi} = 0$).

In a similar manner we insert (8.52) into (8.51), take the curl of the resulting equation, and use the fact that $\mathbf{curl}\ \mathbf{u}_L = 0$. We obtain

$$\mathbf{curl}(c_T^2 \nabla^2 \mathbf{u}_T - \mathbf{u}_{T,tt}) = 0. \tag{8.60}$$

Since the divergence of the expression in parentheses is also zero, we obtain

$$c_T^2 \nabla^2 \mathbf{u}_T = \mathbf{u}_{T,tt}. \tag{8.61}$$

Equation (8.61) is a vector wave equation for $\mathbf{u}_T$ whose solutions yield transverse waves that are rotational but are accompanied by no change in volume (*equivoluminal, transverse, rotational waves*). They propagate with a wave speed c_T.

In this section we deal with an infinite medium. However, when we consider a bounded region, if we formulate such wave propagation problems in terms of the wave equations involving the vector and scalar potentials we must be careful to invoke the required boundary and initial conditions.

Having solved for ϕ and $\boldsymbol{\psi}$ from the appropriate wave equations, the displacement field $\mathbf{u}(x, y, z, t)$ can then be calculated from the system (8.57). For the readers' convenience we write out (8.57) in extended form

$$u = \phi_x + \psi_{3,y} - \psi_{2,z},$$
$$v = \phi_y + \psi_{1,z} - \psi_{3,x},$$
$$w = \phi_z + \psi_{2,x} - \psi_{1,y}, \qquad \psi_{1,x} = \partial\psi_1/\partial x, \quad \text{etc.}$$

Displacement Field Obtained Directly from the Navier Equation

An alternative approach is to determine the displacement field $\mathbf{u}(x, y, z, t)$ directly from the Navier equation without recourse to the scalar and vector potentials. We still consider an infinite region. We construct *plane wave solutions* in terms of progressing and regressing waves of the form

$$u = F(\mathbf{r} \cdot \boldsymbol{\nu} \mp ct),$$
$$v = G(\mathbf{r} \cdot \boldsymbol{\nu} \mp ct), \tag{8.62}$$
$$w = H(\mathbf{r} \cdot \boldsymbol{\nu} \mp ct),$$

where F, G, H are arbitrary functions of the arguments $\mathbf{r} \cdot \boldsymbol{\nu} \mp ct$, and $\mathbf{r} \cdot \boldsymbol{\nu} = l_1 x + l_2 y + l_3 z$, so that (l_1, l_2, l_3) are the direction cosines of the normal $\boldsymbol{\nu}$, where $\boldsymbol{\nu}$ is normal to the wave front. In the arguments the minus sign represents progressing waves and the plus sign represents regressing waves. Inserting (8.62) into the Navier equations (8.43), for the case $\mathbf{F} = 0$, yields the following quadratic equation for c^2:

$$(\rho c^2 - \mu)(\rho c^2 - \lambda - 2\mu) = 0. \tag{8.63}$$

The roots of (8.63) are $c^2 = c_L^2$, c_T^2. This substantiates the fact that there are two plane waves, one traveling with a wave speed c_L and the other traveling with a wave speed c_T.

Separation of Variables

Since the Navier equation consists of a system of PDEs with constant coefficients, we can separate the spatial from the time-dependent part of the solution. We consider a *monochromatic wave*, one characterized by a single frequency given by the angular frequency ω in radians per second. To extend the treatment to a discrete frequency spectrum we can synthesize wave forms by performing Fourier series expansions. For a continuous frequency spectrum we resort to a synthesis by means of the Fourier integral or other appropriate integral transforms. For a monochromatic wave we write the displacement vector in the form

$$\mathbf{u} = \text{Re}[\bar{\mathbf{u}}(x, y, z) \exp(\pm i\omega t)]. \tag{8.64}$$

This method of separating the time-dependent part of $\mathbf{u}$ in the form $\exp(\pm i\omega t)$ means we are looking for *time-harmonic solutions.* Inserting (8.64) into (8.51) we obtain the following PDE for $\mathbf{u}$:

$$c_T\nabla^2\bar{\mathbf{u}}+(c_L-c_T)\,\mathbf{grad}(\operatorname{div}\bar{\mathbf{u}})+\omega^2\bar{\mathbf{u}}=0. \tag{8.65}$$

Inserting (8.52) into (8.65) and using the above technique of taking the divergence and curl of (8.65), respectively, we obtain the following PDEs for $\bar{\mathbf{u}}_L$ and $\bar{\mathbf{u}}_T$:

$$\begin{aligned}\nabla^2\bar{\mathbf{u}}_L+k_L^2\bar{\mathbf{u}}_L&=0,\\ \nabla^2\bar{\mathbf{u}}_T+k_T^2\bar{\mathbf{u}}_T&=0,\end{aligned} \tag{8.66}$$

where

$$k_L^2=\frac{\omega^2}{c_L^2},\qquad k_T^2=\frac{\omega^2}{c_T^2}. \tag{8.67}$$

The equations (8.66) are called the *reduced wave equations* or the *Helmholtz* equations for $\mathbf{u}_L$ and $\mathbf{u}_T$, respectively. It is clear that the system (8.66) can also be obtained by assuming time-harmonic solutions for $\mathbf{u}_L$ and $\mathbf{u}_T$ in the form $\mathbf{u}_L=\bar{\mathbf{u}}_L\exp(-i\omega t)$, $\mathbf{u}_T=\bar{\mathbf{u}}_T\exp(-i\omega t)$ and substituting these expressions into (8.59) and (8.61), respectively.

8.5. Spherically Symmetric Waves

For simplicity, we consider the wave equation for the scalar $f(x, y, z, t)$, where f may stand for ϕ or the components of $\boldsymbol{\psi}$ or $\mathbf{u}$. For spherically symmetric waves we set $r^2=x^2+y^2+z^2$, and neglect the angular dependence. The wave equation for f in spherical coordinates then becomes

$$c^2\left[f_{rr}+\left(\frac{2}{r}\right)f_r\right]=f_{tt}. \tag{8.68}$$

Setting $f(r, t)=\bar{f}(r)\exp(-i\omega t)$ and inserting this expression into (8.68) gives the spherically symmetric reduced wave equation for $\bar{f}$:

$$\bar{f}''+\left(\frac{2}{r}\right)\bar{f}'+k^2\bar{f}=0,\qquad \bar{f}'\equiv\frac{d\bar{f}}{dr},$$

which may be written as

$$(r\bar{f})''+k^2r\bar{f}=0,\qquad \text{where}\quad k^2=\omega^2/c^2. \tag{8.69}$$

The general solution of (8.69) is a linear combination of terms of the form

$$f=\left(\frac{1}{r}\right)\exp(-i\omega t). \tag{8.70}$$

From (8.70) we can write the time-harmonic solution $f(r, t)$ of the spherically symmetric wave equation as a linear combination of terms of the form

$$\left(\frac{1}{r}\right)\exp[ik(r-ct)] \quad \text{and} \quad \left(\frac{1}{r}\right)\exp[-ik(r+ct)]. \tag{8.71}$$

The first expression of (8.71) represents an outgoing attenuated spherical wave. The second expression represents an incoming attenuated wave. Clearly the situation would be reversed if we chose the time-harmonic factor as $\exp(ikct)$ instead of $\exp(-ikct)$ as is customary. These outgoing and incoming waves are analogous to the progressing and regressing waves for one-dimensional wave propagation except that the spherical waves are attenuated as $1/r$. In practical problems we must surround the origin $r=0$ by a spherical surface of small radius, since it is clear that there is a singularity at the origin.

If we relax the condition of time-harmonic dependence we obtain a more general type of spherically symmetric waves. To this end, we write the corresponding wave equation (8.68) as

$$\lambda^2(rf)_{rr}=(rf)_{tt}.$$

If we set $w=rf$ then we obtain the one-dimensional wave equation $\lambda^2 w_{rr}=w_{tt}$. Therefore the general solution for $f(r, t)$ is

$$f(r,t)=\left(\frac{1}{r}\right)F(r-ct)+\left(\frac{1}{r}\right)G(r+ct), \tag{8.72}$$

where F and G are arbitrary functions of their arguments, to be determined from the appropriate initial and boundary conditions.

8.6. Reflection of Plane Waves at a Free Surface

We now study the effect of various boundaries on the propagation of waves in an elastic medium, beginning with the simplest case: the reflection of a plane wave from a free surface. A *free surface* is defined as a surface on which there are no external tractions or surface stresses (a stress-free surface).

To investigate the reflection of plane waves from a free surface we consider the (x, y) plane and let the x axis be the free surface. We can have two types of incident plane waves that reflect off the surface $y=0$:

(1) A longitudinal compression wave that travels with a wave speed c_L, called (in the language of the geophysicist) a P *wave* or pressure wave. It follows that a P wave is irrotational.

(2) A shear or S *wave* which is a transverse wave traveling with a wave speed c_T. An S wave is equivoluminal.

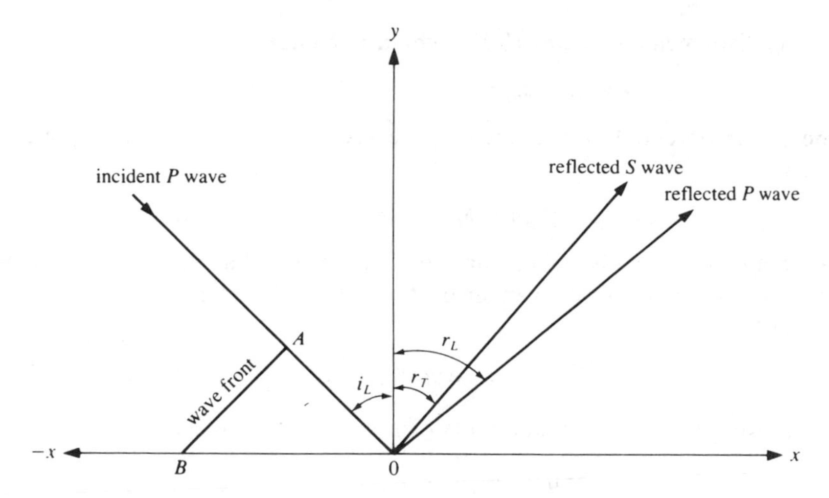

Fig. 8.6. Reflection of P wave into an S and a P wave at a free surface.

We shall show that, for an incident P wave there occur two reflected waves from the boundary: a P and an S wave. Similarly, both a P and an S wave are reflected when an incident S wave impinges on the boundary. To save space we analyze the reflection of an incident P wave; the same type of analysis holds for an incident S wave.

The physical model for the reflection of a P wave is shown in Fig. 8.6 where an incident longitudinal compression wave AO is shown traveling downward from the left with a wave speed c_L and intersecting the free surface at the point O, the origin of the coordinate system. In the figure let the length AO be a portion of the P wave representing the distance traveled by the wave in a unit time. Then $AO = c_L$. At the point A construct a perpendicular to the direction of the incident wave and let it intersect the x axis at B. Then AB represents the *wave front* which is, of course, normal to the direction of the propagating P wave. As this incident wave travels from A to O a corresponding wave will travel along the surface from B to O with a wave speed we shall call c. c is to be determined from the boundary conditions of zero surface stresses which also determine the direction of the appropriate reflected wave.

To analyze this problem we shall make use of the wave equations for the scalar and vector potentials. The problem is two dimensional in the sense that all the dependent variables are functions of x, y, t. Moreover, we consider this as a *plane strain problem*, which means $w = 0$. This gives $\mathbf{u} = (u, v, 0)$, $\boldsymbol{\psi} = (0, 0, \psi)$ where ψ is the z component of the rotation vector or vector potential. The expression for **curl** $\boldsymbol{\psi}$ becomes

$$\mathbf{curl}\ \boldsymbol{\psi} = \mathbf{i}\psi_y,$$

where (in the usual notation) ψ_y is the derivative of the z component of the vector potential with respect to y. We now use the decomposition formula for the displacement vector (8.57) which becomes

$$u = \phi_x + \psi_y, \qquad v = \phi_y - \psi_x, \qquad w = 0. \tag{8.73}$$

The scalar potential ϕ and vector potential ψ satisfy the wave equations given by

$$c_L^2(\phi_{xx} + \phi_{yy}) = \phi_{tt}, \qquad c_T^2(\psi_{xx} + \psi_{yy}) = \psi_{tt}. \tag{8.74}$$

The boundary $y = 0$ is a free surface. This means that the shear stress τ_{xy} and the normal stress τ_{yy} vanish on the x axis. From Hooke's law (8.33) we get

$$\tau_{xy} = 2\mu\varepsilon_{xy}, \qquad \tau_{yy} = \lambda\Theta + 2\mu\varepsilon_{yy}, \qquad \Theta = \varepsilon_{xx} + \varepsilon_{yy}. \tag{8.75}$$

Upon using (8.73) the components of the strain tensor become

$$\begin{aligned} \varepsilon_{xx} &= u_x = \phi_{xx} + \psi_{yx}, \\ \varepsilon_{yy} &= v_y = \phi_{yy} - \psi_{xy}, \\ \varepsilon_{xy} &= \tfrac{1}{2}(v_x + u_y) = \phi_{xy} + \tfrac{1}{2}(-\psi_{xx} + \psi_{yy}). \end{aligned} \tag{8.76}$$

We also have

$$\Theta = \phi_{xx} + \phi_{yy} = 0. \tag{8.77}$$

Inserting (8.76) and (8.77) into (8.75) yields the following boundary conditions for the stress at $y = 0$:

$$\begin{aligned} [\tau_{xy}]_{y=0} &= \mu(2\phi_{xy} - \psi_{xx} + \psi_{yy}) = 0, \\ [\tau_{yy}]_{y=0} &= \lambda\nabla^2\phi + 2\mu(\phi_{yy} - \psi_{xy}) = 0. \end{aligned} \tag{8.78}$$

The problem now is to find time-harmonic solutions for ϕ and ψ that satisfy the wave equations (8.74) and the boundary conditions (8.78). It is reasonable to assume that all solutions are bounded at infinity, since we can imagine the incident wave, which is aimed at the boundary, is "initiated at infinity". It is assumed to be monochromatic. Moreover, the reflected waves are assumed to maintain the same frequency.

Referring to Fig. 8.6, it is easily seen that, since the reflected P wave also travels with a wave speed c_L, we must have $r_L = i_L$ where r_L is the angle of reflection of the reflected P wave and i_L is the angle of incidence of the incident P wave. Therefore the boundary behaves as a mirror with respect to the incident P wave. However, the reflected S wave has a different wave speed from the incident wave, so that we expect its angle of reflection to be different from the angle of incidence of the P wave. We now construct time-harmonic solutions for ϕ and ψ. To this end, we let the reciprocal of the slopes of the incident P wave, the reflected P wave, and the reflected S wave be

$$\gamma_L = \cot i_L = \cot r_L, \qquad \gamma_T = \cot r_T, \tag{8.79}$$

where r_T is the reflected angle of the S wave. Furthermore, we observe from the figure that for the incident P wave x becomes less negative and y decreases as t increases, while for the reflected waves both x and y increase with increasing t. Putting these observations together, we construct the following expressions for the incident and reflected scalar and vector potentials:

$$\begin{aligned}\phi &= A_1 \exp[ik(ct - x + \gamma_L y)] + A_2 \exp[ik(ct - x - \gamma_L y)],\\ \psi &= B_1 \exp[ik(ct - x + \gamma_T y)] + B_2 \exp[ik(ct - x - \gamma_T y)],\end{aligned} \tag{8.80}$$

where A_1 is the amplitude of the incident P wave, A_2 is the amplitude of the reflected P wave, B_1 is the amplitude of the incident S wave (which is zero in this case), B_2 is the amplitude of the reflected S wave, k is the wave number of the surface waves, $k\gamma_L$ is the wave number of the incident and reflected P waves, and $k\gamma_T$ is the wave number of the S wave. Recalling the definition of wave number, we have

$$k = \frac{2\pi}{\lambda} = \frac{\omega}{c},$$

where λ is the wavelength and ω is the frequency. To determine the relations γ_T and γ_L we insert (8.80) into the respective wave equations (8.74) and obtain

$$\gamma_L^2 = \left(\frac{c}{c_L}\right)^2 - 1, \qquad \gamma_T^2 = \left(\frac{c}{c_T}\right)^2 - 1. \tag{8.81}$$

From the definitions of c_L and c_T we get the inequalities $c > c_L > c_T$. For the special case $c = c_L$ the P wave is along the surface $y = 0$. If we now eliminate c from (8.81) we obtain

$$\sin r_T = \left(\frac{c_T}{c_L}\right) \sin i_L. \tag{8.82}$$

Equation (8.82) is called *Snells' law*.† Snells' law was discussed in Chapter 6 in connection with sound waves. This is the law of refraction of a light wave in going from a medium of one refractive index into a medium of another refractive index. In applying (8.82) to our problem we conclude that an incident P wave impinging on the free surface $y = 0$ yields a reflected P wave with the property that $r_L = i_L$ and a reflected S wave which obeys Snell's law (8.82). The reflection of a P wave into a P wave of the same wave speed is analogous to the reflection of a light wave at the surface $y = 0$

† The Dutch astronomer, Willebrord Snell (1591–1626) was the first scientist to discuss the law of refraction given by (8.82), which he applied to light waves. However, the results of his investigations were not published until after his death. The law was discovered independently and first reported by René Descartes (1596–1650).

if it were replaced by a mirror surface. However, the reflection of an incident P wave into a reflected S wave involves a change in wave speed from c_L to c_T, so that the relationship between the incident and reflected angles is given by Snell's law (8.82), which means that the reflection of a P into an S wave (into the same elastic medium) is analogous to the refraction of a light wave from a medium of one wave speed (or refractive index) to a medium of different wave speed. The above analogy with the refraction of light shows us that the reflection of a P wave at a free surface into an S wave is equivalent to the refraction of a stress wave traveling in a medium with the wave speed c_L into a medium traveling with a wave speed c_T.

We now go back to the time-harmonic solutions of (8.74) given by (8.80) and attempt to satisfy the boundary conditions (8.78). Inserting (8.80) into (8.78) and setting $y = B_1 = 0$ yields

$$\begin{aligned} 2\gamma_L(A_1 - A_2) + (\gamma_T^2 - 1)B_2 &= 0, \\ [\lambda + (\lambda + 2\mu)\gamma_L^2](A_1 + A_2) - 2\mu\gamma_T B_2 &= 0. \end{aligned} \tag{8.83}$$

The system (8.83) consists of two equations in the three unknowns A_1, A_2, B_2. However, we can solve for the ratios of the reflected amplitudes to the incident amplitude. To this end, we define

$$\bar{A}_2 = \frac{A_2}{A_1}, \qquad \bar{B}_2 = \frac{B_2}{A_1}. \tag{8.84}$$

Using (8.84), (8.83) becomes

$$\begin{aligned} 2\gamma_L\bar{A}_2 + (\gamma_T^2 - 1)\bar{B}_2 &= 2\gamma_L, \\ [\lambda + (\lambda + 2\mu)\gamma_L^2]\bar{A}_2 - 2\mu\gamma_T\bar{B}_2 &= -[\lambda + (\lambda + 2\mu)\gamma_L^2]. \end{aligned} \tag{8.85}$$

The system (8.85) consists of two equations to be solved for $\bar{A}_2$ and $\bar{B}_2$ in terms of γ_L and γ_T, which are defined by (8.81) and are related by (8.82). Suppose we know the direction of the incident P wave, which means we know i_L. Snell's law allows us to solve for r_T. From the first equation of (8.81) we solve for c and obtain

$$c^2 = c_L \csc^2 i_L, \tag{8.86}$$

which gives the speed c of the incident surface P wave.

As an example we take the approximation $\lambda = \mu$, which gives $c_L/c_T = \sqrt{3}$. The relationship between γ_L and γ_T then becomes $\gamma_T^2 = 3\gamma_L^2 + 2$. The solution for $\bar{A}_2$ and $\bar{B}_2$ becomes

$$\begin{aligned} \bar{A}_2 &= \frac{4\gamma_L\gamma_T - (1 + 3\gamma_L^2)^2}{4\gamma_L\gamma_T + (1 + 3\gamma_L^2)^2}, \\ \bar{B}_2 &= \frac{4\gamma_L(1 + 3\gamma_L^2)}{4\gamma_L\gamma_T + (1 + 3\gamma_L^2)^2}. \end{aligned} \tag{8.87}$$

From the second equation of (8.87) we see that $\bar{B}_2 = 0$ for normal incidence of the P wave ($i_L = 0$) and also for grazing incidence ($i_L = 90°$). This gives us the two conditions for the vanishing of the reflected S wave. The vanishing of the reflected P wave occurs, of course, when $\bar{A}_2 = 0$. Using the relationship $\gamma_T^2 = 3\gamma_L^2 + 2$, the condition for this case becomes

$$4\gamma_L(3\gamma_L^2 + 2)^{1/2} = (1 + 3\gamma_L^2)^2. \tag{8.88}$$

This equation has two roots: $i_L = 77° \, 13'$, $60°$, which means that for these directions there is no reflected P wave. These roots correspond to $c/c_T = 1.776, 2.000$. To compare the approximation $\lambda = \mu$ with physical reality we consider the example of steel where $c_L = 5{,}940$ m/sec, $c_T = 3{,}220$ m/sec, so that we find $c_L/c_T = 1.84$ instead of 1.73.

The reflection of an incident S wave into a reflected P and S wave is treated in the same manner. Since the incident wave is an S rather than a P wave, we set $A_1 = 0$ and divide all terms of the two equations by B_1 to obtain the ratios of the amplitudes of the reflected S and P waves to the amplitude of the incident P wave.

8.7. Surface Waves, Rayleigh Waves

It occurred to Lord Rayleigh to investigate a type of surface wave in which the amplitude of the wave damps off exponentially as one recedes from the surface of the solid. It was anticipated by Lord Rayleigh that solutions of this type might approximate the behavior of seismic waves observed during earthquakes.†

In studying Rayleigh's treatment of surface waves we take the following model: We consider the (x, y) plane bounded by the free surface $y = 0$. Let the upper half plane $y > 0$ be air, and the lower half plane $y < 0$ be the earth which is assumed to be an isotropic elastic medium obeying Hooke's law. We shall suppose that monochromatic progressing waves are propagated in the positive x direction as a result of forces applied at some distance from the free surface (for example, the forces inside the earth that cause earthquakes). Again we consider a plane strain problem so that all the dependent variables are functions of (x, y, t) and $w = 0$. The scalar and vector potentials satisfy their respective wave equations given by (8.74). Since the nature of the disturbing forces is not specified there are infinitely many solutions for the potentials. We shall use Rayleigh's approach which seeks waves that are exponentially damped.

The scalar and vector potentials that have the properties of being progressing waves in the x direction and decaying exponentially in the earth have

† *Proceedings of the London Mathematical Society*, vol. 17 (1887), or *Scientific Papers*, vol. 2, p. 441.

the form

$$\phi = Ae^{-ay}e^{ik(x-ct)}, \qquad \psi = Be^{-by}e^{ik(x-ct)}, \qquad a>0, \quad b>0. \quad (8.89)$$

A is the amplitude of the scalar potential, B is the amplitude of the vector potential, a, b are the positive *decay coefficients* which are determined by the respective wave equations, and y is positive downward. The system (8.89) is the separation of variables representation of the potentials into the products of an exponential decay function in y (with different decay coefficients) and a function whose argument is $ik(x-ct)$, where k is the wave number and c is the velocity of the Rayleigh surface waves, to be determined from the boundary conditions. We recall that $ck=\omega$.

To determine the decay constants a, b, we insert (8.89) into the wave equations (8.74) and obtain

$$\begin{aligned} a^2 - k^2\left[1-\left(\frac{c}{c_L}\right)^2\right] &= 0, \\ b^2 - k^2\left[1-\left(\frac{c}{c_T}\right)^2\right] &= 0. \end{aligned} \quad (8.90)$$

We recall that $y=0$ is a free surface so that the boundary conditions are given by (8.78). We therefore insert (8.89) into (8.78) and obtain the following pair of homogeneous algebraic equations for A and B:

$$\begin{aligned} -2iakA + (b^2+k^2)B &= 0, \\ [2\mu a^2 + \lambda(a^2-k^2)]A + 2i\mu bkB &= 0. \end{aligned} \quad (8.91)$$

We observe that A and B are complex constants and we want the real parts of the potentials ϕ and ψ. The necessary and sufficient condition for nontrivial solutions for A and B is that the determinant of the left-hand sides of (8.91) vanish. This yields

$$4\mu abk^2 - (b^2+k^2)[2\mu a^2 + \lambda(a^2-k^2)] = 0. \quad (8.92)$$

We now eliminate a and b from (8.92) by appealing to (8.90). After a little manipulation (8.92)—which is a cubic equation for $(c/c_T)^2$—can be put in the following form:

$$s^3 - 8s^2 + 8(3-2r)s - 16(1-r) = 0, \quad (8.93)$$

where

$$s = \left(\frac{c}{c_T}\right)^2, \qquad r = \left(\frac{c_T}{c_L}\right)^2 = \frac{\mu}{\lambda+2\mu}. \quad (8.94)$$

Using the approximation $\lambda=\mu$, we get $r=\frac{1}{3}$ and (8.93) becomes

$$(s-4)(3s^2-12s+9) = 0. \quad (8.95)$$

The roots of (8.95) are

$$s = 4, \qquad 2 + \frac{2}{\sqrt{3}}, \qquad 2 - \frac{2}{\sqrt{3}}. \tag{8.96}$$

It is easily seen that the only root that yields positive values for a and b is $s = 2 - 2/\sqrt{3}$. We then obtain the following relationship between c and c_T:

$$c = s^{1/2} c_T = 0.9194 c_T. \tag{8.97}$$

It is of interest to consider the limiting case of an incompressible body so that the dilatation $\Theta = 0$. This gives $r = 0$, so that in this case the velocity c of the Rayleigh wave becomes

$$c = 0.9553 c_T.$$

We have seen that in either case c is slightly less than the velocity c_T of the equivoluminal wave.

We note that c is independent of frequency. This means there is no dispersion—the wave shape is maintained. Having determined c in terms of c_L or c_T we go to (8.90) to calculate the decay constants a and b which determined the rate at which the potentials attenuate with depth. From the definition of k, both a and b are proportional to the frequency ω and $a > b$. This means that for a given frequency, irrotational waves decay faster than equivoluminal waves. We also see that waves of higher frequency are attenuated more rapidly than those of lower frequency.

Seismograph signatures often depict waves similar in structure to Rayleigh waves. However seismograph records of distant earthquakes indicate dispersion (dependence of c on ω), which arises mainly because of the inhomogeneity of the earth, and also because of the viscoelastic properties of the earth. These effects will not be treated here.

CHAPTER 9

Variational Methods in Wave Phenomena

Introduction

In this chapter we treat the subject of wave propagation from the point of view of the calculus of variations. In order to give a self-contained approach we shall discuss in some detail the fundamental principles of the variational methods used. The approach used up to now has made use of partial differential equations (PDEs), specifically the hyperbolic or wave equation in its various forms. Local displacement, velocity fields, etc., were described by solving these PDEs (the field equations) using the appropriate initial and boundary conditions. These equations were derived from the three conservation laws of physics: conservation of mass, conservation of momentum, and the energy equation which supplies the equation of state or constitutive equation defining the medium. In the global approach used here certain *functionals*, which arise from energy considerations, are to be minimized. A functional is a function of a function, and is defined by a certain integral whose integrand involves a class of functions, one of which minimizes the functional.

An example in optics illustrating a functional is the famous *principle of least time* first enunciated by the French mathematician Fermat. This will be discussed in detail in Section 9.1. We give a brief statement of this principle here in order to illustrate the concept of a functional. The actual path of a light beam is the one which minimizes the time (which is expressed as a functional).

We give another example. For a simple harmonic oscillator the equation of motion is $m\ddot{x} = -kx$, where m is the mass, k is the spring constant, and x is the displacement of the mass from equilibrium. Let the velocity of the mass be $\dot{x} = v$. Since $\ddot{x} = v\, dv/dx$, the equation of motion can be written as $v\, dv/dx = -kx$, so that time is eliminated. Integrating with respect to x gives $\frac{1}{2}mv^2 + \frac{1}{2}kx^2 = \text{const}$. This is the energy equation and expresses the conservation of energy for a simple harmonic oscillator. Actually, the appropriate functional is $\int_{t_1}^{t_2} L\, dt$ where $L = T - V$, $T = \frac{1}{2}m\dot{x}^2$, $V = (k/2)x^2$, which will be discussed below in Section 9.2, Example 1.

In general, in mechanics the conservation of energy is a first integral of the equations of motion. In the calculus of variations we start with such functions as expressed by the energy integral. For example, for a system in equilibrium a fundamental variational principle is that the potential energy of the system is a minimum. This minimum principle (and others) lie at the heart of the variational method. Since the calculus of variations is an integral representation of the essential features of a system, it starts, in a sense, with a global picture (as mentioned above). This has the advantage of smearing out local errors from the beginning. We shall show that the calculus of variations leads to the equations of motion of a dynamical system. We start by describing some variational methods which are the foundations of the calculus of variations.

9.1. Principle of Least Time

To show how a functional that we desire to minimize arises in physics, we turn to a classic example in optics that can be formulated as follows: Consider points A and B in the (x, y) plane, for simplicity. (We can easily extend the treatment to three-dimensional space.) We ask the question: What is the path taken by a light ray that passes through these points? Clearly if the velocity of the light ray is constant the answer is a trivial one: a straight line. However, the plane may contain a medium of variable index of refraction so that the velocity of light in the plane can be a variable function of position. Then the question is no longer a trivial one. The answer was given by Pierre Fermat (1601–1665), a French mathematician who enunciated his famous *Principle of Least Time*, also called *Fermat's Principle.* The statement of Fermat's principle is: Of all possible paths taken by a light ray in going from point A to point B (these points being fixed in space), the actual path taken by a light ray is the one that *takes the least time.* If t is time and s is the arc length along any curve $y = y(x)$ joining the points A and B, the time taken for the light beam to travel the portion of the curve between A and B is given by the integral $I(y)$, where

$$
\begin{aligned}
I(y) &= \int_{t_A}^{t_B} dt = \int_A^B \left(\frac{dt}{ds}\right)\left(\frac{ds}{dx}\right) dx \\
&= \int_A^B \frac{\sqrt{(1+\acute{y}^2)}}{c}\, dx = t_B - t_A, \qquad \acute{y} \equiv \frac{dy}{dx},
\end{aligned} \tag{9.1}
$$

where $c = c(x, y(x))$ is the velocity of the light ray. Since the integral $I(y) = t_B - t_A$, it represents the time taken for the light beam to travel from A to B in the plane. The value of $I(y)$ clearly depends on the path given by the curve $y = y(x)$. There is an infinite set of functions $y = y(x)$ which pass through the endpoints A and B; each function represents a possible path of the light beam. Since $I(y)$ depends on this family of functions it is a functional, namely, a function defined on the set of functions y. Of all possible functions y, we want to find one such that along the curve $y = y(x)$, this functional $I(y)$ is a minimum. The integrand of I involves the velocity of light $c = ds/dt$, which is a function of (x, y), which allows for a variable index of refraction, as mentioned above. If c is constant then the minimizing function y is the curve $y = (y_B - y_A)/(x_B - x_A)x$. How do we minimize I to find the required curve $y = y(x)$? The answer is given in the next section.

9.2. One-Dimensional Treatment, Euler's Equation

The integrand in I in (9.1) is of the form $F(x, y, \acute{y})$. We can therefore generalize the principle of least time as follows: Of the set of admissible functions $\{y\}$ passing through the fixed endpoints (x_A, y_A), (x_B, y_B), find that function $y = y(x)$ which minimizes the integral

$$I(y) = \int_{x_A}^{x_B} F(x, y, \acute{y})\, dx, \qquad \acute{y} \equiv \frac{dy}{dx}. \tag{9.2}$$

This is the basic formulation of the calculus of variations in one dimension. The assumption is that a minimum exists. Actually, we settle for a weaker condition, namely, that an *extremum* or *stationary value* exists. By this we mean that $I(y)$ has either a minimum or maximum value. There are problems in physics where we cannot find a curve which gives a stationary value for I. This is a question of the existence of an extremum; we shall not consider such problems here. The reader is referred to such works as [5] for further and deeper studies of this fascinating field.

As mentioned above, the key problem in the calculus of variations is to solve for the minimizing y. Clearly this minimizing y is the curve $y = y(x)$ from P_A to P_B along which $I(y)$ is a minimum. We now show that the necessary condition for the functional $I(y)$ to have a minimum (more generally, an extremum) is that certain PDEs must exist. These are called *Euler's equations* and are the equations of motion of a given system where an extremum exists. For the one-dimensional case we obtain a single Euler equation.

We start by assuming that the set of admissible functions $y(x)$ are continuously differentiable, so that they represent a family of smooth curves passing through the fixed known endpoints $y(x_A) = y_A$, $y(x_B) = y_B$. We also assume that the integrand $F(x, y, \acute{y})$ is continuously differentiable, for all admissible values of y, in some specified region of the (x, y) plane containing the set of curves $\{y(x)\}$. Let $y = \bar{y}(x)$ be any curve in this set such that the functional $I(\bar{y})$ is not necessarily an extremum. Let $y = y(x)$ be the curve in the set such that $I(y)$ is an extremum. Then any $\bar{y}$ in the neighborhood of the minimizing function y can be represented in the form

$$\delta y(x) = \varepsilon\eta(x), \tag{9.3}$$

where

$$\delta y(x) = \bar{y}(x) - y(x). \tag{9.4}$$

$\delta y(x)$ is called the *variation of* $y(x)$. ε is a small real constant that determines the amount of the variation. Clearly if $\varepsilon = 0$ then $\bar{y} = y$. Finally, $\eta(x)$ is any arbitrary, continuously variable, function of x. All curves must pass through the fixed endpoints. This means

$$\eta(x_A) = \eta_A = 0, \qquad \eta(x_B) = \eta_B = 0. \tag{9.5}$$

We now assume that the functional I is never negative and that the extremum is a minimum. Since $y(x)$ minimizes I, we must have $I(\bar{y}) > I(y)$ or

$$I(y + \varepsilon\eta) \geq I(y). \tag{9.6}$$

The left-hand member of this inequality is a continuously differentiable function of the parameter ε, so that the necessary condition for y to minimize the functional I is

$$\left.\frac{dI(y + \varepsilon\eta)}{d\varepsilon}\right|_{\varepsilon=0} = 0.$$

Using (9.2) and differentiation with respect to ε under the integral sign yields

$$\left.\frac{dI(y + \varepsilon\eta)}{d\varepsilon}\right|_{\varepsilon=0} = \int_{x_A}^{x_B} (\eta F_y + F_{\acute{y}}\acute{\eta})\, dx = 0, \qquad \acute{\eta} \equiv \frac{d\eta}{dx}. \tag{9.7}$$

Integrating the second term in the integral by parts gives

$$\int_{x_A}^{x_B} F_{\acute{y}}\acute{\eta}\, dx = F_{\acute{y}}\eta\Big|_{x_A}^{x_B} - \int_{x_A}^{x_B} \eta\left(\frac{dF_{\acute{y}}}{dx}\right) dx = -\int_{x_A}^{x_B} \eta\left(\frac{dF_{\acute{y}}}{dx}\right) dx,$$

where we used the homogeneous end conditions $\eta_A = \eta_B = 0$. We then write (9.7) as

$$\int_{x_A}^{x_B} \left(F_y - \frac{dF_{\acute{y}}}{dx}\right) \eta(x)\, dx = 0. \tag{9.8}$$

This means that the integral of (9.8) vanishes for every admissible $\eta(x)$ which satisfies the homogeneous end conditions. It follows that the integrand must vanish, yielding

$$F_y - \frac{dF_{\acute{y}}}{dx} = 0. \tag{9.9}$$

This conclusion is a lemma due to Lagrange and is given in works on the calculus of variations such as [5]. Equation (9.9) is called the *Euler equation* for the one-dimensional case associated with the variational problem of finding the extremum of the functional $I(y)$. As mentioned above, Euler's equation is a necessary condition that an extremum exists. This means that if there exists a curve $y(x)$ which makes I an extremum then the Euler equation is valid. By expanding the second term in (9.9) and recalling that $F = F(x, y, \acute{y})$, we can put Euler's equation in the form

$$F_{\acute{y}\acute{y}}\acute{\acute{y}} + F_{\acute{y}y}\acute{y} + F_{\acute{y}x} - F_y = 0. \tag{9.10}$$

Equation (9.10) is a second-order ordinary differential equation (ODE) for the minimizing function $y(x)$. It is in general nonlinear. Since the coefficients depend on F they are known functions of $(x, y, \acute{y})$. We now give a few examples.

EXAMPLE 1. We consider an easy example by revisiting the simple harmonic oscillator previously discussed as an example of a conservative system. We introduce the *Lagrangian* L defined by $L = T - V$ where T is the kinetic energy and V is the potential energy of the oscillator. (We shall discuss the Lagrangian function L in some detail below in connection with Hamilton's principle and Lagrange's equations of motion.) Now, if we identify the integrand F with L, we have

$$F = L - V = \tfrac{1}{2}m\acute{y}^2 - \tfrac{1}{2}ky^2,$$

where x is time, y is the displacement of the mass, and $\acute{y} = dy/dx$ is the velocity of the mass. Inserting this expression for F into Euler's equation (9.9) yields

$$m\acute{\acute{y}} + ky = 0,$$

which is the equation of motion for a simple harmonic oscillator. This simple example tells us that by constructing the appropriate F for a dynamical system we can derive the equations of motion for the system by using Euler's equation.

EXAMPLE 2. This example is less trivial. Suppose F is given by

$$F = p(\acute{y})^2 + qy^2 + 2ry,$$

where p, q, and r are prescribed functions of x. Inserting this expression for F into Euler's equation in the form (9.10) gives the following second-order self-adjoint ODE for y:

$$\frac{d(p\acute{y})}{dx} - qy - r = 0.$$

This example arises in certain problems concerning the deflection of bars and strings for the equilibrium case.

9.3. Euler's Equations for the Two-Dimensional Case

We now extend the càlculus of variations to the problem of minimizing (in general, optimizing) the functional I where I is a multiple integral. For simplicity, we define the dependent variable u as a function of the independent variables (x, y), but there is essentially no difficulty in extending the theory to three-dimensional space. We therefore consider the problem of finding the function $u = u(x, y)$ that makes the functional I an extremum, where

$$I(u) = \int\int_{\mathscr{R}} F(x, y, u, p, q)\, dx\, dy, \tag{9.11}$$

where

$$p = u_x, \qquad q = u_y. \tag{9.12}$$

The integration is taken over the region $\mathscr{R}$ bounded by the surface S. $u(x, y)$ is the optimizing function on the set of admissible functions $[u]$ each of which has continuous first derivatives (p, q), and takes on prescribed values $u = f(s)$ on the bounding surface S. We assume F is continuously differentiable with respect to (x, u, p, q).

We now extend the approach to the double integral $I(u)$ that led to Euler's equation for $I(y)$. Of the set of admissible functions $[\bar{u}]$ let $u(x, y)$ be the optimizing function (the function that makes $I(\bar{u})$ an extremum). We choose any function $\bar{u}(xy)$ of the admissible set. Then, for any arbitrary admissible function $\eta(x, y)$, we have

$$\delta u(x, y) = \bar{u}(x, y) - u(x, y) = \varepsilon\eta(x, y). \tag{9.13}$$

Since every $\bar{u} = f(s)$ on the surface S, we must have the following boundary condition on η:

$$\eta(x, y) = 0 \quad \text{on } S. \tag{9.14}$$

We form $I(u + \varepsilon\eta)$ and set the first variation $\delta I(u + \varepsilon\eta) = 0$. We get

$$\delta I = \frac{dI(u + \varepsilon\eta)}{d\varepsilon}\bigg|_{\varepsilon=0} = 0,$$

since $u(x, y)$ optimizes the functional $I(u)$ given by (9.11). From (9.11) we obtain

$$I(u + \varepsilon\eta) = \int\int_{\mathscr{R}} F(x, y, u + \varepsilon\eta, p + \varepsilon\eta_x, q + \varepsilon\eta_y)\, dx\, dy. \tag{9.15}$$

Then the first variation δI becomes

$$\delta I = \int\int_{\mathscr{R}} (F_u\eta + F_p\eta_x + F_q\eta_y)\, dx\, dy. \tag{9.16}$$

We now want to get rid of the terms in η_x and η_y under the integral sign. To this end, we regard the double integral as a repeated integral and integrate one term by parts with respect to x and the other term with respect to y. We take into account the fact that the boundary conditions vanish on S

and obtain

$$\iint_{\mathcal{R}} \left(F_u - \frac{\partial}{\partial x} F_p - \frac{\partial}{\partial y} F_q \right) dx\, dy = 0. \tag{9.17}$$

It can be shown that the lemma due to Lagrange can be extended to more than one dimension, so that the integrand vanishes thus yielding

$$F_u - \frac{\partial}{\partial x} F_p - \frac{\partial}{\partial y} F_q = 0. \tag{9.18}$$

Equation (9.18) is *Euler's equation* for the function $u(x, y)$ which optimizes the functional $I(u)$. Equation (9.18) is a second-order PDE whose solution yields the optimizing function $u = u(x, y)$.

Poisson's Equation

We now give an example of the use of Euler's equation in two dimensions that leads to Poisson's equation. Let F be given by

$$F = p^2 + q^2 + 2gu, \tag{9.19}$$

where g is a prescribed function of (x, y). Let the admissible values of u satisfy the boundary condition

$$u(x, y) = f(s) \quad \text{on } S, \tag{9.20}$$

where f is a prescribed function of arc lengths on the "surface" S which is actually the closed curve C that bounds the region $\mathcal{R}$. The functional is

$$I(u) = \iint_{\mathcal{R}} [p^2 + q^2 + 2gu]\, dx\, dy. \tag{9.21}$$

On substituting $F = p^2 + q^2 + 2gu$ into Euler's equation (9.18) we get

$$\nabla^2 u = g(x, y) \quad \text{in } \mathcal{R}, \qquad u(x, y) = f(s) \quad \text{on } C, \tag{9.22}$$

which is *Poisson's equation in two dimensions*, where f is a prescribed function on the bounding curve C. This tells us that *the optimizing function $u(x, y)$ satisfies Poisson's equation in the (xy) plane.*

We shall now prove that the solution of Poisson's equation (9.22), with the given boundary condition on C, is indeed the minimizing function for the functional given by (9.21). In order to prove this statement we shall show that $I(\bar{u}) > I(u)$ for any $\bar{u}(x, y)$ not equal to the minimizing function $u(x, y)$. To this end, we set

$$\bar{u}(x, y) = u(xy) + \eta(x, y),$$

where $\eta(x, y)$ is an arbitrary function such that $\eta(s)=0$ on C. We have

$$
\begin{aligned}
\Delta I &= I(\bar{u}) - I(u) \\
&= \iint [(p+\eta_x)^2 + (q+\eta_y)^2 \\
&\quad + 2g(u+\eta)]\, dx\, dy - \iint (p^2+q^2+2gu)\, dx\, dy \\
&= 2\iint_{\mathcal{R}} [p\eta_x + q\eta_y + g\eta]\, dx\, dy \\
&\quad + \iint_{\mathcal{R}} [(\eta_x)^2 + (\eta_y)^2]\, dx\, dy.
\end{aligned}
$$

Applying Green's theorem to the first integral gives

$$\iint_{\mathcal{R}} (p\eta_x + q\eta_y)\, dx\, dy = -\iint_{\mathcal{R}} \nabla^2 u\, dx\, dy + \int_C \eta \frac{du}{dv}\, ds.$$

I then becomes

$$
\begin{aligned}
\Delta I &= -2\iint_{\mathcal{R}} \eta(\nabla^2 u - g)\, dx\, dy + \iint_{\mathcal{R}} [(\eta_x)^2 + (\eta_y)^2]\, dx\, dy \\
&= \iint_{\mathcal{R}} [(\eta_x)^2 + (\eta_y)^2]\, dx\, dy,
\end{aligned}
$$

since $\nabla^2 u = g = 0$ in $\mathcal{R}$ and $\eta = 0$ on C. Therefore we have $\Delta I \geq 0$, so that $I(\bar{u}) \geq I(u)$. The equality holds if and only if $\bar{u} = u$, because $\Delta I = 0$ if and only if $\eta_x = \eta_y = 0$, that is, when $\eta = \text{const}$. But $\eta = 0$ and C. We therefore conclude that $\eta = 0$ in $\mathcal{R}$ so that $\bar{u} = u$.

9.4. Generalization to Functionals with More Than One Dependent Variable

The problem of finding the extremum or stationary value of a functional, can be extended to the case where the integrand of the functional depends not on a single dependent variable u but on scalars u_i $(i = 1, 2, \ldots, n)$ which are the components of the vector $\mathbf{u}$. We have $\mathbf{u} = \mathbf{u}(u_1, u_2, \ldots, u_n) = \mathbf{u}(x)$ since each component $u_i = u_i(x)$. We may therefore formulate a typical problem of this type as follows:

Let $F(x, \mathbf{u}(x), \acute{\mathbf{u}}(x))$ be a function of the $2n+1$ arguments $x, u_1, u_2, \ldots, u_n, \acute{u}_1, \acute{u}_2, \ldots, \acute{u}_n$. Let F be a continuously differentiable function of x up to and including the second order in the x interval considered. It is clear that F is a function of x since $\mathbf{u}=\mathbf{u}(x)$. The functional

$$I(\mathbf{u})=\int_{x_1}^{x_2} F(x, \mathbf{u}, \acute{\mathbf{u}})\, dx, \qquad \acute{\mathbf{u}}\equiv\frac{d\mathbf{u}}{dx}, \tag{9.23}$$

over a given interval $x_1 \leq x \leq x_2$ has a definite value determined by the choice of $\mathbf{u}$ (the n functions $u_i(x)$ $(i=1, 2, \ldots, n)$). In the comparison, we regard all functions $\mathbf{u}(x)$ admissible which satisfy the above continuity conditions and for which the boundary values $\mathbf{u}(x_1)$ and $\mathbf{u}(x_2)$ have prescribed values. This means that we consider curves $u_i(x)$ $(i=1, 2, \ldots, n)$, joining two fixed endpoints A and B in the $(n+1)$ dimensional space, in which the coordinates are $(\mathbf{u}, x)=(u_1, u_2, \ldots, u_n, x)$. The variational problem now requires us to find, among all sets of functions $\mathbf{u}(x)$, the optimizing one which is the set that extremizes the functional $I(\mathbf{u})$. We cannot discuss here the actual nature of the extreme value which brings in subtle questions of existence of solutions, etc. We confine ourselves here to inquiring: For what set of $\mathbf{u}(x)$ is the functional $I(\mathbf{u})$ stationary?

We define the concept of a stationary or extreme value of $I(\mathbf{u})$ in exactly the same way we did in the one- and two-dimensional cases. We include the set of functions $\mathbf{u}(x)$ in a one-parameter family of functions depending on the small parameter ε in the following manner: Let $\boldsymbol{\eta}=(\eta_1, \eta_2, \ldots, \eta_n)$ be a set of n arbitrarily chosen functions of x which satisfy the homogeneous boundary conditions $\boldsymbol{\eta}(x_1)=\boldsymbol{\eta}_1=0$, $\boldsymbol{\eta}(x_2)=\boldsymbol{\eta}_2=0$, and which are continuous in the interval $[x_1, x_2]$ and possess continuous first and second derivatives there.

We now consider the family of functions

$$\bar{\mathbf{u}}(x)=\mathbf{u}(x)+\varepsilon\boldsymbol{\eta}(x) \qquad \text{or} \qquad \bar{u}_i=u_i+\varepsilon\eta_i, \qquad i=1, 2, \ldots, n.$$

where $\bar{\mathbf{u}}(x)$ is any admissible member of the family and $\mathbf{u}(x)$ is the optimizing member. The first variation of $\mathbf{u}$ is $\delta\mathbf{u}(x)$, where

$$\delta\mathbf{u}(x)=\bar{\mathbf{u}}(x)-\mathbf{u}(x)=\varepsilon\boldsymbol{\eta}(x). \tag{9.24}$$

Replacing $\mathbf{u}$ by $\mathbf{u}+\varepsilon\boldsymbol{\eta}$ in (9.23) gives the functional I in terms of the parameter ε.

$$I(\mathbf{u}+\varepsilon\boldsymbol{\eta})=\int_{x_1}^{x_2} F(x, \mathbf{u}+\varepsilon\boldsymbol{\eta}, \acute{\mathbf{u}}+\varepsilon\acute{\boldsymbol{\eta}})\, dx. \tag{9.25}$$

Equation (9.25) tells us that $I=I(\varepsilon)$. A necessary condition that we have an extremum is that the first variation of (9.25) be zero at $\varepsilon=0$, or

$$\frac{dI(0)}{d\varepsilon}=0. \tag{9.26}$$

To solve the problem of setting up conditions for an extremum of $I(\mathbf{u})$ we proceed as follows: Pick a particular $\mathbf{u}_j$ and let $\eta_i(x)=0$ for all the $(n-1)$ i's such that $i\neq j$. Then u_j is the only component of $\mathbf{u}$ that varies so that the condition (9.26), upon using the previous result for the one-dimensional case, is equivalent to

$$F_{u_j} - \frac{dF_{\acute{u}_j}}{dx} = 0. \tag{9.27}$$

which is equivalent to (9.9) where $\acute{y}$ is replaced by u_j. Since the $\dot{u}_j$ is arbitrarily chosen, (9.27) is valid for all $j = 1, 2, \ldots, n$. Then (9.27) becomes the system of n Euler equations. This gives us the following theorem:

A necessary condition that the functional $I(\mathbf{u})$ be stationary (have an extremum) is that the n functions u_j $(j = 1, 2, \ldots, n)$ shall satisfy the system of n Euler equations given by (9.27).

Equation (9.27) is a system of n second-order differential equations for the n components of $\mathbf{u}(x)$. All solutions of this system are said to be *extremals* of the variational problem. Therefore the problem of finding the stationary values of the functional reduces to the problem of solving these Euler equations, which can be considered to be the equations of motion of a dynamical system.

Special Case

Since the possibility of giving a general solution of the system of Euler's equations is rather remote we consider the following special case: If the integrand F does not contain x explicitly, so that $F = F(\mathbf{u}, \acute{\mathbf{u}})$, then the expression

$$E = F(\mathbf{u}, \acute{\mathbf{u}}) - \sum_{i=1}^{n} \acute{u}_i F_{\acute{u}_2} = \text{const.} \tag{9.28}$$

is an integral of Euler's system of equations (9.28). The truth of (9.28) follows immediately if we differentiate (9.28) with respect to x and use (9.27) for all the n j's. The pivotal importance of this first integral (9.28) will be seen when we investigate Lagrange's equations of motion. It will be shown that in this case F is identified with the Lagrangian $L = T - V$ where T is the kinetic energy, V is the potential energy, $\mathbf{u}$ are the generalized coordinates, and $\acute{\mathbf{u}}$ are the generalized velocities; then E is the total energy of a conservative system of n particles. But we shall explain these important concepts in the appropriate section below.

EXAMPLE. We close this section by giving, as an example, the *brachistochrone problem* in three dimensions. Problems of this type were first solved by John Bernoulli in 1696. To formulate this problem let the point $A(x_1, y_1, z_1)$ be joined to point $B(x_2, y_2, z_2)$ by a smooth curve, in such a way that the time taken by a particle sliding from A to B without friction along the curve is a minimum. The only force acting on this particle is gravity which is assumed to act in the positive y direction. It is clear that the time taken for the particle to go from A to B is a functional of the family of admissible curves passing through the fixed endpoints A and B. In fact, this functional is the three-dimensional generalization of I as given by (9.1). We therefore have

$$I(y, z) = \int_{x_1}^{x_2} \frac{\sqrt{(1+\acute{y}^2+\acute{z}^2)}}{\sqrt{y}}\, dx.\dagger$$

The functional $I(y, z)$ represents the time it takes for the particle to slide down the curve from A to B. We want to determine the optimizing curve from A to B; specifically, we want to determine two functions, $y = y(x)$, $z = z(x)$, such that $I(y, z)$ is a minimum. In Euler's equations (9.27) $j = 1, 2$, and we identify u_1 with y, u_2 with z, and F with $\sqrt{(1+\acute{y}^2+\acute{z}^2)}$. We obtain

$$\left(\frac{\acute{z}}{\sqrt{y}}\right)\left(\frac{1}{\sqrt{(1+\acute{y}^2+\acute{z}^2)}}\right) = a,$$

$$F - \acute{y}F_{\acute{y}} - \acute{z}F_{\acute{z}} = \left(\frac{1}{\sqrt{y}}\right)\left(\frac{1}{\sqrt{(1+\acute{y}^2+\acute{z}^2)}}\right) = b,$$

where a and b are constants. It follows that $\acute{z} = a/b = k$ is also constant. The optimizing curve must therefore lie in a plane for which $z = kx + c$, where c is a constant. We integrate the differential equation

$$\left(\frac{1}{\sqrt{y}}\right)\left(\frac{1}{\sqrt{(1+k^2+\acute{y}^2)}}\right) = b,$$

and obtain

$$x = k \int \frac{dy}{\sqrt{((1/b^2 y)-1)}} + \text{const.}$$

Upon integration we obtain a *cycloid* as the optimizing curve.

9.5. Hamilton's Variational Principle

Introduction

The calculus of variations is a reliable guide both in formulating and in treating wave propagation phenomena. Wave phenomena involve the equations of motion with inertia terms which, of course, are unsteady. These are best formulated in terms of *Hamilton's variational principle*, which we describe in this section. This leads to the fundamental laws of motion: *Lagrange's equations of motion* expressed as a second-order system in terms of the generalized coordinates and velocities, or the *Hamilton canonical*

---

† The velocity $ds/dt = \sqrt{2g(y-y_1)}$. We neglect the multiplying factor $1/\sqrt{2g}$.

equations which is a system consisting of a pair of first-order equations in terms of the generalized coordinates and momenta. These concepts will be discussed fully below.

Even though the main aim in this book has been to describe wave phenomena in continuous media, for the sake of logical exposition we shall concentrate on discrete systems of particles in this section. A discrete system is one having a finite number of *degrees of freedom*. The number of degrees of freedom is defined as the number of *independent* coordinates necessary to describe the position of all the elements (particles) of the system. A system composed of N particles has $3N$ independent coordinates in three-dimensional space. We shall call these independent coordinates *generalized coordinates*, which we define by the vector $\mathbf{q} = (q_1, q_2, \ldots, q_n)$ for a system of particles with $n = 3N$ degrees of freedom. In defining the generalized coordinates, the operant word is *independent*. If a system of N particles has k constraints then the number of generalized coordinates is $3N - k$. (We shall discuss constraint below.) The *generalized velocities* are given by the vector $\dot{\mathbf{q}} = (\dot{q}_1, \dot{q}_2, \ldots, \dot{q}_n) \equiv d\mathbf{q}/dt$. If the generalized coordinates and generalized velocities are known functions of time for each particle in the system, then we have a complete dynamical description of the system.

Hamilton's Principle

From an energy point of view, we assume that the mechanical properties of our system are determined by the kinetic energy T and the potential energy V. In general, we assume that $T = T(q_1, q_2, \ldots, q_n, \dot{q}_1, \dot{q}_2, \ldots, \dot{q}_n)$. Specifically, we take T as a quadratic function of $\dot{\mathbf{q}}$ of the form

$$T = \sum_{i,j} P_{ij}(q_1, q_2, \ldots, q_n, t)\dot{q}_i\dot{q}_j. \tag{9.29}$$

This allows T to depend arbitrarily on $\mathbf{q}$. We also assume that $V = V(q_1, q_2, \ldots, q_n, t)$ is a quadratic in $\mathbf{q}$ of the form

$$V = \sum_{i,j} b_{ij} q_i q_j, \tag{9.30}$$

where the b_{ij}'s are given constants. Using these definitions, Hamilton's principle states: Between any two instants of time t_0 and t_1, the motion of a system of particles defined by their trajectories $\mathbf{q} = \mathbf{q}(t)$ proceeds in such a way as to make the integral

$$I = \int_{t_0}^{t_1} (T - V)\, dt, \tag{9.31}$$

an extremum (stationary) with respect to the neighboring trajectories given by $\bar{\mathbf{q}}(t)$, subject to the fixed endpoint conditions: $\bar{\mathbf{q}}_i(t_0) = \mathbf{q}_i(t_0)$ and $\bar{\mathbf{q}}_i(t_1) = \mathbf{q}_i(t_1)$. This means that the first variation $\delta I = 0$ so that the integrand $T - V = L$ obeys Euler's equations. (This will be elaborated on below.)

Physically, this means that the actual motion is the one that makes the value of the functional I an extremum (usually a minimum) with respect to all neighboring virtual motions that start at t_0 and end at t_1.

The integrand on the right-hand side of (9.31) is $T - V = L$, where L is called the *Lagrangian*, and is defined as the difference between the kinetic energy and the potential energy of a system of particles. Since L depends in general on $\mathbf{q}$, $\dot{\mathbf{q}}$, and t, (9.31) is a generalization to n dependent variables of the functional I (and was discussed in Section 9.4) where L corresponds to F, t to x, $\mathbf{q}$ to $\mathbf{u}$, and $\dot{\mathbf{q}}$ to $\acute{\mathbf{u}}$. An example of L was given in Example 1 in Section 9.2. There it was shown that $T = \frac{1}{2}m\acute{q}^2$, $V = \frac{1}{2}kq^2$, so that $L = T - V = \frac{1}{2}m\acute{q}^2 - \frac{1}{2}kq^2$.

9.6. Lagrange's Equations of Motion

For any T and V of the form (9.29) and (9.30), respectively, as mentioned above, we define the Lagrangian L as

$$L = T(\dot{\mathbf{q}}, \mathbf{q}, t) - V(\mathbf{q}, t). \tag{9.32}$$

Therefore (9.31) can be written as

$$I(\mathbf{q}) = \int_{t_0}^{t_1} L(\dot{\mathbf{q}}, \mathbf{q}, t)\, dt. \tag{9.33}$$

Hamilton's principle now says that the motion of the system of particles, whose trajectories pass through t_0 and t_1, must proceed in such a way as to make the functional defined by (9.33) an extremum with respect to $\mathbf{q}$, thus yielding the actual trajectory $\mathbf{q} = \mathbf{q}(t)$. Equation (9.33) is the formulation of the fundamental problem in the calculus of variations. Therefore, the necessary condition that makes the functional $I(\mathbf{q})$ an extremum is that the Euler's equations of motion must be satisfied. These become

$$\frac{d}{dt}\left(\frac{\partial L}{\partial \dot{q}_i}\right) - \frac{\partial L}{\partial q_i} = 0, \qquad i = 1, 2, \ldots, n. \tag{9.34}$$

This system of n PDEs is called *Lagrange's equations of motion*. This set of equations is an alternate way of representing Newton's law of motion for an n degree-of-freedom system. The way to use (9.34) is: For a given system we construct the appropriate L, insert it into (9.34), and solve the resulting system of n PDEs for the trajectories $q_i = q_i(t)$ $(i = 1, 2, \ldots, n)$.

9.7. Principle of Virtual Work

What we have done up to now is state Hamilton's principle. This was formulated as a variational method which made the functional $I(\mathbf{q})$ given by (9.33) an extremum, thereby leading to the actual motion of a dynamical system given by $\mathbf{q}=\mathbf{q}(t)$. The necessary condition for such an extremum to exist is that Lagrange's equations of motion (9.34) are satisfied for the appropriate Lagrangian. We showed that Hamilton's principle led to Lagrange's equations, but we did not prove Hamilton's principle. In this section we supply the proof using the important *principle of virtual work*, suggested by the Swiss mathematician James Bernoulli and developed by the French mathematician D'Alembert. We shall now discuss the *principle of virtual work* in order to prove Hamilton's principle.

We first discuss the concept of *virtual displacement.* This concept is necessary to develop the principle of virtual work. A virtual displacement of a system of n particles is defined as any arbitrary infinitesimal change of the particle coordinates $\delta\mathbf{r}_i$ consistent with the forces and constraints imposed on the system at a given time t. Note that the radius vector $\mathbf{r}_i$ for the ith particle is not necessarily the ith generalized coordinate q_i, since the generalized coordinates are independent variables and the $\mathbf{r}_i$'s may involve constraints (which will be discussed below). The displacement of each particle is called *virtual* to distinguish it from an actual displacement of the particle which may occur over a time interval dt during which the forces and constraints may be changing. From elementary mechanics, the work W_i done by a force $\mathbf{F}_i$, acting on the ith particle in displacing it by an amount $\delta\mathbf{r}_i$, is given by $\delta W_i=\mathbf{F}_i\cdot\delta\mathbf{r}_i$. W_i is therefore the work done due to $\mathbf{F}_i$ acting over the virtual displacement $\delta\mathbf{r}_i$ (at a fixed t). If the system of n particles is in equilibrium then $\mathbf{F}_i=\mathbf{0}$ $(i=1,2,\ldots,n)$. Therefore the condition for the system to be in equilibrium is that the virtual work on the system shall vanish. This is given by

$$\sum_{i=1}^{n}\mathbf{F}_i\cdot\delta\mathbf{r}_i=0. \tag{9.35}$$

We now suppose that the ith particle is acted on by the applied force $\acute{\mathbf{F}}_i$ and a constraining force $\mathbf{f}_i$, so that the total force is given by

$$\mathbf{F}_i=\acute{\mathbf{F}}_i+\mathbf{f}_i. \tag{9.36}$$

Then the condition for the virtual work of the system in equilibrium to vanish is

$$\sum_{i=1}^{n}\acute{\mathbf{F}}_i\cdot\delta\mathbf{r}_i+\sum_i\mathbf{f}_i\cdot\delta\mathbf{r}_i=0. \tag{9.37}$$

We now restrict our system to one where the virtual work of the forces of constraint is zero. This will not be true for a system undergoing frictional forces. For the present we exclude this case. Then the condition for the system to be in equilibrium is that the virtual work of the applied forces on the system must vanish. This gives

$$\sum_i\acute{\mathbf{F}}_i\cdot\delta\mathbf{r}=0. \tag{9.38}$$

Equation (9.38) is often called the *principle of work* due to the applied forces $\acute{\mathbf{F}}_i$. It says that the virtual work due to the applied forces summed over all the particles is equal to zero.

The $\mathbf{r}_i$'s are not independent, due to the constraints on the system (as implied above, the $\mathbf{r}$ space is not a linearly independent space). It follows that, in general, we cannot equate the coefficient of the virtual displacement of each of the n particles to zero. This means that not all the $\acute{\mathbf{F}}_i$'s can vanish in (9.38). In order to be able to equate each coefficient $\acute{\mathbf{F}}_i$ to zero we must transform the $\mathbf{r}$ space to the linearly independent generalized coordinate space $\mathbf{q}$. Only then can the coefficient of each virtual displacement δq_i be set equal to zero. It is clear that if $\acute{\mathbf{F}}_i = 0$ then the ith particle is in equilibrium if acted on only by the applied force $\acute{\mathbf{F}}_i$. This is the situation we want, and this is the situation that we shall get if we use the generalized coordinate space $\mathbf{q}$. With this in mind, we shall investigate this transformation from $\mathbf{r}$ space to $\mathbf{q}$ space.

However, we have another problem. The principle of virtual work expressed by (9.38) is based on the condition of equilibrium, which means that we are dealing with problems in statics. Since we wish to consider dynamical systems we would like to generalize the principle given by (9.38) to problems in dynamics (where inertial forces are considered). This is just what D'Alembert did. He ingeneously wrote the equations of motion in the following form:

$$\mathbf{F}_i - \dot{\mathbf{p}}_i = \mathbf{O}, \qquad i = 1, 2, \ldots, n. \tag{9.39}$$

$\mathbf{p}_i = m_i \mathbf{v}_i$, where $\mathbf{p}_i$ is the momentum of the ith particle, m_i is its mass, and $\mathbf{v}_i$ is its velocity. The inertial force on the ith particle is $d\mathbf{p}_i/dt \equiv \dot{\mathbf{p}}_i$. The right-hand side of (9.39) is the zero force vector. This seemingly innocuous device of putting all the forces on the left-hand side has the advantage of treating the inertial force $\dot{\mathbf{p}}_i$ in the same manner as the external force $\mathbf{F}_i$. Therefore (9.39) says that the sum of the inertial and external force on the ith particle (which consists of all the forces on the particle) vanishes, so that we have the same situation as a particle in equilibrium. This means that D'Alembert, by this simple algebraic trick, transformed the system of n particles from a problem in dynamics to an apparent problem in statics, so that the principle of virtual work can be applied to the system under the action of the $\mathbf{F}_i - \dot{\mathbf{p}}_i$ force system. This device expressed by (9.39) is aptly called *D'Alembert's principle.* Another way of stating this principle is: The system of n particles is in equilibrium under the action of the resultant of the external forces $\mathbf{F}_i$ and the "reversed inertial forces" $-\dot{\mathbf{p}}_i$. The principle of virtual work can now be written as

$$\sum_i (\mathbf{F}_i - \dot{\mathbf{p}}_i) \cdot \delta\mathbf{r} = 0. \tag{9.40}$$

We have again restricted ourselves to systems for which the virtual work due to the forces of constraint $\mathbf{f}_i$ vanish. Equation (9.40) is the dynamical analogue of (9.38).

Constraints

We now discuss a system of n particles subject to k constraints. These constraints will be classified and examples of each type of constraint will be given. It is important to understand the nature of constraints on a system in setting up the Lagrange's equations of motion for a system with constraints.

We first discuss a simple example of the nature of constraints, by considering a single particle which is free to move anywhere on a given surface but cannot leave it. Then the particle has two degrees of freedom and one constraint, or $n=2$, $k=1$. The equation defining the surface on which the particle moves ties up one of the three coordinates so that there are only two independent coordinates ($n=2$). This is an example of a *holonomic constraint*, which is defined as one expressed by an equation connecting the coordinates. Another example of a holonomic constraint is a rigid body composed of n particles. The condition of constraint is given by

$$(\mathbf{r}_i - \mathbf{r}_j)^2 = \mathbf{c}_{ij}, \qquad i, j = 1, 2, \ldots, n, \quad i \neq j,$$

where the $\mathbf{c}_{ij}$'s are constants. Clearly this means that the defining property of a rigid body is characterized by no relative motion between any two particles. In general, the motion of a system of n particles with k constraints is governed by the k constraining equations of the form

$$f_i(\mathbf{r}_1, \mathbf{r}_2, \ldots, \mathbf{r}_n) = 0, \qquad i = 1, 2, \ldots, k.$$

Not all constraints are holonomic. An example of a *nonholonomic constraint* is a particle initially placed on the surface of a sphere of radius R and given an initial velocity. In a gravitational field the particle will slide down the surface and eventually fall off. This type of constraint is not governed by an equation, but by the inequality

$$\mathbf{r}^2 - R^2 \geq 0,$$

Constraints are also classified as *scleronomic* (time independent) or *rheonomic* (containing time explicitly). A bead sliding on a stationary wire is an example of a scleronomic constraint. If the bead slides on a moving wire, then the constraint is rheonomic.

Regardless of the type of constraint, if our system of n particles has k constraints expressed by k equations and/or inequalities, then the number of independent or generalized coordinates is $n-k$. This means that the n Lagrange's equations given by (9.34) are not independent but depend upon the k constraints. The method of undetermined multipliers developed by Lagrange is used to eliminate constraints and construct the appropriate equations of motion. An excellent treatment of this subject is given in [17, Chap. 2]. We shall not discuss this matter here.

9.8. Transformation to Generalized Coordinates

We now carry out our program of transforming the coordinate system $\mathbf{r}$ (where the $\mathbf{r}_i$'s are not linearly independent due to the constraints) to the generalized coordinate system $\mathbf{q}$, so that the coefficient of the ith virtual displacement $\mathbf{q}_i$ can vanish for all i. In other words, we seek a coordinate transformation

$$\mathbf{r}_i = \mathbf{r}_i(q_1, q_2, \ldots, q_n), \qquad i = 1, 2, \ldots, n, \tag{9.41}$$

such that the virtual work on the dynamical system expressed by (9.40) will be transformed to the corresponding system in order that each $\mathbf{F}_i - \dot{\mathbf{p}}_i$ shall $\mathbf{F}_i - \dot{\mathbf{p}}_i = \mathbf{O}$. In this case $\mathbf{p}_i$ is the *generalized momentum* for the ith particle.

Transformation of Coordinates for a Single Particle

Before we see how this works for the system of n particles (which is more complicated mathematically but not conceptually), we discuss the motion of a system consisting of a single particle in three-dimensional space $\mathbf{r} = (x, y, z)$. (The subscript i defining the ith particle is obviously suppressed.) The virtual displacements are $\delta\mathbf{r} = (\delta x, \delta y, \delta z)$ and are not linearly independent since they are tied together by k constraints, where $k < 3$. (Obviously for $k = 3$ there would be no motion.) The number of degrees of freedom is $3 - k$. The essential ideas of the transformation from $\mathbf{r} = (x, y, z)$ to $\mathbf{q} = (q_1, q_2, q_3)$ space are embedded in this simplified system. The generalized coordinates $\mathbf{q}$ may be curvilinear or any other set of three linearly independent variables.

The transformation given by (9.41) becomes

$$x = x(q_1, q_2, q_3), \qquad y = y(q_1, q_2, q_3), \qquad z = z(q_1, q_2, q_3). \tag{9.42}$$

The relations given by (9.42) are continuously differentiable functions of time, so that we can calculate the components of the particle velocity $\mathbf{v}$ by expanding the operator d/dt according to the chain rule. We get

$$\dot{x} = \frac{\partial x}{\partial q_1}\dot{q}_1 + \frac{\partial x}{\partial q_2}\dot{q}_2 + \frac{\partial x}{\partial q_3}\dot{q}_3, \tag{9.43}$$

and similar expansions for $\dot{y}$ and $\dot{z}$. It is clear from (9.43) that the coefficients of the $\dot{q}_i$'s are functions of the q_i's. This gives

$$\dot{x} = \dot{x}(q_1, q_2, q_3, \dot{q}_1, \dot{q}_2, \dot{q}_3)$$

and similar functional relations for $\dot{y}$ and $\dot{z}$. We can calculate the following partial derivatives from these functional relations:

$$\frac{\partial \dot{x}}{\partial q_1}, \quad \frac{\partial \dot{x}}{\partial q_2}, \quad \frac{\partial \dot{x}}{\partial q_3}, \quad \frac{\partial \dot{x}}{\partial \dot{q}_1}, \quad \frac{\partial \dot{x}}{\partial \dot{q}_2}, \quad \frac{\partial \dot{x}}{\partial \dot{q}_3}, \quad \text{etc.}$$

From (9.43) we get

$$\frac{\partial \dot{x}}{\partial \dot{q}_i} = \frac{\partial x}{\partial q_i}, \qquad \frac{\partial \dot{y}}{\partial \dot{q}_i} = \frac{\partial y}{\partial q_i}, \qquad \frac{\partial \dot{z}}{\partial \dot{q}_i} = \frac{\partial z}{\partial q_i}, \qquad i = 1, 2, 3. \tag{9.44}$$

The partial derivatives $\partial x/\partial q_i$, etc., are functions of q_1, q_2, and q_3. We follow the motion of a particle so that the term $\partial/\partial t$ in the operator d/dt vanishes. The above expressions lead to

$$\frac{d}{dt}\frac{\partial x}{\partial q_i} = \frac{\partial \dot{x}}{\partial q_i}, \qquad \frac{d}{dt}\frac{\partial y}{\partial q_i} = \frac{\partial \dot{y}}{\partial q_i}, \qquad \frac{d}{dt}\frac{\partial z}{\partial q_i} = \frac{\partial \dot{z}}{\partial q_i}. \tag{9.45}$$

Equations (9.44) and (9.45) are important to the development of Lagrange's equations for a single particle.

We now turn to the kinetic energy T for a particle,

$$T = \left(\frac{m}{2}\right)(\dot{x}^2 + \dot{y}^2 + \dot{z}^2). \tag{9.46}$$

Since $(\dot{x}, \dot{y}, \dot{z})$ are functions of $\dot{q}_1, \ldots, \dot{q}_3$ we can write T as a quadratic form in $(\dot{q}_1, \dot{q}_2, \dot{q}_3)$.

$$T = \tfrac{1}{2}(a_{11}\dot{q}_1^2 + a_{12}\dot{q}_1\dot{q}_2 + a_{13}\dot{q}_1\dot{q}_3 + a_{22}\dot{q}_2^2 + a_{23}\dot{q}_2\dot{q}_3 + a_{33}\dot{q}_3^2), \tag{9.47}$$

where the coefficients a_{ij} are functions of (q_1, q_2, q_3) and $a_{ij} = a_{ji}$. We now obtain the three equations

$$\frac{\partial T}{\partial \dot{q}_i} = \frac{\partial T}{\partial \dot{x}}\frac{\partial \dot{x}}{\partial \dot{q}_i} + \frac{\partial T}{\partial \dot{y}}\frac{\partial \dot{y}}{\partial \dot{q}_i} + \frac{\partial T}{\partial \dot{z}}\frac{\partial \dot{z}}{\partial \dot{q}_i} = m\dot{x}\frac{\partial x}{\partial q_i} + m\dot{y}\frac{\partial y}{\partial q_i} + m\dot{z}\frac{\partial z}{\partial q_i}. \tag{9.48}$$

From (9.44), (9.45), (9.46), and (9.48) we obtain

$$\frac{d}{dt}\frac{\partial T}{\partial \dot{q}_i} = m\ddot{x}\frac{\partial x}{\partial q_i} + m\ddot{y}\frac{\partial y}{\partial q_i} + m\ddot{z}\frac{\partial z}{\partial q_i} + m\dot{x}\frac{\partial \dot{x}}{\partial q_i} + m\dot{y}\frac{\partial \dot{y}}{\partial q_i} + m\dot{z}\frac{\partial \dot{z}}{\partial q_i}, \tag{9.49}$$

$$\frac{\partial T}{\partial q_i} = m\dot{x}\frac{\partial \dot{x}}{\partial q_i} + m\dot{y}\frac{\partial \dot{y}}{\partial q_i} + m\dot{z}\frac{\partial \dot{z}}{\partial q_i}.$$

The components of the equations of motion $m\ddot{\mathbf{r}} = \mathbf{F}$ are

$$m\ddot{x} = X, \qquad m\ddot{y} = Y, \qquad m\ddot{z} = Z. \tag{9.50}$$

Subtracting the second of (9.49) from the first and using the equations of motion yields the system of three equations

$$\frac{d}{dt}\frac{\partial T}{\partial \dot{q}_i} - \frac{\partial T}{\partial q_i} = X\frac{\partial x}{\partial q_i} + Y\frac{\partial y}{\partial q_i} + Z\frac{\partial z}{\partial q_i}. \tag{9.51}$$

We now introduce the virtual work δW done by (X, Y, Z) due to the virtual displacements $(\delta x, \delta y, \delta z)$. We get

$$\delta W = X\delta x + Y\delta y + Z\delta z = Q_1\delta q_1 + Q_2\delta q_2 + Q_3\delta q_3, \tag{9.52}$$

where $\mathbf{Q} = (Q_1, Q_2, Q_3)$ is defined as the *generalized force* whose components are given by

$$Q_i = X\frac{\partial x}{\partial q_i} + Y\frac{\partial y}{\partial q_i} + Z\frac{\partial z}{\partial q_i}. \tag{9.53}$$

Upon using (9.53), (9.51) becomes

$$\frac{d}{dt}\frac{\partial T}{\partial \dot{q}_i} - \frac{\partial T}{\partial q_i} = Q_i. \tag{9.54}$$

The set of three equations given by (9.54) are Lagrange's equations of motion for a particle in terms of T and $\mathbf{Q}$.

Conservative Force

A conservative force is defined as an external force $\mathbf{F}$ that is *derivable from a potential function.* The potential energy $V = V(\mathbf{q})$ is such a potential function. Then a conservative $\mathbf{F}$ has the property that $\mathbf{F} = \mathbf{grad}\ V$. In the generalized coordinate system the Q_i's are given by

$$Q_i = -\frac{\partial V}{\partial q_i}. \tag{9.55}$$

If we introduce the Lagrangian $L = T - V$ (9.54) then becomes

$$\frac{d}{dt}\frac{\partial L}{\partial \dot{q}_i} - \frac{\partial L}{\partial q_i} = 0, \tag{9.56}$$

which are the three Lagrange's equations of motion for a particle acted upon by a conservative force. It is important to realize that $\partial L/\partial \dot{q}_i$ is the ith component of the generalized momentum and $\partial L/\partial q_i$ is the ith component of the generalized force derivable from a potential. Therefore (9.56) is the ith component of Newton's equations of motion, namely, $\dot{p}_i = F_i$. If F_i is *not derivable from a potential* then the generalized force $\mathbf{Q}$ is not given by (9.55) since it depends on $\dot{\mathbf{q}}$ as well as possibly $\mathbf{q}$. Using $L = T - V$ (9.54) becomes

$$\frac{d}{dt}\frac{\partial L}{\partial \dot{q}_i} - \frac{\partial L}{\partial q_i} = Q_i + \frac{\partial V}{\partial q_i} - \frac{d}{dt}\frac{\partial V}{\partial \dot{q}_i}. \tag{9.57}$$

Equation (9.57) tells us that the generalized force Q_i for a nonconservative system is given by

$$Q_i = -\frac{\partial V}{\partial q_i} + \frac{d}{dt}\frac{\partial V}{\partial \dot{q}_i}. \tag{9.58}$$

For a conservative force system (9.58) reduces to (9.55), since

$$\frac{\partial V}{\partial \dot{q}_i} = 0.$$

Another important observation is that $\partial T/\partial \dot{q}_i = p_i$ and $\partial T/\partial q_i$ can be absorbed into the generalized force so that we again get Newton's equation of motion $\dot{\mathbf{p}} = \mathbf{Q}$.

The *n* Particle System

We now generalize our discussion to the case of a system of n particles. The physical concepts are the same as for a single particle. The only difference is that the mathematical technique is a little more complicated since we must sum over all the particles. We start with the radius vector of the ith particle $\mathbf{r}_i$ which involves constraints, so that the virtual displacement vectors $\mathbf{r}_i$ are not linearly independent. The set of transformations to the n generalized coordinates $\mathbf{q}$ is given by (9.41). The velocity of the ith particle is $\mathbf{v}_i = d\mathbf{r}_i/dt$. We express $\mathbf{v}_i$ in terms of the generalized velocity components q_i by using the chain rule and obtain

$$\mathbf{v}_i = \sum_{j=1}^{n} \frac{\partial \mathbf{r}_i}{\partial q_j} \dot{q}_j + \frac{\partial \mathbf{r}_i}{\partial t}. \tag{9.59}$$

Similarly, the virtual displacement of the ith particle is connected with $\mathbf{q}$ by the following expansion:

$$\delta \mathbf{r}_i = \sum_j \frac{\partial \mathbf{r}_i}{\partial q_j} \delta \dot{q}_j. \tag{9.60}$$

In terms of $\mathbf{q}$ the virtual work δW summed over all particles due to the $\mathbf{F}_i$'s is

$$\delta W = \sum_i \mathbf{F}_i \cdot \delta \mathbf{r}_i = \sum_{i,j} \mathbf{F}_i \cdot \frac{\partial \mathbf{r}_i}{\partial q_j} \delta q_j = \sum_j Q_j \delta q_j, \tag{9.61}$$

where Q_j is the generalized force of the jth particle, and is defined by

$$Q_j = \sum_i \mathbf{F}_i \cdot \frac{\partial \mathbf{r}_i}{\partial q_j}. \tag{9.62}$$

We now go back to the principle of virtual work given by (9.40). We sum over all the particles, use the definition of $\dot{\mathbf{p}}_i$, and consider the term

$$\sum_i \dot{\mathbf{p}}_i \cdot \delta \mathbf{r}_i = \sum_i m_i \ddot{\mathbf{r}}_i \cdot \delta \mathbf{r}_i = \sum_i m_i \ddot{\mathbf{r}}_i \cdot \frac{\partial \mathbf{r}_i}{\partial q_j} \delta q_j.$$

We express the right-hand side of this equation in the form

$$\sum_i m_i \ddot{\mathbf{r}}_i \cdot \frac{\partial \mathbf{r}_i}{\partial q_j} = \sum_i \left[\frac{d}{dt} \left(m_i \dot{\mathbf{r}}_i \cdot \frac{\partial \mathbf{r}_i}{\partial q_j} \right) - m_i \dot{\mathbf{r}}_i \cdot \frac{d}{dt} \left(\frac{\partial \mathbf{r}_i}{\partial q_j} \right) \right].$$

In the last term of the right-hand side we can interchange the operators d/dt and $\partial/\partial q_j$. We get

$$\frac{d}{dt} \frac{\partial \mathbf{r}_i}{\partial q_j} = \sum_k \frac{\partial^2 \mathbf{r}_i}{\partial q_j \, \partial q_k} \dot{q}_k + \frac{\partial^2 \mathbf{r}_i}{\partial q_j \, \partial t} = \frac{\partial \mathbf{v}_i}{\partial q_j}.$$

From these expressions we obtain

$$\sum_i m_i \ddot{\mathbf{r}}_i \cdot \frac{\partial \mathbf{r}_i}{\partial q_j} = \sum_i \left[\frac{d}{dt}\left(m_i \mathbf{v}_i \cdot \frac{\partial \mathbf{v}_i}{\partial q_j} \right) - m_i \mathbf{v}_i \cdot \frac{\partial \mathbf{v}_i}{\partial q_j} \right].$$

This gives

$$\sum_i \dot{\mathbf{p}}_i \cdot \delta r_i = \sum_j \left[\frac{d}{dt} \frac{\partial}{\partial \dot{q}_j} \left(\sum_i \tfrac{1}{2} m_i v_i^2 \right) - \frac{\partial}{\partial q_j} (\tfrac{1}{2} m_i v_i^2) \right] \delta q_j.$$

The kinetic energy of the system $T = \frac{1}{2} \sum_i m_i v_i^2$, so that D'Alembert's principle (9.40) becomes

$$\sum_j \left(\frac{d}{dt} \frac{\partial T}{\partial \dot{q}_j} - \frac{\partial T}{\partial q_j} - Q_j \right) \delta q_j = 0. \tag{9.63}$$

Since all the virtual displacements δq_j are independent, the coefficients of δq_j must vanish, yielding

$$\frac{d}{dt} \frac{\partial T}{\partial \dot{q}_j} - \frac{\partial T}{\partial q_j} = Q_j. \tag{9.64}$$

The system of equations (9.64) are the Lagrange equations for the kinetic energy T, and generalized forces Q_j in terms of the generalized coordinates and velocities.

For a conservative system the potential energy V is a quadratic function of $\mathbf{q}$. Therefore the force $\mathbf{F}_i$ on the ith particle is derivable from the potential V, so that for each particle we have $\mathbf{F}_i = -\mathbf{grad}_i V$, and the generalized forces become

$$Q_j = \sum_i \mathbf{F}_i \cdot \frac{\partial \mathbf{r}_i}{\partial q_j} = -\sum_i \mathbf{grad}_i V \cdot \frac{\partial \mathbf{r}_i}{\partial q_j} = -\frac{\partial V}{\partial q_j}. \tag{9.65}$$

Recalling that T is a function of $\mathbf{q}$ and $\dot{\mathbf{q}}$ and V is a function of $\mathbf{q}$, we introduce the Lagrangian $L = T - V$ so that (9.64) becomes

$$\frac{d}{dt} \frac{\partial L}{\partial \dot{q}_j} - \frac{\partial L}{\partial q_j} = 0, \qquad j = 1, 2, \ldots, n. \tag{9.66}$$

These are Lagrange's equations of motion for a conservative system of n particles.

9.9. Rayleigh's Dissipation Function

Suppose the system of n particles is acted upon by frictional or viscous forces so that the system is not conservative, which means that the external forces are not derivable from a potential. Physically, this means that energy is dissipated in the form of heat. Even for this case Lagrange's equation can still be put in the form of (9.66) if we introduce a *generalized potential*

$U(\mathbf{q}, \dot{\mathbf{q}})$ such that

$$Q_j = -\frac{\partial U}{\partial q_j} + \frac{d}{dt}\frac{\partial U}{\partial \dot{q}_j}. \tag{9.67}$$

Clearly if U does not depend on $\dot{\mathbf{q}}$ then the last term in the right-hand side of (9.67) vanishes, so that $U = V(\mathbf{q})$. If the Lagrangian is now defined as

$$L = T - U, \tag{9.68}$$

then (9.64) becomes

$$\frac{d}{dt}\frac{\partial(L+U)}{\partial \dot{q}_j} - \frac{\partial(L+U)}{\partial q_j} = Q_j. \tag{9.69}$$

These equations become Lagrange's equations (9.66), upon using (9.67), where L is given by (9.68).

If only some of the generalized forces acting on the system are derivable from a potential (either V or the more general U), then Lagrange's equations can always be written as

$$\frac{d}{dt}\frac{\partial L}{\partial \dot{q}_j} - \frac{\partial L}{\partial q_j} = Q_j, \tag{9.70}$$

where L contains V or U, and Q_j represents the force *not* derivable from V or U. Such a situation arises for the case of frictional or viscous forces. If the motion is small enough, the viscous force is proportional to the particle velocity. (As an aside; for aircraft or rockets that fly at greater velocities than those that occur in the materials sciences, the viscous drag is proportional to the square of the velocity.) If the frictional force $\mathbf{f}_i$ on the ith particle is proportional to the velocity, we have

$$\mathbf{f}_i = -\mathbf{k}\mathbf{v}_i, \tag{9.71}$$

where $\mathbf{k} = (k_x, k_y, k_z) > 0$, and the particle velocity $\mathbf{v}_i = (u_i, v_i, w_i)$. In [32], Lord Rayleigh introduced the scalar quadratic function of the velocity $\mathscr{F}$ called the *Rayleigh dissipation function.*† $\mathscr{F}$ is defined by

$$\mathscr{F} = \tfrac{1}{2}\sum_i (k_x u_i^2 + k_y v_i^2 + k_z w_i^2). \tag{9.72}$$

From this definition we obtain

$$\mathbf{f} = -\mathbf{grad}_v \mathscr{F} \qquad \text{where} \qquad \mathbf{grad}_v = \mathbf{i}\frac{\partial}{\partial u} + \mathbf{j}\frac{\partial}{\partial v} + \mathbf{k}\frac{\partial}{\partial w}, \tag{9.73}$$

† The Rayleigh dissipation function is described in the above-cited reference [32, Vol. I, Chap. IV]. Lord Rayleigh mentions that this function first appeared in an article written by him in a paper entitled "On General Theorems Relating to Vibrations" published in the *Proceedings of the London Mathematical Society*, June, 1873.

and $(\mathbf{i}, \mathbf{j}, \mathbf{k})$ are the usual unit vectors. We may give a physical interpretation of the dissipation function $\mathscr{F}$ by considering a differential amount of work dW_f done by the system against the frictional force $\mathbf{f}$. This is given by

$$dW_f = -\mathbf{f} \cdot d\mathbf{r} = -\mathbf{f} \cdot \mathbf{v}\, dt = (k_x u^2 + k_y v^2 + k_z w^2).$$

This means, from (9.72), that $\mathscr{F}$ is half the rate of energy dissipated by friction. Let Q_j be the jth component of the generalized force due to the frictional force. This is given by

$$\begin{aligned} Q_j &= \sum_i \mathbf{f}_i \cdot \frac{\partial \mathbf{r}}{\partial q_j} = -\sum_i \mathbf{grad}_v \mathscr{F} \cdot \frac{\partial \dot{\mathbf{r}}_i}{\partial \dot{q}_j} \\ &= -\frac{\partial \mathscr{F}}{\partial \dot{q}_j}. \end{aligned} \tag{9.74}$$

Lagrange's equations now become

$$\frac{d}{dt}\frac{\partial L}{\partial \dot{q}_j} - \frac{\partial L}{\partial q_j} = -\frac{\partial \mathscr{F}}{\partial \dot{q}_j}. \tag{9.75}$$

We see from (9.75) that the two scalars L and $\mathscr{F}$ must be prescribed in order to solve for the motion of a system subjected to a viscous force.

9.10. Hamilton's Equations of Motion

For a system of n particles we recall that Lagrange's equations involved n generalized coordinates $\mathbf{q}$ and n generalized velocities $\dot{\mathbf{q}}$, so that there are n second-order PDEs for $L(\mathbf{q}, \dot{\mathbf{q}})$, thus requiring n initial values of the q_i's and n of the $\dot{q}_i$'s. Thus the Lagrange formulation can be considered as a description of a dynamical situation, in terms of trajectories defined by the generalized coordinates and velocities as functions of time. These are the same requirements on the initial condition demanded by Newton's equations.

Hamilton developed an alternative formulation of the equations of motion. This was based on using the n generalized coordinates $\mathbf{q}$, but substituting *n generalized momenta p* for the generalized velocities $\dot{\mathbf{q}}$. Associated with each q_i he introduced a generalized p_i called the *canonical momentum* or *conjugate momentum* (conjugate to the q_i). Clearly each $p_i = m_i\dot{q}_i$. Specifically, Hamilton defined p_i, by using the Lagrangian, as

$$p_i = \frac{\partial L}{\partial \dot{q}_i}.$$

By this definition the Lagrange equation for the ith particle for a conservative force system is transformed into

$$\frac{d}{dt} p_i - \frac{\partial V}{\partial q_i} = 0 \qquad \text{or} \qquad \dot{p}_i = \frac{\partial V}{\partial q_i}, \tag{9.76}$$

which is a first-order system for p_i and V. But this form of the transformed Lagrange's equations is not used, since we seek a first-order system of PDEs where the independent variables are the q_i's and p_i's. We shall see below that the appropriate dependent variable turns out to be the total energy.

We recall that in Lagrange's equations the relationship between the momentum and the Lagrangian is

$$p_i = \frac{\partial L}{\partial \dot{q}_i}. \tag{9.77}$$

We observe that the dimensions of p_i are (ML/T) provided q_i is given as length. We note that the Lagrangian has the dimensions of energy so that the right-hand side of (9.77) is dimensionally correct. However, a particular q_i may have dimensions other than length; for example, it may be given in terms of an angle or a Fourier component (which is a linearly independent variable), in which case the corresponding p_i is not in units of ML/T, but is still in units of (mass) $\times$ (generalized velocity), whatever the units of q_i are.

The motivation for the definition of momentum given by (9.77) is seen by considering the following example: Consider a system of n particles under no constraints, under the influence of a conservative force (derivable from a potential V). In the Cartesian reference frame the generalized coordinates and velocities are $\mathbf{q} = (x, y, z)$, $\dot{\mathbf{q}} = (\dot{x}, \dot{y}, \dot{z}) = (u, v, w)$. This gives

$$L = T - V = \tfrac{1}{2}\sum_i m_i[(u_i)^2 + (v_i)^2 + (w_i)^2] - \tfrac{1}{2}k\sum_i [(x_i)^2 + (y_i)^2 + (z_i)^2].$$

We differentiate L with respect to u_i, v_i, and w_i, and obtain the x, y, and z components of p_i, the momentum for the ith particle.

$$\frac{\partial L}{\partial u_i} = m_i u_i = (p_i)_x, \qquad \frac{\partial L}{\partial v_i} = m_i v_i = (p_i)_y, \qquad \frac{\partial L}{\partial w_i} = m_i w_i = (p_i)_z.$$

Hamilton defined a function, aptly called the *Hamiltonian H*, which gives the total energy of a conservative system and is a function of $\mathbf{p}$ and $\mathbf{q}$.

$$H = H(\mathbf{p}, \mathbf{q}) = T + V. \tag{9.78}$$

We may obtain some of the properties of the Hamiltonian from the Lagrangian of a conservative system. For a conservative system $\mathbf{F} = -\mathbf{grad}\ V$, where V is a function of $\mathbf{q}$ (as mentioned above). Suppose the constraints of the system are independent of time (scleronomic). Then expanding $d/dt\, L(\mathbf{q}, \dot{\mathbf{q}})$ gives

$$\frac{d}{dt} L = \sum_i \frac{\partial L}{\partial q_i}\frac{dq_i}{dt} + \sum_i \frac{\partial L}{\partial \dot{q}_i}\frac{d\dot{q}_i}{dt}.$$

From Lagrange's equations we get

$$\frac{d}{dt} L = \sum_i \frac{d}{dt}\frac{\partial L}{\partial \dot{q}_i}\dot{q}_i + \sum \frac{\partial L}{\partial \dot{q}_i}\frac{d}{dt}\dot{q}_i = \sum_i \frac{d}{dt}\left(\dot{q}_i \frac{\partial L}{\partial \dot{q}_i}\right).$$

This yields

$$\frac{d}{dt}\left[\sum_i \dot{q}_i \frac{\partial L}{\partial \dot{q}_i} - L\right] = 0.$$

This means that the expression in the brackets must be constant. Setting this expression equal to H gives

$$\sum_i \dot{q}_i \frac{\partial L}{\partial \dot{q}_i} - L = H = \text{const.}$$

The Hamiltonian then becomes

$$H = \sum_i \dot{q}_i p_i - L. \tag{9.79}$$

From this important equation we now show that H is indeed the total energy $T+V$ for a conservative system. Using the definition of L and recognizing that V is independent of $\dot{\mathbf{q}}$, (9.79) gives

$$p_i = \frac{\partial T}{\partial \dot{q}_i}.$$

Since T is a homogeneous quadratic function of the $\dot{q}_i$'s, the first term of the right-hand side of (9.79) becomes

$$\sum_i \dot{q}_i \frac{\partial T}{\partial \dot{q}_i} = 2T.$$

Equation (9.79) now yields $H = 2T - (T - V) = T + V$, which is the total energy for a conservative system of n particles. Note that this is the case where H is independent of t.

We shall now obtain a pair of first-order differential equations for H, where we consider $H = H(\mathbf{p}, \mathbf{q}, t)$. To this end, we first expand dH and obtain

$$dH = \sum_i \frac{\partial H}{\partial q_i} dq_i + \sum_i \frac{\partial H}{\partial p_i} dp_i + \frac{\partial H}{\partial t} dt. \tag{9.80}$$

Next we form dH obtaining

$$dH = \sum_i \dot{q}_i \, dp_i + \sum_i p_i \, d\dot{q}_i - \sum_i \frac{\partial L}{\partial q_i} dq_i - \sum_i \frac{\partial L}{\partial \dot{q}_i} d\dot{q}_i - \frac{\partial L}{\partial t} dt. \tag{9.81}$$

The coefficients of the $d\dot{q}_i$'s are zero, because of the definition of the generalized momenta given by (9.77). We also have (since T is independent of $\mathbf{q}$)

$$\frac{\partial L}{\partial q_i} = \dot{p}_i.$$

Therefore (9.81) reduces to

$$dH = \sum_i \dot{q}_i \, dp_i - \sum_i p_i \, d\dot{q}_i - \frac{\partial L}{\partial t} \, dt. \tag{9.82}$$

Comparing (9.82) with (9.80) yields the following system of $2n+1$ equations:

$$\dot{q}_i = \frac{\partial H}{\partial p_i}, \qquad \dot{p}_i = -\frac{\partial H}{\partial q_i}, \qquad i = 1, 2, \ldots, n, \tag{9.83}$$

$$-\frac{\partial L}{\partial t} = \frac{\partial H}{\partial t}. \tag{9.84}$$

Equations (9.83) are called *Hamilton's canonical equations of motion.* They constitute the system of $2n$ first-order PDEs for $H(\mathbf{p}, \mathbf{q}, t)$, which replaces the Lagrange system of n second-order equations for $L(\mathbf{q}, \dot{\mathbf{q}}, t)$. Equation (9.84) relates L to H for the time-varying case.

As an example of the use of Hamilton's equations (9.83) we consider a system of n simple harmonic oscillators, where $\mathbf{q} = (q_1, q_2, \ldots, q_n)$ is interpreted as a set of normal coordinates so that each mass ocillates in an uncoupled manner and obeys the equation of motion

$$m_i \ddot{q}_i = -kq_i,$$

where the spring constant k is assumed to be the same for all the particles. Since the momentum of the ith particle is $m_i \dot{q}_i = p_i$, the kinetic energy of the system becomes

$$T = \tfrac{1}{2} \sum_i \frac{(p_i)^2}{m_i}.$$

The potential energy is

$$V = \tfrac{1}{2} k \sum_i (q_i)^2.$$

The Hamiltonian becomes

$$H = \tfrac{1}{2} \sum_i \frac{(p_i)^2}{m_i} + \tfrac{1}{2} k \sum_i (q_i)^2 = \tfrac{1}{2} \sum \mathbf{p} \cdot \frac{\mathbf{p}}{m_i} + \tfrac{1}{2} k \sum \mathbf{q} \cdot \mathbf{q} = H(\mathbf{p}, \mathbf{q}). \tag{9.85}$$

From this we obtain

$$\frac{\partial H}{\partial q_i} = kq_i, \qquad \frac{\partial H}{\partial p_i} = m_i p_i.$$

Upon using Hamilton's equations these equations yield

$$kq_i = -\dot{p}_i, \qquad \frac{p_i}{m_i} = \dot{q}_i.$$

The second equation is merely the definition of the momentum of the ith particle. The first equation is the equation of motion for the simple harmonic motion of the ith particle, which is what we set out to demonstrate.

9.11. Cyclic Coordinates

If the Lagrangian of a system does not contain a given generalized coordinate q_i (although it may contain the corresponding velocity $\dot{q}_i$), then that q_i is said to be a *cyclic coordinate.* If q_i is a cyclic coordinate then

$$\frac{\partial L}{\partial q_i} = 0, \tag{9.86}$$

so that the Lagrange equation of motion for the ith particle becomes

$$\frac{d}{dt}\frac{\partial L}{\partial \dot{q}_i} = \frac{dp_i}{dt} \equiv \dot{p}_i = 0. \tag{9.87}$$

This gives the important result that

$$p_i = \text{const.} \qquad \text{for } q_i \text{ cyclic.} \tag{9.88}$$

From this we get the following generalized conservation theorem:

The generalized momentum conjugate to a cyclic coordinate is conserved.

Looking at (9.79) we see that if q_i is cyclic, then $H = -L$ (for that coordinate) so that the generalized conservation theorem can be readily transformed to the Hamilton formulation.

There is another point that makes the Hamiltonian procedure particularly adaptable to the treatment of problems involving cyclic coordinates. This is shown by recognizing that the momentum p_i conjugate to a cyclic coordinate q_i yields $dp_i/dt = 0$. The second of Hamilton's equations (9.84) immediately tells us that H does not contain that q_i. This gives the following principle:

A cyclic coordinate q_i will be absent from the Hamiltonian H. Conversely, if a generalized coordinate q_i does not occur in H then the conjugate momentum p_i is conserved (is constant) and the q_i is cyclic.

This last statement is easily seen from the fact that $\partial H/\partial q_i = 0$.

Cyclic coordinates lead to certain *symmetry properties.* For example, suppose a particular cyclic coordinate corresponds to a pure rotation about the x axis. Then the conjugate momentum is the angular momentum, and it is conserved. If the system is spherically symmetric, then all the components of the angular momentum are conserved. Another simple example: If a coordinate corresponding to the displacement of the center of gravity of the system in a given direction is cyclic, then the system is invariant with respect to a pure translation in that direction, and the conjugate linear momentum is conserved.

The physical significance of H was seen above where we showed that if L is not an explicit function of t, then the external forces are derivable from a potential $V(\mathbf{q})$ so that the energy is conserved and H is constant. We may prove this in another way by directly expanding dH/dt and using Hamilton's canonical equations. We get

$$\frac{dH}{dt}=\sum_i\left(\frac{\partial H}{\partial q_i}\dot{q}_i+\frac{\partial H}{\partial p_i}\dot{p}_i+\frac{\partial H}{\partial t}\right).$$

Upon using Hamilton's equations we eliminate the $\dot{q}_i$'s and $\dot{p}_i$'s in the above expression, which now becomes

$$\frac{dH}{dt}=\sum_i\left(\frac{\partial H}{\partial q_i}\frac{\partial H}{\partial p_i}-\frac{\partial H}{\partial p_i}\frac{\partial H}{\partial q_i}\right)+\frac{\partial H}{\partial t},$$

which yields

$$\frac{dH}{dt}=\frac{\partial H}{\partial t}=-\frac{\partial L}{\partial t}. \tag{9.89}$$

Clearly if L does not explicitly depend on t, then $dH/dt=0$ so that H is constant. Note that H may be constant and not the energy integral if, for example, a dissipative force exists yielding a velocity-dependent potential. We saw above that H is the energy integral only if V is velocity independent.

Another advantage of using the Hamiltonian formulation is as follows: Suppose we label the n generalized coordinates so that q_n is cyclic. We can write the Lagrangian as

$$L=L(q_1, q_2, \ldots, q_{n-1}, \dot{q}_1, \dot{q}_2, \ldots, \dot{q}_n, t).$$

All the n generalized velocities still occur in the Lagrangian so that there are still n Lagrange's equations of motion. This means that we must still solve an n degree-of-freedom system in obtaining the trajectories for the n q_i's from the equations of motion (even though we have the simplification that the conjugate p_n is constant). On the other hand, if we use the Hamilton formulation, the cyclic coordinate q_n truly deserves its alternative description as an *ignorable coordinate*, since the Hamiltonian contains $n-1$ degrees of freedom. Indeed, the conjugate $p_n=\alpha=$ const. so that

$$H=H(q_i, q_2, \ldots, q_{n-1}, p_1, p_2, \ldots, p_{n-1}, \alpha, t).$$

This tells us that, in contrast to the Lagrangian, the Hamiltonian actually describes an n particle system involving $n-1$ degrees of freedom if one of the coordinates is cyclic. This means that we need only solve for $n-1$ trajectories from Hamilton's equations which are thus valid for $i=1, 2, \ldots, n-1$. We completely ignore the cyclic coordinate q_n and recognize that the conjugate $p_n=\alpha$, which is the constant of integration arising from the conservation of p_n. α is given as an initial condition. The trajectory of the q_n can be determined by integrating the corresponding equation of motion

$$\dot{q}_n = \frac{\partial H}{\partial \alpha}. \tag{9.90}$$

We shall capitalize on the use of cyclic coordinates in Hamilton's formulation when we treat Hamilton–Jacobi theory below. For the present we switch gears and treat a *continuum* first by the Lagrangian, then by the Hamiltonian formulation.

9.12. Lagrange's Equations of Motion for a Continuum

Introduction

We now have the mathematical equipment and physical ideas necessary to discuss the relationship between variational principles and wave propagation phenomena. We therefore extend the discussion to consider continuous media, since we are interested in wave propagation in such media. We start by developing Lagrange's equations for a continuum.

One-Dimensional Case

In this section we use a one-dimensional bar as a model to develop the Lagrangian for a continuous distribution of matter. This will be used to derive the wave equation from the Lagrange equation for a continuum.

We start by considering a discrete model of a bar of elastic material. To this end, we approximate a finite bar by a coupled system of N particles which are represented by a uniform distribution of points on the x axis. Let x_i be an interior point representing the ith particle of the bar, and let $u_i(t)$ be the displacement of the particle from its equilibrium configuration. Then the velocity of the ith particle is $\dot{u}_i = du_i/dt$. The kinetic energy of the system is

$$T = \tfrac{1}{2}\sum_i \Delta m(\dot{u}_i)^2 = \tfrac{1}{2}\sum_i \rho A \Delta x(\dot{u}_i)^2, \tag{9.91}$$

where Δm is the particle mass, $\Delta x = x_{i+1} - x_i = \text{const.}$, A is the cross-sectional area of the bar, and ρ is its density. The potential energy of the system is

$$V = \tfrac{1}{2}\sum_i \frac{EA}{\Delta x}(u_{i+1} - u_i)^2, \tag{9.92}$$

where E is the Young's modulus of the bar and $EA/\Delta x = K$ is the spring force per unit length. We may justify (9.92) by deriving the elastic or spring force F_i on the ith particle and comparing it with the force derivable from the potential V. The force due to the spring on the right of the particle is $K(u_{i+1} - u_i)$ while the force on the left of the particle is $-K(u_i - u_{i-1})$, so that the total force is $F_i = K(u_{i+1} - 2u_i + u_{i-1})$. This agrees with $F_i = -\partial V/\partial u_i$ as obtained from (9.92).

We now construct the Lagrangian from (9.91) and (9.92) and obtain

$$L = T - V = \tfrac{1}{2}A \sum_i \left[\rho(\dot{u}_i)^2 - E\left(\frac{u_{i+1} - u_i}{\Delta x}\right)^2\right] \Delta x. \tag{9.93}$$

We take the limit of the right-hand side of (9.93) as $N \to \infty$, $\Delta x \to 0$, so that in the limit we have $u_i = u(x, t)$, the longitudinal displacement of a continuous particle from its equilibrium position, and $\dot{u}_i = u_t$. The sum becomes the integral from $x = 0$ to $x = L$, the length of the bar, and

$$\lim_{\Delta x + 0} \left(\frac{1}{\Delta x}\right)(u_{i+1} - u_i) = u_x.$$

Then the Lagrangian for the continuous bar becomes

$$L = \int_0^L \mathscr{L}\, dx, \tag{9.94}$$

where $\mathscr{L}$ is called the *Lagrangian density function* and is given by

$$\mathscr{L} = \tfrac{1}{2}A[\rho(u_t)^2 - E(u_x)^2] = \mathscr{L}(u_t, u_x). \tag{9.95}$$

We see that L is a functional whose integrand $\mathscr{L}$ depends on u_t and u_x.

The Euler equation, which is a necessary condition for the functional L to be stationary, is given by

$$\frac{d}{dt}\frac{\partial \mathscr{L}}{\partial u_t} - \frac{\partial \mathscr{L}}{\partial u} = 0. \tag{9.96}$$

We must interpret the operator d/dt as the partial derivative with respect to t, since before we performed the limiting process each u_i was a differentiable function of t. Inserting (9.95) into (9.96) yields the one-dimensional wave equation for wave propagation in the bar for a wave speed c,

$$c^2 u_{xx} - u_{tt} = 0, \qquad c^2 = \frac{E}{\rho}. \tag{9.97}$$

If an external force $f(x, t)$ is applied to the bar then the term $f(x, t)u(x, t)\, dx$ must be added to the potential energy. The Euler equation then yields the nonhomogeneous wave equation

$$Eu_{xx} - \rho u_{tt} = f(x, t). \tag{9.98}$$

Two-Dimensional Case

We now extend Lagrange's method to the (x, y) plane by using the membrane as an example. $u(x, y, t)$ is interpreted as the lateral displacement of the membrane (in the z direction) whose equilibrium configuration is the plane

$z=0$. A membrane is defined as a portion of a surface with the potential energy proportional to a change in surface area. The proportionality factor is known as the tension of the membrane. Suppose the membrane at rest is in a region $\mathcal{R}$ in the (x, y) plane. Let u be small in the sense that second-order terms in u, u_x, u_y, and u_t are neglected.

First we consider the equilibrium case where $u=u(x, y)$. Suppose $\mathcal{R}$ is a finite region bounded by the curve C, and suppose we prescribe the boundary condition $u=f(s)$ on C, where s is the arc length along C. The required potential energy that we desire to minimize depends on the first-order approximation to the surface area of the membrane, when subject to the prescribed boundary condition. The appropriate functional to minimize is (apart from a constant)

$$\frac{1}{2}\int_{\mathcal{R}} [(u_x)^2+(u_y)^2]\, dx\, dy. \tag{9.99}$$

This expression is the Lagrangian for the two-dimensional steady-state case. The term in brackets is the corresponding Lagrange density function. The Euler equation that minimizes this functional (the Lagrangian) is

$$\nabla^2 u = u_{xx}+u_{yy}=0, \qquad u=f(s) \quad \text{on } C. \tag{9.100}$$

Thus the problem of finding the equilibrium configuration of the membrane subject to the given boundary condition is equivalent to the following classical problem in potential theory: Find the solution to the two-dimensional Laplace equation in $\mathcal{R}$ subject to the prescribed boundary condition $u=f(s)$ on C.

Next we consider the case of a vibrating membrane occupying the same region but subject to the homogeneous boundary condition $u=f(s)=0$ on C. In addition, we suppose that the membrane is acted on by a prescribed lateral force $f(x, y, t)$. The expression for the potential energy is the same as above. The kinetic energy is

$$T=\tfrac{1}{2}\rho(u_t)^2,$$

so that the Lagrangian functional is of the form

$$L=\int_{\mathcal{R}} \mathcal{L}(u_t, u_x, u_y, f(x, y, t)u)\, dx\, dy, \tag{9.101}$$

where the Lagrange density function is given by

$$\mathcal{L}=\tfrac{1}{2}\rho(u_t)^2-\tfrac{1}{2}\mu[(u_x)^2+(u_y)^2]-f(x, y, t)u. \tag{9.102}$$

μ is the elastic modulus of the membrane. The Euler equation for this problem is

$$\mu\nabla^2 u-\rho u_{tt}=f(x, y, t), \tag{9.103}$$

where ∇^2 is the two-dimensional Laplacian. This is the two-dimensional nonhomogeneous wave equation. For specific problems we require two initial conditions as well as appropriate boundary conditions.

Three-Dimensional Case

As we have observed in investigating wave propagation in elastic media, the situation is complicated by the fact that traveling waves exhibit various modes of vibration: lateral and longitudinal. To develop the appropriate equations of motion for these more complicated cases that arise in continuous media, we need to extend Hamilton's principle and thereby extend the Lagrange formulation. To this end, we start by considering the displacement vector $\mathbf{u} = \mathbf{u}(x, y, z, t)$. We now replace this notation by a symmetric one given by $\mathbf{u} = \mathbf{u}(u_1, u_2, u_3)$, where $u_i = u_i(x_1, x_2, x_3, t)$ $(i = 1, 2, 3)$. Let the particle velocity $\mathbf{u}_t = (u_{1,t}, u_{2,t}, u_{3,t})$. Using this notation and the tensor summation convention, the kinetic energy becomes

$$T = \tfrac{1}{2}\rho u_{i,t} u_{i,t} \qquad \text{where} \quad u_{i,t} \equiv \frac{\partial u_i}{\partial t}.$$

The potential energy becomes

$$V = \tfrac{1}{2}\mu u_{i,j} u_{i,j} \qquad \text{where} \quad u_{i,j} \equiv \frac{\partial u_j}{\partial x_j}.$$

The Lagrangian functional is of the form

$$L = \int_{\mathscr{R}} \mathscr{L}(\mathbf{u}_{\mathbf{x}}, \mathbf{u}_t)\, dV, \tag{9.104}$$

where $dV = dx\, dy\, dz$. Upon using the above expressions for T and V, the Lagrange density function becomes

$$\mathscr{L} = T - V = \tfrac{1}{2}\rho u_{i,t} u_{i,t} - \tfrac{1}{2}\rho u_{i,j} u_{i,j}, \tag{9.105}$$

where we sum over i and j. The generalization of Hamilton's principle for a three-dimensional continuum for the various modes of wave propagation described by $\mathbf{u}(\mathbf{x}, t)$ can be formulated as follows: Find that function $\mathbf{u} = \mathbf{u}(\mathbf{x}, t)$ which makes the functional $I(\mathbf{u})$ an extremum, or which makes the first variation $\delta I(\mathbf{u}) = 0$. We introduce the virtual displacement vector $\delta\mathbf{u}$, which is the generalization for a continuum of the generalized virtual displacement vector $\delta\mathbf{q}$. Consequently, Hamilton's principle takes on the form

$$\delta I(\mathbf{u}) = \delta \int_{t_A}^{t_B} \int_{\mathscr{R}} \mathscr{L}\, dV\, dt = \int_{t_A}^{t_B} \int_{\mathscr{R}} \delta\mathscr{L}\, dV\, dt = 0. \tag{9.106}$$

Note that we integrate over $\mathscr{R}$ in space and over t from t_A to t_B. The operator δ operates on the u_i's and $u_{i,t}$'s but does not operate on the x_i's or t. Used here, the operator δ is a generalization of the same operator used in developing the principle of virtual work for a discrete system of particles, where t was fixed while performing the variation δ. Since $\mathscr{L}$ is a function of the u_i's and $u_{i,t}$'s, upon expanding we obtain

$$\delta\mathscr{L} = \frac{\partial\mathscr{L}}{\partial u_i}\,\delta u_i + \frac{\partial\mathscr{L}}{\partial u_{i,t}}\,\delta u_{i,t} + \frac{\partial\mathscr{L}}{\partial u_{i,j}}\,\delta u_{i,j}. \tag{9.107}$$

We now insert (9.107) into (9.106) and first integrate by parts with respect to t, obtaining

$$\int_{t_A}^{t_B} \frac{\partial\mathscr{L}}{\partial u_{i,t}}\,\delta u_{i,t}\,dt = -\int_{t_A}^{t_B} \frac{d}{dt}\,\frac{\partial\mathscr{L}}{\partial u_{i,t}}\,\delta u_{i,t}\,dt.$$

Next we work on the integrals involving the spatial derivatives of the u_i's by interchanging the $\partial/\partial x_j$'s and the δ variations. We obtain

$$\int \frac{\partial\mathscr{L}}{\partial u_{i,j}}\,\delta u_{i,j}\,dx_j = \int \frac{\partial\mathscr{L}}{\partial u_{i,j}}\left(\frac{\partial\,\delta u_i}{\partial x_j}\right)dx_j.$$

An integration by parts converts the integral on the right-hand side to the following expression:

$$\frac{\partial\mathscr{L}}{\partial u_{i,j}}\,\delta u_i - \int \frac{d}{dx_j}\left(\frac{\partial L}{\partial u_{i,j}}\right)\delta u_i\,dx_j.$$

Since the u_i's vanish at the endpoints t_A, t_B, the integrated term also vanishes. Collecting these results yields the following form for Hamilton's principle:

$$\delta I(\mathbf{u}) = \int_{t_A}^{t_B}\int_{\mathscr{R}} \delta u_i\left[\frac{\partial\mathscr{L}}{\partial u_i} - \frac{d}{dt}\left(\frac{\partial\mathscr{L}}{\partial u_{i,t}}\right) - \frac{d}{dx_j}\left(\frac{\partial\mathscr{L}}{\partial u_{i,j}}\right)\right]dV\,dt = 0. \tag{9.108}$$

The four-dimensional integral can vanish only if the coefficient of each of the linearly independent virtual displacements δu_i vanishes. This gives the following three Euler equations of motion:

$$\frac{d}{dt}\left(\frac{\partial\mathscr{L}}{\partial u_{i,t}}\right) + \frac{d}{dx_j}\left(\frac{\partial\mathscr{L}}{\partial u_{i,j}}\right) - \frac{\partial\mathscr{L}}{\partial u_i} = 0. \tag{9.109}$$

(Reminder: In each of these equations we sum over j.) For $\mathscr{L}$ given by (9.105) we get the scalar wave equations for each component of $\mathbf{u}$.

$$c^2\nabla^2 u_i - u_{i,tt} = 0, \tag{9.110}$$

where the wave speed c is c_L for longitudinal waves and c_T for transverse waves. For an external force $f(\mathbf{x}, t)$ we add the terms $-fu_i$, so that Euler's equations yield the nonhomogeneous wave equations $\mu\nabla^2 u_i - u_{i,tt} = f(x_i, t) = f_i$. For the equilibrium case, Euler's equations reduce to Laplace's equation for $\mathbf{u}$ or $\nabla^2 u_i = 0$.

Summary

We developed the continuum formulation of Lagrange's equations by generalizing Hamilton's principle. To this end we introduced the Lagrange density function. We then showed that the three-dimensional wave equations were obtained from Euler's equations. We made use of the fact that the generalized coordinates $\mathbf{q}$ and virtual displacements $\delta\mathbf{q}$ were replaced by $\mathbf{u}$ and $\delta\mathbf{u}$ for a continuum. We could have used another approach: that of decomposing $\mathbf{u}$ into the gradient of a scalar potential and the curl of a vector potential. We would then have arrived at the appropriate wave equations for ϕ and $\boldsymbol{\psi}$. In addition, we could also extend the Lagrange approach to include nonconservative forces such as the Rayleigh dissipation function to take account of viscous damping; this would lead to damped wave equations (which involves the $\mathbf{u}_t$ term expressing the fact that the damping force is proportional to the velocity).

Whichever way we look at Lagrange's equations we must remember that they are based on energy concepts, since kinetic and potential energy and the velocity potential—Rayleigh's dissipation function for a dissipative system—were involved in constructing the appropriate Lagrange density function. As mentioned previously, energy, being a first integral of the equations of motion, has the advantage of giving a global approach and also of smoothing out errors. Based on energy considerations, Lagrange's formulation also has the advantage of being invariant with respect to the choice of coordinate systems, since the Lagrange density function only involves physical quantities that are scalar invariants. This aspect of invariance is important in other fields of physics such as relativity theory, quantum mechanics, and electromagnetic theory.

It was observed previously that Lagrange's equations are second-order PDEs. We also learned in treating the wave equation that this second-order equation can be reformulated as a pair of first-order PDEs by defining new dependent variables such as strain and particle velocity. Similarly, in the use of variational approaches to obtain equations of motion, we saw in the discrete model that Hamilton developed a pair of first-order equations for the Hamiltonian in terms of $\mathbf{p}$ and $\mathbf{q}$, and we noted the advantages over the Lagrange formulation.

In accordance with our program of relating wave propagation phenomena to the variational treatment in a continuum we extend Hamilton's formulation to a continuum in the next section.

9.13. Hamilton's Equations of Motion for a Continuum

One-Dimensional Case

In this section we use the same one-dimensional model as for the corresponding Lagrange case, i.e., we start by approximating longitudinal wave propagation in a bar by a discrete model. The Lagrangian L can be rewritten as

$$L = \Delta x \sum_i L_i, \qquad L_i = \tfrac{1}{2}A\left[\rho(u_i)^2 - E\left(\frac{u_{i+1} - u_i}{\Delta x}\right)^2\right]. \tag{9.111}$$

p_i is the momentum conjugate to u_i and is given by

$$p_i = \frac{\partial L}{\partial u_i} = \Delta x \frac{\partial L_i}{\partial \dot{u}_i}. \tag{9.112}$$

We can rewrite the Hamiltonian as

$$H = \sum_i (p_i \dot{u}_i - \Delta x L_i) = \sum_i \left(\dot{u}_i \frac{\partial L_i}{\partial u_i} - L_i \right) \Delta x. \tag{9.113}$$

To obtain the continuous bar we pass to the limit as $\Delta x \to 0$, $u_i(t) \to u(x, t)$, $\dot{u}_i \to u_t$, and the sum tends to an integral with respect to x. Then H tends to

$$H = \int_0^L \left(\frac{\partial \mathscr{L}}{\partial u_t} u_t - \mathscr{L} \right) dx, \tag{9.114}$$

where $\mathscr{L}$ is the Lagrange density function and is given by (9.95) for the continuous bar. As seen from (9.112) the canonical momentum p_i for each mass vanishes in the limit for the continuum, since each p_i is defined in terms of L for the discrete case. However, we can get around this difficulty by defining the *momentum density* π in terms of the Lagrange density function for the continuous case. We obtain

$$\pi = \frac{\partial \mathscr{L}}{\partial u_t}. \tag{9.115}$$

Equation (9.115) implies that π does not depend on the individual masses but is defined for a continuous bar. By analogy with the definition of $\mathscr{L}$ we define the *Hamiltonian density function* $\mathscr{H}$ as the integrand of (9.114). Using (9.115) we have

$$\mathscr{H} = \pi u_t - \mathscr{L}, \tag{9.116}$$

so that we get

$$H = \int_0^L \mathscr{H} \, dx. \tag{9.117}$$

Three-Dimensional Case

In this section we reinvestigate wave propagation in an extended three-dimensional continuum from the point of view of the Hamilton formulation for a continuum. As for the corresponding Lagrange case we have $\mathbf{u} = (u_1, u_2, u_3)$ where $u_i = u_i(\mathbf{x}, t)$. In three dimensions the momentum density function is the vector $\boldsymbol{\pi}$ whose ith component is

$$\pi_i = \frac{\partial \mathscr{L}}{\partial u_{i,t}}. \tag{9.118}$$

The three-dimensional generalization for the Hamiltonian is

$$H = \int_V \mathscr{H}\, dV = \int_V (\pi_i u_{i,t} - \mathscr{L})\, dV, \qquad \text{summed over } i, \tag{9.119}$$

For three-dimensional wave propagation $\mathscr{H}$ is a given function of $\mathbf{u}_x$, $\mathbf{u}_t$, and $\boldsymbol{\pi}$.

We now obtain Hamilton's canonical equations of motion for a continuum or the *field equations*. To do this we first expand dH. (As usual, we use the tensor notation.)

$$dH = \int_V \left(\frac{\partial \mathscr{H}}{\partial \pi_i} d\pi_i + \frac{\partial \mathscr{H}}{\partial u_{i,j}} du_{i,j} + \frac{\partial \mathscr{H}}{\partial u_{i,t}} du_{i,t} \right) dV.$$

We now perform an integration by parts on the integrals of the form

$$\int \frac{\partial \mathscr{H}}{\partial u_{i,j}} du_{i,j}\, dV.$$

The integrated term can be made to vanish by making the volume of integration so large that the values of $\mathbf{u}$ and $\mathscr{H}$ can be made to vanish on the surface at infinity. As a result we write dH as

$$dH = \int_V \left(\frac{\partial \mathscr{H}}{\partial \pi_i} d\pi_i - \frac{d}{dx_j} \frac{\partial \mathscr{H}}{\partial u_{i,j}} du_i \right) dV. \tag{9.120}$$

Forming the differential of (9.119) gives

$$dH = \int_V \left[\pi_i du_{i,t} + u_{i,t}\, d\pi_i - \frac{\partial \mathscr{L}}{\partial u_i} du_i - \frac{\partial \mathscr{L}}{\partial u_{i,t}} du_{i,t} \right] dV. \tag{9.121}$$

Using the equations of motion for $\mathscr{L}$ given by (9.109) and equating like coefficients from (9.120) and (9.121) gives the analogue of Hamilton's canonical equations for a three-dimensional continuum in terms of the Hamilton density function. They can be written as

$$\frac{\partial \mathscr{H}}{\partial u_i} - \frac{d}{dx_j} \frac{\partial \mathscr{H}}{\partial u_{i,j}} = -\pi_{i,t}, \qquad \frac{\partial \mathscr{H}}{\partial \pi_i} = u_{i,t}. \tag{9.122}$$

For the general case where H is an explicit function of t we have the additional equation

$$\frac{\partial \mathscr{H}}{\partial t} = -\frac{\partial \mathscr{L}}{\partial t}. \tag{9.123}$$

We now apply the equations of motion for $\mathscr{H}$ given by (9.122) to wave propagation in three dimensions. Using the above results we obtain

$$\mathscr{H} = \pi_i u_{i,t} - \tfrac{1}{2}\rho u_{i,t} + \tfrac{1}{2}\mu u_{i,j} u_{i,j}. \tag{9.124}$$

The second equation of (9.122) becomes an identity. Since $\mathscr{H}$ is independent of the u_i's, $\partial\mathscr{H}/\partial u_i = 0$. Also $\partial\mathscr{H}/\partial u_{i,j} = -\mu u_{i,j}$. Using these expressions

(9.122) becomes

$$\frac{d}{dx_j} u_{i,j} = \mu u_{i,jj} = \pi_{i,t} = \rho u_{i,tt},$$

which gives the wave equation for the ith component of the displacement as

$$c^2 u_{i,jj} - c^2 \nabla^2 u_i = u_{i,tt}, \qquad c^2 = \frac{\mu}{\rho}. \tag{9.125}$$

Propagation of Sound Waves in a Gas

As an illustration of the use of the Hamilton formulation, we investigate the production of sound waves in a gas with a view toward obtaining the equations of motion for the longitudinal vibrations of an adiabatic gas. This will turn out to be the wave equation that yields the sound field. We recall that a sound wave is a surface in three-space. It is the limiting form of a shock wave of infinitesimal strength. This means that the pressure p and density ρ differ but slightly from their undisturbed states p_0, ρ_0, in the sense that second-order terms are neglected.

In order to form the Hamilton density function $\mathcal{H}$ we need to first construct the Lagrange density function $\mathcal{L}$ which is given in terms of the kinetic energy density function $\mathcal{T}$ and the potential energy density function $\mathcal{V}$:

$$\mathcal{L} = \mathcal{T} - \mathcal{V}. \tag{9.126}$$

Note that H, T, and V are the volume integrals of their respective density functions. $\mathcal{T}$ is given by

$$\mathcal{T} = \tfrac{1}{2}\rho[(u_{1,t})^2 + (u_{2,t})^2 + (u_{3,t})^2]. \tag{9.127}$$

To obtain $\mathcal{V}$ requires a little more work. The potential energy of the gas is a measure of the work the gas can do in expanding against the pressure. For example, suppose we have a semi-infinite tube of gas with a piston at the front end. If the piston is pushed in, the pressure is increased and the gas does work against the piston. The work done by the piston in maintaining that pressure is the negative of the work done by the gas. Consider a mass m of gas occupying a volume V_0 so that $V_0 = m/\rho_0$. $\mathcal{V}V_0$ is the potential energy of the gas occupying the volume V_0. A sound wave changes the density and the volume from V_0 to $V_0 + \Delta V$. The force exerted on an element of surface dA of V_0 is $p\,dA$ and is directed normal to the surface inward. In expanding, the surface moves a differential distance dx in the direction of the outward normal, so that the external work done is $-pA\,dx = -p\,dV$. Therefore the potential energy corresponding to the volume change ΔV is

$$\mathcal{V}V_0 = -\int_{V_0}^{V_0+\Delta V} p\,dV.$$

Consider a typical pressure–volume curve where p is plotted against V. In going along the curve from the point A: (p_0, V_0) to B: $(p_0+\Delta p, V_0+\Delta V)$ we approximate the portion of the curve between A and B by a straight line. Since $p=p(V)$ (and is given by the adiabatic equation of state) we have $p=(\partial p/\partial V)_0\Delta V$. We can therefore easily evaluate the integral in the above expression and obtain

$$\mathcal{V}V_0=-\left[p_0\Delta V+\frac{1}{2}\left(\frac{\partial p}{\partial V}\right)_0(\Delta V)^2\right]. \tag{9.128}$$

To evaluate $\partial p/\partial V$ we need the equation of state relating p and V which we said was the adiabatic equation of state. The equation of state for an adiabatic gas is

$$pV^\gamma=p_0V_0^\gamma=\text{const.}, \tag{9.129}$$

where γ is the ratio of specific heat at constant pressure to that at constant volume. For air under normal conditions $\gamma=1.4$.

The classical experiments made in the eighteenth century on measuring the velocity of sound in air, demonstrated that an adiabatic equation of state (no heat exchange with the environment) gave a theoretical value of the sound speed, that was closer to the experimental value than the isothermal equation of state (which was used by Newton, thus yielding erroneous results). In 1816, Laplace used the hypothesis of the adiabatic equation of state and from this he calculated the sound speed.

For a derivation of this expression see, for example, [30, Chapter XX]. We now express the change in volume in terms of the corresponding change in density. If s is the fractional change in density we have

$$\rho=\rho_0(1+s). \tag{9.130}$$

Then $\mathcal{V}$ becomes

$$\mathcal{V}=p_0s+\tfrac{1}{2}p_0s^2, \tag{9.131}$$

where we have used the fact that

$$\left(\frac{\partial p}{\partial V}\right)_0=-\frac{\gamma p_0}{V_0},$$

which arises from the adiabatic equation of state. We need to express s in terms of $\mathbf{u}$. To do this we apply Gauss's divergence theorem to the mass flux of gas flowing across the closed surface S surrounding the finite volume V of gas due to the disturbance caused by the sound wave. The mass flux must equal the negative of the volume integral of the change in density. The reason for the minus sign is that a positive divergence of mass flow must decrease the density of the gas in V_0. The mathematical relationship is therefore

$$\rho_0\oint_S\mathbf{u}\cdot d\mathbf{A}=-\rho_0\int_{V_c}s\,dV.$$

Applying the divergence theorem to the surface integral gives

$$\int_{V_0} \nabla \cdot \mathbf{u}\, dV = -\int_{V_0} s\, dV.$$

Since this equality holds for an arbitrary volume, we get

$$\nabla \cdot \mathbf{u} = -s. \tag{9.132}$$

Using the above results, the final form of the potential energy density function becomes

$$\mathcal{V} = -p_0 \nabla \cdot \mathbf{u} + \tfrac{1}{2}\gamma p_0 (\nabla \cdot \mathbf{u})^2. \tag{9.133}$$

The Lagrange density function then becomes

$$\mathcal{L} = \tfrac{1}{2}[\rho_0 \mathbf{u}_t \cdot \mathbf{u}_t + 2p_0 \nabla \cdot \mathbf{u} - \gamma p_0 (\nabla \cdot \mathbf{u})^2]. \tag{9.134}$$

Using the momentum density in the form $\boldsymbol{\pi} = \rho_0 \mathbf{u}_t$ the Hamiltonian density function becomes

$$\mathcal{H} = \boldsymbol{\pi} \cdot \mathbf{u}_t - \mathcal{L} = \tfrac{1}{2}\rho_0 \boldsymbol{\pi} \cdot \boldsymbol{\pi} + \tfrac{1}{2}\gamma p_0 (\nabla \cdot \mathbf{u})^2. \tag{9.135}$$

Clearly the Hamiltonian density function is the total energy density, or $\mathcal{H} = \mathcal{T} + \mathcal{V}$. Hamilton's canonical equations then yield the identity $\mathbf{u}_t = \boldsymbol{\pi}/\rho_0$ and

$$\pi_{i,t} = \frac{d}{dx_i}(\gamma p_0 \nabla \cdot \mathbf{u}),$$

for the ith component of the equations of motion for sound wave propagation. Using the definition of π_i, these three equations can be combined into one vector equation as

$$\rho_0 \mathbf{u}_{tt} - \gamma p_0 \nabla \nabla \cdot \mathbf{u} = 0. \tag{9.136}$$

We may transform this into the three-dimensional wave equation for s. We obtain

$$\nabla^2 s - \left(\frac{\rho_0}{\gamma p_0}\right) s_{tt} = 0. \tag{9.137}$$

The solution of (9.137) yields the density field due to the propagation of a sound wave in a three-dimensional gas. The wave speed is

$$c = \sqrt{\frac{\gamma p_0}{\rho_0}}, \tag{9.138}$$

which is based on the adiabatic equation of state. It is easily shown that $\mathbf{u}$ and p satisfy the same wave equation. Since sound waves in gas are longitudinal waves, a potential $\phi = \phi(\mathbf{x}, t)$ exists such that $\mathbf{u} = \mathbf{grad}\, \phi$, and ϕ also satisfies the wave equation. It is clear that the same wave equation can also be derived from the Lagrange formulation.

9.14. Hamilton–Jacobi Theory

In this section we discuss an important approach to the treatment of the PDEs that occur in wave propagation phenomena. This is the so-called *Hamilton-Jacobi theory*. Aside from its use in the study of wave propagation phenomena, it is most important in adding to our fundamental knowledge of the theory of PDEs.

We start by sketching a little of the history of this theory. It was developed by Hamilton in the early part of the nineteenth century, primarily for problems in particle dynamics (commonly called "analytical dynamics"). Hamilton went further than this by attempting to tie in the fields of optics and mechanics through the variational principle he developed in dynamics and Fermat's principle in optics. He therefore seized upon the analogy between the variational principles of geometric optics and analytical dynamics and used them as guides in the development of optical and dynamical theory. Hamilton's canonical equations of motion were extended by the German mathematician Jacobi to first-order PDEs which are basic in continuum dynamics and wave propagation. Thus this theory is aptly called the Hamilton-Jacobi theory.

The mathematical methods of analytical dynamics essentially consist of finding the properties of systems of ODEs with constraints, since systems of particles are involved. When we consider wave propagation in continuous media we are concerned with PDEs. We showed several examples where systems of ODEs that characterize discrete models converge to a single PDE when the number of particles becomes infinite, in such a way as to have the discrete mediun tend to a continuum. The relationship between PDEs and ODEs was brought into sharp focus by using characteristic theory to describe hyperbolic systems. Characteristic theory yields a set of ODEs, the characteristic equations, which are fundamental to the solution of the corresponding system of hyperbolic PDEs. This is the motivation for the Hamilton-Jacobi theory.

In order to understand the Hamilton-Jacobi theory we again invoke the concept of cyclic coordinates. We recall that a cyclic coordinate was defined as one that does not occur in the Lagrangian. This leads to the conservation of the generalized momentum conjugate to the cyclic coordinate. The use of cyclic coordinates is not merely academic. Indeed, if a canonical transformation can be found that transforms *all* the generalized coordinates into cyclic coordinates, then Hamilton's canonical equations of motion lead to particularly simple solutions. These solutions arise from the property that all the conjugate momenta are conserved. Such a transformation can always be found for a conservative system.

Contact Transformations

We seek a canonical transformation from the coordinates and momenta $(\mathbf{q}, \mathbf{p})$ at any time t to a new set of constant quantities $(\mathbf{q}_0, \mathbf{p}_0)$, which can be interpreted as the $2n$ initial conditions. The transformation must preserve the form of Hamilton's equations. Such a canonical transformation is called a *contact transformation* and clearly gives cyclic coordinates, since the initial momentum vector $\mathbf{p}_0$ is constant. *The method of finding such a class of transformations is the basis of the Hamilton–Jacobi theory.*

We now consider the canonical or contact transformation described above, where the transformed momentum vector is given by the initial $\mathbf{p}_0 = (\alpha_1, \alpha_2, \ldots, \alpha_n)$, so that the transformed $\mathbf{q}_0$ are the initial values of $\mathbf{q}$ and are cyclic. This means that H is independent of $\mathbf{q}$ so that $H = H(\alpha_1, \alpha_2, \ldots, \alpha_n)$. Let the transformed or cyclic coordinates be the Q_i's and P_i's. The transformation equations take the form

$$\begin{aligned} Q_i &= Q_i(q_1, \ldots, q_n, p_1, \ldots, p_n, t), \\ P_i &= P_i(q_1, \ldots, q_n, p_1, \ldots, p_n, t) = \alpha_i, \qquad i = 1, 2, \ldots, n, \end{aligned} \tag{9.139}$$

where Q_i, P_i are the initial values of q_i, p_i. This system is a canonical transformation provided there exists a function $K(\mathbf{Q}, \mathbf{P}, t)$ such that Hamilton's canonical equations are valid in the transformed coordinate system with respect to K. We have

$$\frac{\partial K}{\partial P_i} = \dot{Q}_i, \qquad \frac{\partial K}{\partial Q_i} = -\dot{P}_i. \tag{9.140}$$

The original canonical coordinates $(\mathbf{q}, \mathbf{p})$ must satisfy Hamilton's principle which may be put in the following form:

$$\delta \int_{t_A}^{t_B} [\dot{\mathbf{q}} \cdot \mathbf{p} - H(\mathbf{q}, \mathbf{p}, t)] \, dt = 0. \tag{9.141}$$

The transformed coordinates $(\mathbf{Q}, \mathbf{P})$ must also satisfy Hamilton's principle which is

$$\delta \int_{t_A}^{t_B} [\dot{\mathbf{Q}} \cdot \mathbf{P} - K(\mathbf{Q}, \mathbf{P}, t)] \, dt = 0. \tag{9.142}$$

Equation (9.142) is the image of (9.141) in the transformed coordinate system. In order for both (9.141) and (9.142) to be valid, their integrands must be connected by the following relationship:

$$K(\mathbf{Q}, \mathbf{P}, t) - H(\mathbf{q}, \mathbf{p}, t) = \frac{dS}{dt} + \dot{\mathbf{Q}} \cdot \mathbf{P} - \dot{q} \cdot p, \tag{9.143}$$

where S is an arbitrary function of the old and new canonical coordinates obeying the canonical transformations, such that the variation of S at the endpoints of the variation of the integral is zero. S may be a function of $(\mathbf{q}, \mathbf{Q}, t)$, $(\mathbf{p}, \mathbf{P}, t)$, $(\mathbf{Q}, \mathbf{p}, t)$, or $(\mathbf{q}, \mathbf{P}, t)$. Taking into account the transformation equations we see that these are the only possible functional relations involving the new coordinates. We choose the fourth possibility and thus set $S = S(\mathbf{q}, \mathbf{P}, t)$. The motivation for this choice is: since $\mathbf{P}$ (the initial momentum vector) is constant, if we choose the initial momenta such that $P_i = 0$ for all i, then $S = S(\mathbf{q}, t)$ so that

$$\frac{dS}{dt} = \frac{\partial S}{\partial q_i}\dot{q}_i + \frac{\partial S}{\partial t} \quad \text{(summed over } i\text{)}. \tag{9.144}$$

This equation is also valid in the more general case where $S = S(\mathbf{q}, \mathbf{P}, t)$ since $\mathbf{P}$ is independent of t. We now let S be of the form

$$S(\mathbf{q}, \mathbf{P}, t) = G(\mathbf{q}, \mathbf{P}, t) - \mathbf{P} \cdot \mathbf{Q}. \tag{9.145}$$

The reason for subtracting off the term $\mathbf{P} \cdot \mathbf{Q}$ will become apparent. We obtain

$$S_t = G_t, \qquad \dot{S} = \dot{G} - P_i \dot{Q}_i \tag{9.146}$$

We then get

$$K - H = -p_i \dot{q}_i + \frac{\partial G}{\partial q_i}\dot{q}_i + \frac{\partial S}{\partial t}. \tag{9.147}$$

Since the q_i's are linearly independent coordinates the coefficient of each $\dot{q}_i$ must vanish. This gives the following system:

$$\frac{\partial G}{\partial q_i} = \frac{\partial S}{\partial q_i} = p_i, \tag{9.148}$$

$$K - H = \frac{\partial S}{\partial t}. \tag{9.149}$$

Since we have chosen $\mathbf{Q}$ and $\mathbf{P}$ as the initial coordinates, $\dot{\mathbf{Q}} = \dot{\mathbf{P}} = 0$. From the canonical equations for K (9.140) we see that K is at most a function of t. If we further demand that $K = 0$, then a fortiori $\dot{Q}_i = \dot{P}_i = 0$, and we get $H + \partial S/\partial t = 0$. Since $H = H(\mathbf{q}, \mathbf{p}, t)$ we obtain the important PDE

$$H\left(q_1, \ldots, q_n, \frac{\partial S}{\partial q_1}, \ldots, \frac{\partial S}{\partial q_n}, t\right) + \frac{\partial S}{\partial t} = 0. \tag{9.150}$$

Equation (9.150) is the *Hamilton–Jacobi equation*. It is a first-order PDE for $S = S(\mathbf{q}, \mathbf{P}, t)$, which is nonlinear, in general. S is called *Hamilton's principle function*. It is a *generating function* which serves the important purpose of allowing us to transform from $(\mathbf{q}, \mathbf{p}, t)$ to $(\mathbf{Q}, \mathbf{P}, t)$ space. We see that (9.150) is a PDE in the $n+1$ independent variables $(q_1, q_2, \ldots, q_n, t)$ for S. Consequently a complete solution must involve $n+1$ constants of integration. We observe that S does not appear in (9.150); only its partial derivatives with respect to the q_i's and t are involved. Therefore, if S is a solution to the Hamilton–Jacobi equation, then $S + \alpha_{n+1}$ is also a solution where α_{n+1} is an additive constant. The complete solution of this equation can then be written as

$$S = S(q_1, q_2, \ldots, q_n, \alpha_1, \alpha_2, \ldots, \alpha_n) + \alpha_{n+1},$$

where

$$\alpha_i = P_i, \qquad i = 1, 2, \ldots, n,$$

and we have chosen α_i to be the ith component of the initial momentum $\mathbf{P}$. Each p_i can then be determined from (9.148). The cyclic coordinate conjugate to P_i is given by

$$Q_i = \frac{\partial S}{\partial \alpha_i} = \beta_i. \tag{9.151}$$

If the Jacobian of the mapping from $(\mathbf{q}, \mathbf{p})$ to $(\mathbf{Q}, \mathbf{P})$ is not zero then we obtain the trajectories $\mathbf{q} = \mathbf{q}(\alpha_1, \alpha_2, \ldots, \alpha_n, \beta_1, \beta_2, \ldots, \beta_n, t)$ which solves the problem by giving the generalized coordinates as functions of time and the $2n$ initial conditions $(\alpha_1, \ldots, \alpha_n, \beta_1, \ldots, \beta_n)$.

We may gain further insight into the nature of Hamilton's principle function S by showing that it is equal to the time integral of the Lagrangian. This is done by expanding dS/dt, obtaining

$$\frac{dS}{dt} = \frac{\partial S}{\partial q_i}\dot{q}_i + \frac{\partial S}{\partial t} = p_i\dot{q}_i - H = L, \tag{9.152}$$

Integrating from t_A to t_B gives

$$S = \int_{t_A}^{t_B} L\, dt, \tag{9.153}$$

which gives the important result that S is the functional which satisfies Hamilton's principle. Hamilton recognized that S is a solution to the Hamilton–Jacobi equation, and Jacobi realized that S could be considered a generating function used to generate the proper canonical transformations, and thereby solve problems in dynamics.

Harmonic Oscillator

Before we go any further, by showing the relationship between the Hamilton–Jacobi PDE and Hamilton's canonical equations of motion, we give a simple example which illustrates the Hamilton–Jacobi theory and shows how the generating function S is used in solving a problem. We consider the simple harmonic oscillator. The generalized coordinates are (q, p). The Hamiltonian is

$$H = \frac{p^2}{2m} + \frac{kq^2}{2}.$$

By setting $p = \partial S/\partial q$, the Hamilton–Jacobi equation (9.150) becomes

$$\frac{1}{2m}\left(\frac{\partial S}{\partial q}\right)^2 + \frac{kq^2}{2} + \frac{\partial S}{\partial t} = 0,$$

where $S = S(q, \alpha, t)$ and the cyclic coordinate α is the initial value of the momentum. Recognizing that $H = E$, the total energy of the conservative system, we may separate out the time-dependent part of S by introducing a function $W = W(q, \alpha)$ such that

$$S(q, \alpha, t) = W(q, \alpha) - Et.$$

Inserting this expression into the Hamilton–Jacobi equation gives the following equation for W:

$$\frac{1}{2m}\left(\frac{\partial W}{\partial q}\right)^2 + \frac{kq^2}{2} = E.$$

Since α is constant we integrate and obtain

$$W = \sqrt{mk}\int_{Q=\beta}^{q} \sqrt{\left(\frac{2E}{k}\right) - q^2}\, dq = S + Et.$$

We are not interested in W as such, but in its partial derivatives with respect to the parameters α and E. Differentiating the above expression with respect to E and recognizing that S is independent of E, gives

$$\frac{\partial W}{\partial E} = \sqrt{\frac{m}{k}}\int \frac{dq}{\sqrt{(2E/k) - q^2}} = -\sqrt{\frac{m}{k}}\cos^{-1} q\sqrt{\frac{k}{2E}} + \tau = t,$$

where τ is a constant having the dimension of time. We solve this expression for q and obtain

$$q = \sqrt{\frac{2E}{k}}\cos \omega(t - \tau),$$

where it is clear that $\omega\tau$ is the phase angle. The momentum conjugate to q then becomes

$$p = -\sqrt{2Em}\sin \omega(t - \tau).$$

This gives the cyclic coordinate $P = \alpha = \sqrt{2Em}\sin \omega\tau$. Note that we used the expression $\omega = \sqrt{k/m}$ for the natural frequency of the oscillator. We can then easily calculate L and then show that S is the time integral of L.

This example shows that it is possible to integrate the Hamilton–Jacobi equation because S was separated into a time-independent part $W(q, \alpha)$ and a part proportional to time. The decomposition of S is always possible if H is independent of time, since the only time-dependent part of the Hamilton–Jacobi equation is the S_t term. We again point out that $H = \text{const.}$ does not necessarily mean energy is conserved unless the constant is the total energy E. If H involves an additive function of t then the same type of decomposition of S can be made.

2*n* Degree-of-Freedom Case

We now indicate how the Hamilton–Jacobi equation (9.150) can be solved for the $2n$ degree-of-freedom case, which means that $\mathbf{q}$ and $\mathbf{p}$ are n dimensional vectors. We consider the case where $H=\alpha_{n+1}=\text{const.}$ and decompose S into a time-independent and a time-dependent part.

$$S(\mathbf{q}, \boldsymbol{\alpha}, t) = W(\mathbf{q}, \boldsymbol{\alpha}) - \alpha_{n+1}t, \tag{9.154}$$

where $\boldsymbol{\alpha}=(\alpha_1, \ldots, \alpha_n)$ is the initial or transformed momentum vector. Inserting (9.154) into (9.150) gives

$$H\left(q_1, q_2, \ldots, q_n, \frac{\partial W}{\partial q_1}, \frac{\partial W}{\partial q_2}, \ldots, \frac{\partial W}{\partial q_n}\right) = \alpha_{n+1}, \tag{9.155}$$

where we set

$$p_i = \frac{\partial W}{\partial q_i}. \tag{9.156}$$

Equation (9.155) is called the Hamilton–Jacobi equation for W. W is called *Hamilton's characteristic function.* We can decompose W as follows:

$$W(q_1, \ldots, q_n, \alpha_1, \ldots, \alpha_n) = \sum_{i=1}^{n} W_i(q_i, \alpha_1, \ldots, \alpha_n), \tag{9.157}$$

where each W_i is a function of only one q_i (and, of course, $\boldsymbol{\alpha}$). If this decomposition is valid, then the q_i's occurring in (9.155) are said to be *separable*, and (9.155) can then be split into n equations, the ith equation being

$$H_i\left(q_i, \frac{\partial W_i}{\partial q_i}\right) = \alpha_i.$$

Now each of these n equations involves only one q_i and the conjugate momentum $\partial W_i/\partial q_i$ (which is actually a total derivative). The system (9.150) then becomes a set of n ODEs of a simple form. Because each q_i and p_i appears as a square (H is a quadratic function of $\mathbf{q}$ and $\mathbf{p}$), each ODE is first order and second degree in $\partial W_i/\partial q_i$ and q_i. It is always possible to reduce each equation to quadratures by solving for $\partial W_i/\partial q_i$ and integrating with respect to q_i.

A separation of variables in Hamilton's characteristic function of the form (9.157) can be accomplished when all but one of the q_i's are cyclic. Suppose q_1 is the only noncyclic coordinate. Since the p_i's conjugate to the q_i's (for $i \neq 1$) are cyclic, we must have

$$\frac{\partial W_i}{\partial q_i} = p_i = \alpha_i, \qquad W_i = \alpha_i q_i, \qquad i \neq 1. \tag{9.158}$$

The Hamilton–Jacobi equation for W_1 then reduces to

$$H\left(q_1, \frac{\partial W_1}{\partial q_1}, \alpha_2, \ldots, \alpha_n\right) = \alpha_1,$$

which is an ODE of the above-mentioned type and is therefore immediately soluble. Then W can be written as

$$W = W_1 + \sum_{i=2}^{n} \alpha_i q_i,$$

so that W is decomposed into separable functions of the q_i's. We can perform a similar transformation that allows q_2 to be noncyclic and demands that the q_i's ($i = 3, \ldots, n$) be cyclic. We get

$$W = W_1 + W_2 + \sum_{i=3}^{n} \alpha_i q_i$$

and the Hamilton–Jacobi equation becomes

$$H\left(q_2, \frac{\partial W}{\partial q_2}, \alpha_1, \alpha_3, \ldots, \alpha_n\right) = \alpha_2.$$

We now repeat this process until all the q_i's are separable.

n Harmonic Oscillators

We now consider another case where we can separate the q_i's in W, namely, n uncoupled simple harmonic oscillators, all having the same mass and spring constant (for simplicity). The Hamiltonian or energy equation is

$$H = \frac{1}{2m} \sum_{i=1}^{n} (p_i)^2 + \frac{k}{2} \sum_{i=1}^{n} (q_i)^2 = E.$$

Using (9.156), this expression for H becomes

$$\frac{1}{2m} \sum_i \left(\frac{\partial W}{\partial q_i}\right)^2 + \frac{k}{2} \sum_i (q_i)^2 = E. \tag{9.159}$$

Note that (9.159) is a special case of (9.155) where $\alpha_{n+1} = E$. We now insert (9.159) into (9.157) and obtain

$$\frac{1}{2m} \sum_i \left(\frac{dW_i}{dq_i}\right)^2 + \frac{k}{2} \sum_i (q_i)^2 = E.$$

If we put this expression in the form

$$\frac{1}{2m} \left(\frac{dW_1}{dq_1}\right)^2 + \frac{k}{2}(q_1)^2 = F(q_2, \ldots, q_n),$$

we see that the left-hand side depends on q_1 and the right-hand side is independent of q_1. This means that $F = \text{const.}$ since the q_i's are independent.

We therefore obtain

$$\frac{1}{2m}\left(\frac{dW_1}{dq_1}\right)^2+\frac{k}{2}(q_1)^2=H_1=\text{const.}=\alpha_1.$$

We perform the same process on F by writing F explicitly in the form

$$\frac{1}{2m}\left(\frac{dW_2}{dq_2}\right)^2+\frac{k}{2}(q_2)^2=H_2=\text{const.}=\alpha_2,$$

by the same argument. By continuing this process we finally arrive at

$$\frac{1}{2m}\left(\frac{dW_n}{dq_n}\right)^2+\frac{k}{2}(q_n)^2=H_n=\alpha_n.$$

This procedure shows that the decomposition given by (9.157) is valid, and the Hamiltonian can be split into a sequence of n Hamiltonians given by (9.158).

9.15. Characteristic Theory in Relation to Hamilton–Jacobi Theory

We previously mentioned that the Hamiltonian canonical equations of motion are the characteristics of the Hamilton-Jacobi PDE. In this section we explore this correspondence between the Hamilton-Jacobi PDE and the corresponding characteristic ODE by developing the theory of characteristics in the setting of the Hamilton-Jacobi theory.

First we observe that if Hamilton's principle function $S(\mathbf{q}, \mathbf{P})$ is any solution of (9.155), then the generalized coordinates $(\mathbf{q}, \mathbf{P})$ satisfy the appropriate Hamilton canonical equations. This is easily seen.

We return to our investigation of characteristic theory vis-à-vis Hamilton-Jacobi theory. Our starting point is (9.150). For ease of comparison with the standard theory of PDEs, we change the notation to conform with PDE theory. We revert to using subscripts for partial differentiation and use the following correspondence between the notation of Hamilton-Jacobi theory and that of PDEs.

$$\begin{aligned} &S\sim u-t, \qquad W\sim u, \qquad q_i\sim x_i, \\ &\frac{\partial S}{\partial q_i}\equiv S_q\sim u_x=p_i, \qquad \frac{\partial S}{\partial t}\equiv S_t=-E. \end{aligned} \tag{9.160}$$

The Hamilton-Jacobi equation (9.150) can then be written in the general form

$$F(x_1,\ldots,x_n,p_1,\ldots,p_n,t,\alpha_1,\ldots,\alpha_n)=0. \tag{9.161}$$

Note that F does not depend explicitly on u. This does not change the form of the characteristic equations we shall derive, since, as will be seen, these

ODEs only depend on the principle part of the PDE which are the partial derivatives. We now address ourselves to the problem of obtaining solutions of (9.161). For simplicity, we start with a two degree-of-freedom system.

Two Degree-of-Freedom System

Let $q_1 = x$, $q_2 = y$, $S = u$, $S_q = u_x = p$, $S_q = u_y = q$ (do not confuse this q with a generalized coordinate). Then (9.161) becomes

$$F(x, y, u, p, q) = 0. \tag{9.162}$$

Note that in the Hamilton–Jacobi PDE u only appears explicitly in terms of its partial derivatives p and q. Nevertheless, we still need to solve for $u = u(x, y)$ as well as $p = p(x, y)$ and $q = q(x, y)$. For example, for a two degree-of-freedom oscillating system (two uncoupled simple harmonic oscillators) (9.162) becomes

$$H - E = \frac{1}{2m}(p^2 + q^2) + \frac{k}{2}(x^2 + y^2) - E = 0$$

We now consider (9.162) and derive) the characteristic ODEs associated with this PDE. Equation (9.162) is a first-order PDE but is not quasilinear since the partial derivatives p and q do not appear linearly—in fact they appear, in general, in quadratic form.

To give *a brief review of the quasilinear case*:† Suppose we have a quasilinear PDE of the form $ap + bq = c$ where (a, b, c) are given functions of (x, y, u). We pick a point P_0 in the (x, y) plane. At P_0 we have $a(x_0, y_0, u_0) = a_0$, $b(x_0, y_0, u_0) = b_0$, and $c(x_0, y_0, u_0) = c_0$, where (a_0, b_0, c_0) are known. Then the PDE tells us that the normal $(p_0, q_0, -1)$ to every integral surface through P_0 is perpendicular to the direction given by (a_0, b_0, c_0). This means that the scalar product $(p_0, q_0, -1) \cdot (a_0, b_0, c_0) = 0$. In other words, *every integral surface through P_0 is tangent to the fixed direction* (a_0, b_0, c_0). This gives the following situation for the quasilinear case: At every point in the (x, y) plane we have a one-parameter family of integral surfaces, all going through the axis (the *Monge axis*) whose direction is given by (a_0, b_0, c_0). This family of surfaces is tangent to a surface called an *envelope*. The envelope formed by this family of integral surfaces surrounding the Monge axis is called the *Monge pencil*. The characteristic curves lie on the integral surfaces or Monge pencil and thus have the direction of the Monge axis at each point P_0.

For the fully nonlinear case given by the PDE (9.162) the situation is more complicated. The normals to the integral surfaces through P_0 are not perpendicular to the Monge axis (as they are in the quasilinear case). These normals form a one-parameter family of lines all going through P_0, namely,

---

† The reader is referred to Chapter 3, Section 3.2 for a more detailed discussion of first-order quasilinear PDEs.

the family of directions given by $(p_0, q_0, -1)$, where (p_0, q_0) are connected by the PDE $F(x_0, y_0, p_0, q_0) = 0$. This means that the one-parameter family of integral surfaces through P_0 (which are perpendicular to this family of normals) form an envelope which is a conical surface called the *Monge cone*. The Monge cone whose vertex is P_0 is therefore an element of surface that is the envelope of the tangent planes to the integral surfaces. The generator of the Monge cone (along which the integral surface is tangent) is a characteristic. Therefore, in this fully nonlinear case there is a whole *cone of characteristics through each field point* P_0.

Suppose we have a solution field for (9.162). This means that every field point in the region of the (x, y) plane, where the solution exists, is the vertex of a Monge cone which is the envelope of integral surfaces stemming from that point. The problem of solving (9.162), or obtaining the solution field, is thereby reduced to the problem of finding the Monge cone through each field point P_0. This means the problem is now one of finding the family of characteristic curves or characteristic field. In the theory of ODEs the situation is simpler, in that the problem reduces to that of finding the family of isoclines which consist of elements of the tangent to the solution curve at each field point.

We now give a simple example that illustrates how to get an envelope. Let $f(x, y, u, s) = x^2 + y^2 + (u - s)^2 - 1 = 0$. This equation represents a one-parameter family of integral surfaces in (x, y, u) space—the parameter being s. Specifically, it represents a family of spheres of unit radius whose centers are on the u axis. The envelope is obtained by eliminating s between the above equation and the equation $f_s = 0$ which is $u - s = 0$. The equation of the envelope then becomes the cylindrical surface $x^2 + y^2 - 1 = 0$ of unit radius whose axis is the u axis. It is also intuitively obvious that the envelope of a family of spheres whose centers are on a line is a cylindrical surface whose axis is that line. We return to this example after developing the characteristic equations.

Returning to the main problem, our first task is to find the generator of the Monge cone that yields the characteristic field. First we need the equation of the tangent plane to the integral surface u going through P_0 whose coordinates we take as (x, y, u). This is

$$p(x - \xi) + q(y - \eta) - (u - \zeta) = 0, \qquad (9.163)$$

where (ξ, η, ζ) are the running coordinates on the tangent plane. Since p and q are connected by the PDE (9.162) we may use this equation to solve for q as a function of p. Inserting $q(p)$ into (9.163) tells us that this equation represents a one-parameter of tangent planes that are the integral surfaces through (x, y, u), the parameter, of course, being p. To obtain the envelope we first differentiate (9.163) with respect to p. This gives

$$(x - \xi) + \acute{q}(p)(y - \eta) = 0, \qquad \acute{q} \equiv \frac{dq}{dp}. \qquad (9.164)$$

Then we would eliminate the parameter p between (9.163) and (9.164). This, in principle, would give us the envelope which is the desired Monge cone. In general, this cannot be done explicitly. However, what we really want is the equations of the characteristics which are the generators of the Monge cone. Clearly this family of characteristics is a set of curves lying on the Monge cone—which, we recall, is tangent to the family of integral surfaces. These are given implicitly by (9.163) and (9.164). In order to obtain the explicit representation of the characteristic curves we first go to the PDE $F(x, y, p, q)=0$, and differentiate it with respect to p, obtaining

$$F_p + \acute{q} F_q = 0.$$

Eliminating $\acute{q}(p)$ from this equation and (9.164) and using the above equations yields the following ratios:

$$\frac{\xi - x}{F_p} = \frac{\eta - y}{F_q} = \frac{\zeta - u}{pF_p + qF_q}. \tag{9.165}$$

Both the fixed point (x, y, u) and the variable point (ξ, η, ζ) lie in the tangent plane. Since the Monge cone was constructed as an envelope of the tangent planes the three equations given by (9.165) are also direction numbers of a characteristic curve lying on the Monge cone. This is what we set out to find. Since this analysis took place in the neighborhood of the fixed point (x, y, u) ("in the small") we may replace $\xi - x$, $\eta - y$, and $\zeta - z$ by dx, dy, and dz, respectively. We then get the following system of ODEs:

$$\frac{dx}{dt} = F_p, \qquad \frac{dy}{dt} = F_q, \qquad du/dt = pF_p + qF_q. \tag{9.166}$$

These are the ODEs whose solution yields the family of characteristic curves for the PDE (9.162). The system (9.166) can be shown to be the same equations as in the quasilinear case, which were derived in a different way in Chapter 3, taking advantage of the linearity of p and q. They depend on the solution u as well as x and y. However, unlike the quasilinear case, the fully nonlinear PDE (9.162) involves p and q nonlinearly. The system (9.166) consists of three equations. But there are five unknowns, namely, (x, y, u, p, q). Therefore we need two additional equations.

These two equations are obtained by expanding dp/dt and dq/dt by the chain rule and using the first two equations of (9.166). We get

$$\frac{dp}{dt} = p_x \frac{dx}{dt} + p_y\, dy/dt = p_x F_p + p_y F_q,$$

$$\frac{dq}{dt} = q_x \frac{dy}{dt} + q_y \frac{dy}{dt} = q_x F_p + q_y F_q.$$

Differentiating the PDE $F(x, y, p, q)=0$ gives

$$F_x + pF_u + p_x F_p + q_x F_q = 0,$$

$$F_y + qF_u + p_y F_p + q_y F_q = 0.$$

Using the fact that $q_x = p_y$ and inserting the expressions for dp/dt and dq/dt into the above equations yields the two additional equations we set out to obtain

$$\frac{dp}{dt} = -(F_x + pF_p),$$
$$\frac{dq}{dt} = -(F_y + qF_q). \tag{9.167}$$

Equations (9.166) and (9.167) are the complete set of characteristic equations for the fully nonlinear case.

As mentioned above, for the quasilinear case we only need the three equations given by (9.166). We easily see how to obtain the characteristics for this case by setting $F = ap + bq - c = 0$, where the coefficients are known functions of (x, y, u). (Note the linear relation between p and q.) By using this quasilinear PDE and (9.166) we get

$$\frac{dx}{dt} = F_p = a, \qquad \frac{dy}{dt} = F_q = b, \qquad \frac{du}{dt} = pF_p + qF_q = ap + bq = c,$$

which are the same as the characteristic equations obtained in Chapter 3, given by (3.17) and (3.18).

We revisit the example give above as an illustration of the use of the characteristic equations in solving a nonlinear PDE. From the family of integral surfaces given by $f = x^2 + y^2 + (u - s)^2 - 1 = 0$ we have $f_x = 0$, $f_y = 0$, which yields $py - xq = 0$, upon eliminating $u - s$. This gives the equation of the envelope as the cylindrical surface $x^2 + y^2 - 1 = 0$, which was obtained above. From the definition of $f(x, y, u, s)$ and $f_x = 0$ which gives $x^2 = p^2(u - s)^2$ and $f_y = 0$ which gives $y^2 = q^2(u - s)^2$, we obtain the corresponding PDE which is $F(u, p, q, s) = (u - s)^2(p^2 + q^2 + 1) - 1 = 0$. Upon using the definition of F, the characteristic equations (9.166) and (9.167) become

$$\frac{dx}{dt} = 2p(u - s)^2,$$

$$\frac{dy}{dt} = 2q(u - s)^2,$$

$$\frac{du}{dt} = 2[1 - (u - s)^2],$$

$$\frac{dp}{dt} = -\frac{2p}{u - s},$$

$$\frac{dq}{dt} = -\frac{2q}{u - s}.$$

The last two equations give $dq/dp = q/p$ from which we get $p = kq$ where k is a constant. Manipulating these equations we get

$$q^2 d\left(\frac{p}{q}\right) = q\,dp - p\,dq = d[x + p(u-s)] = d[y + q(u-s)] = 0,$$

which gives $x + p(u-s) = \text{const.} = 0$, $y + q(u-s) = 0$. From these solutions of the characteristic equations and $F = 0$ we obtain $x^2 + y^2 + (u-s)^2 - 1 = 0$, the integral surfaces, and the envelope $x^2 + y^2 - 1 = 0$ arising from $p = kq$.

Two Degree-of-Freedom Oscillator

Another example more in keeping with Hamilton–Jacobi theory is the two degree-of-freedom oscillator (two uncoupled oscillators) mentioned above. From the Hamiltonian we obtain the PDE for this system as

$$H - E = F(x, y, p, q, E) = \frac{1}{2m}(p^2 + q^2) + \frac{k}{2}(x^2 + y^2) - E = 0. \quad (9.168)$$

The characteristic ODEs (9.166) and (9.167) become

$$\begin{aligned} \frac{dx}{dt} &= \frac{p}{m}, \\ \frac{dy}{dt} &= \frac{q}{m}, \\ \frac{du}{dt} &= \frac{1}{m}(p^2 + q^2), \\ \frac{dp}{dt} &= -kx \\ \frac{dq}{dt} &= -ky. \end{aligned} \quad (9.169)$$

The first two equations of (9.169) give the definition of the x and y components of the momentum. The last two equations are the equations of motion for the two degree-of-freedom simple harmonic oscillator. Manipulating the system (9.169) gives

$$\frac{1}{m}\left(p\frac{dp}{dt} + q\frac{dq}{dt}\right) + k\left(x\frac{dx}{dt} + y\frac{dy}{dt}\right) = 0.$$

Integrating this equation yields (9.168) where the constant of integration E is interpreted as the total energy for the conservative system. The third equation of (9.169) tells us that $du/dt = 2T$. The principle of decomposition of the Hamiltonian, as given by (9.131), applies here. Setting $F = F_1 + F_2$

we have

$$F_1 = \left(\frac{1}{2m}\right)p^2 + \left(\frac{k}{2}\right)x^2, \qquad F_2 = \left(\frac{1}{2m}\right)q^2 + \left(\frac{k}{2}\right)y^2.$$

We then set $u = u_1 + u_2 = W_1(x) + W_2(y)$, so that $p = u_{1,x}$, $q = u_{2,y}$. This reduces the two degree-of-freedom problem to two one degree-of-freedom simple harmonic oscillator problems.

n Degree-of-Freedom System

We now generalize characteristic theory in relation to Hamilton–Jacobi theory to an n degree-of-freedom system. We still use the notation of PDEs. Let $\mathbf{x} = (x_1, \ldots, x_n)$ and $\mathbf{p} = (p_1, \ldots, p_n)$ be the generalized coordinate and generalized momentum vectors, respectively, where $p_i = \partial u/\partial x_i = u_{x_i}$. Note that for the two-dimensional case $p_1 = p = u_x$, $p_2 = q = u_y$, and $p_i = 0$ for $i = 3, \ldots, n$. Then the fully nonlinear PDE given by (9.162) is generalized to

$$F(\mathbf{x}, u, \mathbf{p}) = 0. \tag{9.170}$$

The characteristic equations (9.166) and the first of (9.167) are generalized to

$$\begin{aligned} \frac{dx_i}{dt} &= F_p, \qquad i = 1, 2, \ldots, n, \\ \frac{du}{dt} &= \sum_{i=1}^{n} p_i F_{p_i}, \\ \frac{dp_i}{dt} &= -F_{x_i} - p_i F_u. \end{aligned} \tag{9.171}$$

The system (9.171) is a set of $2n+1$ equations for the $2n+1$ variables $(x_1, \ldots, x_n, u, p_1, \ldots, p_n)$.

We now revert to the physical notation (the notation of Hamilton–Jacobi theory) by using the correspondence given by (9.160). The Hamilton–Jacobi PDE is given by (9.150), where $\partial S/\partial q_i = p_i$. The characteristic ODEs (9.171) now become

$$\begin{aligned} \frac{dq_i}{dt} &= H_{p_i}, \\ \frac{dp_i}{dt} &= -H_{q_i}. \end{aligned} \tag{9.172}$$

9.16. Principle of Least Action

It is appropriate at this stage to introduce another variational principle also associated with the Hamilton canonical equations. It is known as the principle of *Least Action* and will be useful to us when we investigate the Hamilton–Jacobi theory in relation to wave propagation in the next section.

In mechanics (more particularly, in quantum mechanics) there is an integral called the *action integral* A defined by

$$A = \int_{t_A}^{t_B} \sum_i p_i \dot{q}_i \, dt. \tag{9.173}$$

The Principle of Least Action states that in a conservative system, which is one for which the Hamiltonian H is equal to the constant total energy E, we have the condition that

$$\Delta A = \int_{t_A}^{t_B} p_i \dot{q}_i \, dt = 0, \tag{9.174}$$

where the variational operator Δ represents a new type of variation of the path of the particles in configuration space from t_A to t_B, as follows: Recall that in our discussion of virtual work the variational operator δ was shown to correspond to virtual displacements, which were defined as those displacements in which time is held fixed and the coordinates are varied subject to the constraints imposed on the system. Such a virtual displacement does not always coincide to an actual physical displacement occurring in the course of the motion, for example, when the constraints depend on time. Consequently, the Hamiltonian H may not be conserved in a conservative system when we vary the path using the δ operator. In contrast, the Δ operator is defined as dealing with displacements which do involve a change in time, so that the varied path is obtained by a succession of displacements, each of which includes a differential change in time dt. We also require that the required path be consistent with the physical motion, meaning that in a conservative system H is conserved on the varying path as well as the actual path. The points in configuration space obtained by using the δ operator all travel with the same speed, while it is clear that the time of transit of the points along varying paths, by using the Δ operator, need no longer be constant so that the same point may slow down or speed up in order to keep H constant. As a result the Δ operator allows for a variation of t even at the endpoints, although the variation of the q_i's at the endpoints must still remain zero. We point out that it is possible that the actual path varied in configuration space may be the same with respect to the δ and Δ operators, except that the system point travels along the path with different speeds in each case. In the δ variation the speed of the system point is such as to make the time of transit the same for the varied path as for the actual path (H is not necessarily constant along the varied path). In the Δ variation the times of transit are not the same but *H is constant along every path.*

To describe mathematically the Δ variation process we start by tagging each curve (system path in configuration space) with a parameter α. Since the variation includes the time associated with each point, therefore t must also be a function of α. It follows that, for the configuration space $\mathbf{q}$, each component $q_i = q_i(\alpha, t)$. We then expand Δq_i in the limit as

$$\Delta q_i \to d\alpha\left(\frac{dq_i}{d\alpha}\right) = d\alpha\left(\frac{\partial q_i}{\partial \alpha}\right) + d\alpha\left(\frac{dt}{d\alpha}\right)\frac{dq_i}{dt}. \tag{9.175}$$

The first term $d\alpha(\partial q_i/\partial\alpha) = \delta q_i$ while the coefficient of dq_i/dt represents the change in t occurring as a result of the Δ variation and can therefore be designated as Δt. Consequently (9.149) can be written as

$$\Delta q_i = \delta q_i + \dot{q}_i \Delta t. \tag{9.176}$$

This relation between the operators δ and Δ as shown by (9.176) holds for any function $f(\mathbf{q}, t)$ so that

$$f(\mathbf{q}, t) = \delta f + \dot{f}\Delta t = \sum_i \left(\frac{\partial f}{\partial q_i}\right)\delta q_i + \left[\sum_i \left(\frac{\partial f}{\partial q_i}\right)\dot{q}_i + \frac{\partial f}{\partial t}\right]\Delta t. \tag{9.177}$$

It is clear that the first summation term is the expansion of δf, while the terms in the brackets represent the total time derivative of f, namely, $\dot{f} = df/dt$.

The action integral A can be written as

$$A = \int_{t_A}^{t_B} \sum_i p_i q_i \, dt = \int_{t_A}^{t_B} (L+H)\, dt = \int_{t_A}^{t_B} L\, dt + H(t_B - t_A), \tag{9.178}$$

since H is constant. We apply Δ to A and obtain

$$\Delta A = \Delta \int_{t_A}^{t_B} L\, dt + H(\Delta t_B - \Delta t_A).$$

In applying Δ to the functional $I = \int_{t_A}^{t_B} L\, dt$ we must remember that the limits of the integral are also subject to the variation. We therefore obtain

$$\begin{aligned}\Delta \int_{t_A}^{t_B} L\, dt &= \Delta I(t_B) - \Delta I(t_B) = \delta I(t_B) - \delta I(t_A) + \dot{I}(t_B)\Delta t_B - \dot{I}(t_A)\Delta t_A,\\ &= \delta \int_{t_A}^{t_B} L\, dt + L\Delta t\Big|_{t_A}^{t_B},\end{aligned} \tag{9.179}$$

upon using the fact that $\Delta f = \delta f + \dot{f}\Delta t$. We note the important point that the variation of the functional, $\delta \int_{t_A}^{r_B} L\, dt$, does not obey Hamilton's principle and hence does not vanish, because in the Δ process used here the variations δq_i do not vanish at the endpoints (the Δq_i do however, but they are not used here). We now calculate $\delta \int_{t_A}^{t_B} L\, dt$.

$$\delta \int_{t_A}^{t_B} L\, dt = \int_{t_A}^{t_B} \delta L\, dt = \sum_i \int_{t_A}^{t_B} \left[\left(\frac{\partial L}{\partial q_i}\right)\delta q_i + \left(\frac{\partial L}{\partial \dot{q}_i}\right)\delta \dot{q}_i\right] dt,$$

which, by using Lagrange's equations, becomes

$$\delta \int_{t_A}^{t_B} L\,dt = \sum_i \int \left[\frac{d}{dt}\left(\frac{\partial L}{\partial \dot{q}_i}\right) \delta q_i + \left(\frac{\partial L}{\partial \dot{q}_i}\right)\left(\frac{d}{dt}\right) \delta q_i \right] dt$$

$$= \sum_i \int \left[\frac{d}{dt}\left(\frac{\partial L}{\partial \dot{q}_i} \delta q_i\right)\right] dt.$$

Applying (9.176) this last equation becomes

$$\delta \int_{t_A}^{t_B} L\,dt = \sum_i \int_{t_A}^{t_B} \frac{d}{dt}\left[\left(\frac{\partial L}{\partial \dot{q}_i}\right)\Delta q_i - \left(\frac{\partial L}{\partial \dot{q}_i}\right)\dot{q}_i \Delta t\right] dt$$

$$= \sum_i \left[\left(\frac{\partial L}{\partial \dot{q}_i}\right)\Delta q_i - \left(\frac{\partial L}{\partial \dot{q}_i}\right)\dot{q}_i \Delta t\right].$$

The Δq_i's vanish at the endpoints, but Δt does not vanish because the time of transit is not constant. Consequently, upon using the fact that $\partial L/\partial \dot{q}_i = p_i$, we write the last expression as

$$\delta \int_{t_A}^{t_B} L\,dt = -\sum_i p_i \dot{q}_i \Delta t \Big|_{t_A}^{t_B}. \tag{9.180}$$

Combining the above results, the Δ variation of the action integral becomes

$$\Delta A = \left[-\sum_i p_i \dot{q}_i + L + H \right] \Delta t \Big|_{t_A}^{t_B} = 0.$$

This completes the principle of least action which states that the Δ variation of the action integral A as given by (9.173) shall vanish between the limits t_A and t_B.

We can exhibit the principle of least action in a variety of forms. For example, $\sum_i p_i \dot{q}_i = H + L = 2T$, since the system is assumed to be conservative. Therefore, the principle of least action can be written as

$$\Delta \int_{t_A}^{t_B} T\,dt = 0. \tag{9.181}$$

We first consider a single particle. The kinetic energy $T = (m/2)|dr/dt|^2$. Setting $dr^2 = ds^2$, where ds is the element of arc length, and solving for dt, the principle of least action given by (9.181) becomes

$$\Delta \int_{t_A}^{t_B} 2T\,dt = \Delta \int_{t_A}^{t_B} \sqrt{2mT}\,ds = \Delta \int_{t_A}^{t_B} \sqrt{2m(H-V)}\,ds. \tag{9.182}$$

We now extend the form of the least action principle given by (9.182) to a system of n particles. We seek a more general expression for dt in terms of an arc length. To this end, we use a set of generalized coordinates and write the kinetic energy as a quadratic function of the generalized velocities.

We have (using the tensor summation convention)

$$T = \tfrac{1}{2}\sum_{ij} m_{ij}\dot{q}_i\dot{q}_j = \tfrac{1}{2}\sum_{ij} \frac{m_{ij}\,dq_i\,dq_j}{(dt)^2}. \tag{9.183}$$

We now define a new differential $d\rho$ as follows:

$$(d\rho)^2 = \sum_{ij} m_{ij}\,dq_i\,dq_j. \tag{9.184}$$

The differential $d\rho$ as given by (9.184) is well known from differential geometry. It represents the most general form of the element of path length for a trajectory in *configuration space q*. In the application given here the coefficients m_{ij} are the elements of the *metric tensor.* In a Cartesian coordinate system the elements of the metric tensor reduce to $m_{ij} = \delta_{ij}$, as can be seen by comparing the expression $(ds)^2 = (dx)^2 + (dy)^2 + (dz)^2$ with (9.184). Using (9.184) we may put the kinetic energy in the form

$$T = \frac{1}{2}\left(\frac{d\rho}{dt}\right)^2 \quad \text{or} \quad dt = \frac{d\rho}{\sqrt{2T}}. \tag{9.185}$$

Using this relation for dt the least action principle is put in the form

$$\Delta \int_{t_A}^{t_B} T\,dt = \Delta \int_{t_A}^{t_B} \sqrt{T}\,d\rho = \Delta \int_{t_A}^{t_B} \sqrt{H - V(\mathbf{q})}\,d\rho = 0. \tag{9.186}$$

Equation (9.186) is sometimes called *Jacobi's form of the least action principle.* It is the generalization of (9.182) to a system of n particles. It should be emphasized that the Jacobi form of the least action principle is concerned with the path of the system point in configuration space, rather than its motion in time. In fact, the time nowhere appears in the integrand of the right-hand side of (9.186), since $\mathbf{q}$ is independent of time. We shall use the above arguments concerning the principle of least action in the next section on the Hamilton–Jacobi theory in its relation to wave propagation.

9.17. Hamilton–Jacobi Theory and Wave Propagation

It is of interest to show that Hamilton–Jacobi theory can be applied to wave propagation in continuous media. Physicists have applied this theory to geometric optics and quantum mechanics. We continue to consider only those systems that are conservative, so that the Hamiltonian is a constant of the motion and is identified with the total energy. Hamilton's principle function S and the characteristic function W are related by

$$S(\mathbf{q}, \mathbf{P}, t) = W(\mathbf{q}, \mathbf{P}) - Et. \tag{9.187}$$

We now consider the n-dimensional space *configuration space* given by $\mathbf{q} = (q_1, \ldots, q_n)$. Both W and S are embedded in this space. Since W is independent of t, the surfaces of constant W are fixed in the configuration space. However, a surface of constant S moves with time according to (9.187).

Suppose at some time t the surface of constant S corresponds to the surface $W = \text{const.}$ in configuration space. At time $t + dt$ that surface coincides with the surface for which $W = S + E\,dt$. During the time interval dt, the surface $S = \text{const.}$ travels from W to a new surface given by $W + dW$. The surface $S = \text{const.}$ is interpreted as the *wave front*, so that surfaces of constant S may be considered as wave fronts propagating in configuration space. The outward drawn normal at each point on the surface $S = \text{const.}$ gives the direction of the wave or phase velocity. If S is a planar surface then the wave velocity is constant. However, in general, the wave velocity is not constant in space so that the surface $S = \text{const.}$ is not planar. In any case, at each point on the surface, the wave velocity c is given by the normal distance the wave front or surface S moves divided by dt: $c = ds/dt$. From the definition of dW and (9.187) we get

$$\frac{dW}{dt} = \mathbf{grad}\ W \cdot c = E,$$

so that

$$c = \frac{ds}{dt} = \frac{E}{|\mathbf{grad}\ W|}. \tag{9.188}$$

For the case of a single particle without constraints the configuration space $\mathbf{q}$ reduces to three dimensions so that $\mathbf{q} = (x_1, x_2, x_3)$, where we may use rectangular coordinates. $\mathbf{grad}\ W$ is furnished by specializing the Hamilton–Jacobi equation (9.155) to a single particle. We get

$$\begin{aligned} \left(\frac{1}{2m}\right)[(W_x)^2 + (w_y)^2 + (W_z)^2] &= \left(\frac{1}{2m}\right) \mathbf{grad}\ W \cdot \mathbf{grad}\ W \\ &= \left(\frac{1}{2m}\right)(\mathbf{grad}\ W)^2 = E - V. \end{aligned} \tag{9.189}$$

From (9.188) and (9.189) we get the following expressions for c:

$$c = \frac{E}{\sqrt{2m(E - V)}} = \frac{E}{\sqrt{2mT}} = \frac{E}{p}. \tag{9.190}$$

Since, in general $p_i = \partial W/\partial q_i$ we have

$$\mathbf{p} = \mathbf{grad}\ W, \tag{9.191}$$

which clearly means that $\mathbf{grad}\ W$ determines the normal to the surfaces of constant S or W, and thus gives the direction of wave propagation. Any family of surfaces of constant W gives a set of trajectories of possible motion which are always normal to these surfaces. As the particle moves along one of the trajectories the surfaces of S generating the motion will also travel through our one-dimensional configuration space, but the two motions do not keep in step. In fact, when the surfaces move faster the particle slows down, and conversely. This is seen from the reciprocal relation between c and p, as shown by (9.190). Thus far we have specialized to a system consisting of a single particle. But these results also hold for a system of particles, as shown below.

n Particles

Generalizing to a system of n particles, we use the n-dimensional configuration space $\mathbf{q}$ and appeal to the principle of least action. For an ensemble of n particles the system point (consisting of the n particles) moves in a prescribed trajectory in configuration space, so that at each time t the system point on the trajectory represents the state of the system at that time. In the section on the principle of least action we introduced the metric tensor given by (9.184) in connection with configuration space. In terms of the kinetic energy, the metric tensor is such that the arc length $d\rho$ is given by (9.185). In configuration space the wave velocity c of the moving surfaces becomes $c = d\rho/dt$ (using the metric tensor $d\rho$) rather than $c = ds/dt$ for a single particle. Therefore, instead of obtaining (9.190) for the relationship between c and T, we get

$$c = \frac{E}{\sqrt{2(E-V)}} = \frac{E}{\sqrt{2T}}. \tag{9.192}$$

We know that the velocity of the system point is proportional to $\sqrt{T}$, since this velocity depends on the kinetic energy of the system. Thus the reciprocal relation between the wave velocity and the velocity of the system point is preserved in configuration space as shown by (9.192). Also, all the trajectories of the system point are found to be normal to the surfaces of constant S. We then see that the transition from a single particle to an n-particle system introduces no new physical results.

Light Waves

We shall show that the Hamilton–Jacobi equation tells us that the geometrical optics limit of wave motion corresponds to classical mechanics, in the sense that the light rays that are orthogonal to the wave fronts correspond to surfaces of constant S. This is the reason for this excursion into the realm of optics.

The surfaces of constant S were characterized as wave fronts because they propagate in space in the same manner as wave surfaces. We now go further and investigate the phenomena of wave motion using light waves as our model. Light waves are electromagnetic waves which are transverse waves; they therefore involve both the scalar and vector potentials. For simplicity, we consider the scalar wave equation

$$\nabla^2\phi - \frac{n^2}{c_0^2}\phi_{tt} = 0, \tag{9.193}$$

where ϕ is a scalar quantity such as the electromagnetic potential, c_0 is the velocity of light in a vacuum, and n is the index of refraction which is defined as the ratio of c_0 to the velocity of light c in the optical medium. In general, n will be a function of space since it depends on the optical density of the medium. If n is constant then (9.193) is satisfied by the plane wave solution

$$\phi = \phi_0 \exp[i(\mathbf{k}\cdot\mathbf{r} - \omega t)], \tag{9.194}$$

where ϕ_0 is the amplitude which is complex and we take the real part of the right-hand side. Recall that (9.194) represents a progressing wave and that $\mathbf{k}$ is the wave number, sometimes called the "wave vector" (see Chapter 6, (6.35)). We have

$$k = \frac{2\pi}{\lambda} = \frac{n\omega}{c_0} = \frac{\omega}{c}, \tag{9.195}$$

where ω is the frequency and λ is the wavelength. Recall that $\mathbf{r}$ is the radius vector to a point on the wave surface. $\mathbf{k}\cdot\mathbf{r}$ is the projection of $\mathbf{r}$ in the direction of the propagating wave. If $\mathbf{k}$ is independent of $\mathbf{r}$ then the term $\mathbf{k}\cdot\mathbf{r}$ in the exponent of (9.194) means we have plane waves. Since $\mathbf{k}$ depends on n this means the refractive index is constant.

Suppose n is spatially dependent so that $n = n(\mathbf{r})$. At this point we are interested in the *geometrical optics case* where n varies very slowly with $\mathbf{r}$. The spatial variation of n will distort the wave form so that we can no longer obtain plane wave solutions. However, we seek a solution to (9.193) which resembles (9.194) as closely as possible, taking into account the slow variation of n. For simplicity, we take the direction of $\mathbf{k}$ to be along the x axis. Let $k = k_0$ be the wave number in the vacuum where $n = 1$. The plane wave solution (9.194) becomes

$$\phi = \phi_0 \exp[ik_0(nx - c_0 t)]. \tag{9.196}$$

Since we desire a solution to (9.193) closely resembling that given by (9.196) we replace $k_0 nx$ by $k_0\psi(\mathbf{r})$, and ϕ_0 by $e^{A(\mathbf{r})}$, where the amplitude A and ψ are slowly varying functions of $\mathbf{r}$. For the case $n =$ const., ψ reduces to nx. ψ is therefore called the *optical path length* or phase of the wave. ψ is also called the *eiconal.* Then for an "almost planar wave" (9.196) is generalized to

$$\phi = \exp[A(\mathbf{r}) + ik_0(\psi(\mathbf{r}) - c_0 t)]. \tag{9.197}$$

We now insert (9.197) into the wave equation (9.193). To this end, we successively apply the **grad** operator to (9.197) and obtain

$$\mathbf{grad}\ \phi = \phi\ \mathbf{grad}(A + ik_0\psi),$$

$$\nabla^2\phi = \phi[\nabla^2(A + ik_0\psi) + (\mathbf{grad}(A + ik_0\psi))^2]$$

$$= \phi[\nabla^2 A + ik_0\nabla^2\psi + (\mathbf{grad}\ A)^2 - k_0^2(\mathbf{grad}\ \psi)^2 + 2ik_0\ \mathbf{grad}\ A \cdot \mathbf{grad}\ \psi].$$

The wave equation becomes

$$ik_0[2\ \mathbf{grad}\ A \cdot \mathbf{grad}\ \psi + \nabla^2\psi]\phi + [\nabla^2 A + (\mathbf{grad}\ A)^2 - k_0^2(\mathbf{grad}\ \psi)^2 + n^2k_0^2]\phi = 0. \tag{9.198}$$

Since both A and ψ are real, the real and imaginary parts of (9.198) must be set equal to zero. This gives

$$\begin{aligned} \nabla^2 A + (\mathbf{grad}\ A)^2 + k^2(n^2 - (\mathbf{grad}\ \psi)^2) &= 0, \\ 2\ \mathbf{grad}\ A \cdot \mathbf{grad}\ \psi + \nabla^2\psi &= 0. \end{aligned} \tag{9.199}$$

Both equations of (9.199) are exact since no approximations have as yet been made. We now invoke the hypothesis that n varies very slowly with distance. If the distance is of the order of the wavelength λ which means that we consider the case where λ is small or the wave number k_0 is large, by using (9.195) in the form $k_0 = 2\pi/\lambda_0$; it also means the high frequency case since $k_0 = n\omega/c_0$. The small wavelength or high frequency case is just the assumption used in geometrical optics. The second equation of (9.199) remains unchanged. Letting k_0 become very large in the first equation yields

$$(\mathbf{grad}\ \psi)^2 = n^2. \tag{9.200}$$

Equation (9.200) is a very important equation in optics, and in wave phenomena in other media as well. It is called the *eiconal equation. The surfaces of constant optical phase determined by* (9.200) *define the wave fronts.* This important statement allows us to relate geometrical optics to classical mechanics by relating (9.200) to the Hamilton–Jacobi equation for the characteristic function W which is given by (9.189). Thus the eiconal or optical phase in geometrical optics plays the same role as the characteristic function W in classical mechanics, and also n plays the same role as $\sqrt{2m(E - V)}$. The Hamilton–Jacobi equation therefore tells us that classical mechanics corresponds to the geometrical optics† limit of wave motion (high frequency), in which the particle trajectories, orthogonal to the surfaces of constant S, correspond to the light rays which are orthogonal to the wave front or surfaces of constant eiconal. We can also see the correspondence between classical mechanics and geometrical optics, by relating the

† Strictly speaking, geometrical optics deals with light phenomena as light rays, neglecting its wave properties such as diffraction effects.

principle of least action given in the form

$$\Delta \int \sqrt{2mT}\, ds = 0,$$

to Fermat's principle of least time of geometrical optics given by (9.1). This is easily seen since $\sqrt{2mT} = \sqrt{2m(E - V)}$ is analogous to n, and the principle of least action can thus be written as

$$\Delta \int n\, ds = \Delta \int \frac{ds}{c} = \Delta \int \frac{ds}{\lambda(\omega, \mathbf{r})} = 0, \tag{9.201}$$

Equations (9.201) are well-known variations of Fermat's principle of least time. The local phase velocity of light c is a function of the frequency ω and the spatial coordinates $\mathbf{r}$. As the frequency is treated as a constant in varying the functional, and as the local wavelength λ is equal to $2\pi c/\omega$, we may substitute λ for c in the statement of the principle of least time and hence obtain the last equation of (9.201).

9.18. Application to Quantum Mechanics

Quantum mechanics is concerned with the *duality of particles and waves.* This is brought out to some extent in classical mechanics by connecting it with geometrical optics by way of the Hamilton–Jacobi equation, as shown above. However, there is a big hole in relating Hamilton–Jacobi theory to optics. All that we did above was determine that the wavelength must be very much smaller than the spatial portion of the potential, in order to obtain the correspondence between geometrical optics and classical mechanics. We cannot go any further than this in the realm of classical mechanics. This means that we cannot establish the frequencies and wavelengths of the waves associated with motion in classical mechanics. In relating classical mechanics to optics we can only relate it to geometrical optics wherein the wave phenomena of interference and diffraction cannot occur—in geometrical optics a light wave is treated as a ray.

However, if we speculate about what form the wave equation for the potential would take, for which the Hamilton–Jacobi equation represents the short wavelength limit, we arrive at the astonishing conclusion that the optical wave equation for the potential becomes the *Schrödinger equation* of quantum mechanics. In order to reach this conclusion we use the following argument: The similarity of the eiconal equation (9.200) with the Hamilton–Jacobi equation (9.189) tells us that the optical path length or eiconal ψ is proportional to the characteristic function W. If W corresponds to ψ, then $S = W - Et$ must be equal to a constant times the "total phase" of the light wave, which is given by the imaginary part of the exponent in (9.197) so that we have

$$W - Er = (2\pi h) k_0(\psi(\mathbf{r}) - c_0 t),$$

where h is a constant. (The factor 2π is given for convenience.) This tells us that the total energy $E = (2\pi h) k_0 c_0$. Now $k_0 c_0 = \omega = \nu/2\pi$, where ν is the frequency in cycles/sec. This yields

$$E = h\nu, \tag{9.202}$$

Equation (9.202) is of fundamental importance in quantum mechanics. It applies on the molecular level and is the basis for the modern theory of atomic structure. It tells us that the total energy of an atomic or molecular system is "quantized", in the sense that the energy depends on the frequency in a discrete manner. For example, if the electron of a hydrogen atom is excited by an external energy source then it emits a quantum of energy equal to $h\nu$. h is called *Planck's constant.* It is a universal constant (independent of the coordinate system) and is equal to 6.625×10^{-22} joul·sec, having units of (energy) × (time). The frequency ν, wavelength λ, and the wave speed c are connected by

$$c = \lambda\nu.$$

According to Hamilton–Jacobi theory the relationship between p, E, and c is given by (9.190). We therefore get another fundamental equation in quantum mechanics, namely,

$$\lambda = \frac{h}{p}. \tag{9.203}$$

The optical wave equation for the potential given by (9.193) can be put in the form

$$c^2\nabla^2\phi - \phi_{tt} = 0, \tag{9.204}$$

where $c^2 = c_0^2/n^2$ (using the definition that $n = c_0/c$). If we set

$$\phi(\mathbf{r}, t) = u(\mathbf{r})\, e^{i\omega t} = u\, e^{2\pi i c t/\lambda}, \tag{9.205}$$

and insert (9.205) into (9.204), we obtain

$$\nabla^2 u + \left(\frac{4\pi^2}{\lambda^2}\right) u = 0. \tag{9.206}$$

Equation (9.206) is called the *reduced wave equation* corresponding to (9.204). Since $p = \sqrt{2m(E - V)}$, (9.203) becomes $\lambda = h/\sqrt{2m(E - V)}$. Inserting this expression for the wavelength into (9.206) yields

$$\nabla^2 u + \left(\frac{8\pi^2 m}{h^2}\right)(E - V)u = 0. \tag{9.207}$$

Equation (9.207) is the *Schrödinger equation* of quantum mechanics. The function u is called the amplitude or "space factor" of the *wave function* which is a fundamental variable in quantum mechanics.

When (9.203) is applied to bodies of macroscopic dimensions the wavelengths that are obtained are exceedingly small. For example, for a golf ball weighing 47 grams and traveling with a speed of 1 meter/sec the wavelength $\lambda = 1.41 \times 10^{-20}$ meter. This means that diffractions effects are hopelessly beyond the reach of experiment in the case of large-scale bodies. On the other hand, the wavelength becomes appreciable if we apply (9.203) to atomic scale problems. For example, an oxygen molecule with a speed corresponding to the mean thermal energy of 300 K has a wavelength of approximately 1.5×10^{-8} cm, which is a dimension of the order of magnitude of an atomic diameter and X-ray wavelengths. It is therefore clear that diffraction effects play an important role in atomic physics.

The above discussion merely gives a correspondence between classical mechanics, as formulated by the Hamilton–Jacobi equations, and optics as given by the optical wave equation for the potential. It does not shed any light on why quantized energy levels are crucial to quantum mechanics. The real reason for the introduction of quantized energy levels stems from the pioneering work of Max Planck on blackbody radiation, and extended by Einstein, which gives both an experimental and theoretical foundation for quantum mechanics. We cannot go any further in this book on this fascinating subject. The author is planning a book on wave propagation in electromagnetic media (to be published by Springer-Verlag), which will incorporate the subject of wave propagation on an atomic scale where the Schrödinger equation and the corresponding theory will be investigated in some detail.

9.19. Asymptotic Phenomena

We now sketch an approach to the relationship between the classical mechanics of Hamilton–Jacobi theory and the wave phenomena of optics. The technique to be used here is an asymptotic one, which gives us a deeper insight into how we go from geometrical optics to physical optics which allows for wave phenomena such as interference and diffraction. The approach is essentially due to the seminal work of Joseph Keller and his co-workers [22], [23].

It was pointed out that geometrical optics is the limiting case of the optical wave equation, as the wave number k or frequency ω becomes very large. A very powerful method in mathematical physics is that of asymptotic expansions with respect to a physical parameter which is either very large or very small. In our case, the parameter is k and we look for an expansion of time-harmonic solutions of the optical wave equation for the potential ϕ as k becomes very large.

Since we are interested in time-harmonic solutions of the wave equation (9.204), we may focus our attention on the reduced wave equation (9.206), where ϕ can be obtained from u from (9.205). It is more convenient to deal with k than λ so that we rewrite (9.206) in the form

$$\nabla^2 u + k^2 u = 0. \tag{9.208}$$

We assume that $u(\mathbf{r})$ has an asymptotic expansion (for $k \to \infty$) of the form

$$u \sim e^{ik\psi} \sum_{n=0}^{\infty} \frac{v_n(\mathbf{r})}{(ik)^n}. \tag{9.209}$$

By an asymptotic expansion we mean that the ratio of u to the series in the right-hand side of (9.209) tends to unity as $k \to \infty$. The motivation for this form of the series is that we want to separate out the optical path length ψ from the other terms. We attempt to solve for the set of v_n's as functions of space $\mathbf{r}$. v_n is called the nth *expansion coefficient.* It will turn out that we shall obtain a set of *recursion* relationships wherein each v_n is given as a function of v_{n-1}. It is convenient to insert the parameter k in the form $(ik)^{-n}$ in the series since ik appears explicitly as $e^{ik\psi}$.

The procedure is to insert (9.209) into (9.208) and equate coefficients of like powers of $(ik)^{-n}$. To this end, we successively apply the **grad** operator to (9.209) (considered as an equality). It is convenient to rewrite (9.209) in the form

$$u = (e^{ik\psi})v, \tag{9.210}$$

where

$$v = \sum_{n=0}^{\infty} \frac{v_n}{(ik)^n}. \tag{9.211}$$

We obtain the following:

$$\begin{aligned}
\mathbf{grad}\, u &= e^{ik\psi}[\mathbf{grad}\, v + ikv\, \mathbf{grad}\, \psi], \\
\nabla^2 u &= e^{ik\psi}[ik\, \mathbf{grad}\, \psi \cdot (\mathbf{grad}\, v + ikv\, \mathbf{grad}\, \psi) + \nabla^2 v + ik\, \mathbf{grad}(v\, \mathbf{grad}\, \psi)] \\
&= e^{ik\psi}[2ik(\mathbf{grad}\, \psi) \cdot (\mathbf{grad}\, v) - k^2 v(\mathbf{grad}\, \psi)^2 \\
&\quad + \nabla^2 v + ikv(\nabla^2 \psi)].
\end{aligned} \tag{9.212}$$

Inserting (9.212) into (9.208) gives

$$ik(2(\mathbf{grad}\, \psi) \cdot (\mathbf{grad}\, v) + v\nabla^2 \psi) + k^2 v[1 - (\mathbf{grad}\, \psi)^2] + \nabla^2 v = 0. \tag{9.213}$$

Using (9.211) we get

$$\begin{aligned}
\sum_{n=0}^{\infty} (ik)^{-n}[\nabla^2 v_n + ik(2(\mathbf{grad}\, \psi) \cdot (\mathbf{grad}\, v_n) + v_n \nabla^2 \psi) \\
+ k^2 v_n (1 - (\mathbf{grad}\, \psi)^2)] = 0.
\end{aligned} \tag{9.214}$$

In order for (9.214) to hold, the coefficient of each power of $(ik)^{-n}$ must be set equal to zero. This gives

$$\nabla^2 v_n + 2(\mathbf{grad}\ \psi) \cdot (\mathbf{grad}\ v_{n+1}) + v_{n+1}\nabla^2\psi + v_{n+2}[1-(\mathbf{grad}\ \psi)^2] = 0, \qquad v_n = 0 \quad \text{for} \quad n<0. \tag{9.215}$$

Setting $n=-2$ in (9.215) gives

$$v_0[1-(\mathbf{grad}\ \psi)^2] = 0.$$

Since we demand that $v_0 \neq 0$ we obtain

$$(\mathbf{grad}\ \psi)^2 = 1. \tag{9.216}$$

We recognize (9.216) as the eiconal equation (9.200) normalized for a unit index of refraction ($n=1$). Recall that the surfaces of constant optical path length or optical phase define the wave fronts.

Using (9.216), (9.215) becomes

$$v_n\nabla^2\psi + 2(\mathbf{grad}\ \psi)\cdot(\mathbf{grad}\ v_n) = -\nabla^2 v_{n-1}, \qquad n = 0, 1, \ldots; \quad v_{-1} \equiv 0. \tag{9.217}$$

In (9.217) we set the expansion coefficient $v_{-1} \equiv 0$ in order to be able to solve for the expansion coefficient v_0 which becomes

$$v_0\nabla^2\psi + 2(\mathbf{grad}\ \psi)\cdot(\mathbf{grad}\ v_0) = 0. \tag{9.218}$$

Comparing (9.218) with the second equation of (9.199), we see that $v_0 = A$, where $e^A = \phi_0$ the amplitude of the "almost planar wave" given by (9.197).

Returning to (9.217), we observe that

$$(\mathbf{grad}\ \psi)\cdot(\mathbf{grad}\ v_n) = \frac{dv_n}{ds}, \tag{9.219}$$

where s denotes the arc length along a light ray (which is a curve orthogonal to the wave fronts $\psi = \text{const.}$). This means that (9.217) is a recursive system of ODEs along the rays. The solution is easily seen to be

$$v_n(s) = v_n(s_0)\exp\left[-\frac{1}{2}\int_{s_0}^{s}\nabla^2\psi\, ds\right] - \frac{1}{2}\int_{s_0}^{s}\exp\left[-\frac{1}{2}\int_{r}^{s}\nabla^2\psi\, ds'\right]\nabla^2 v_{n-1}(\tau)\, d\tau. \tag{9.220}$$

The first term on the right-hand side of (9.220) is the complementary solution—we see that it only depends on ψ. The second term is a particular integral—which depends on the previous iterate v_{n-1}. Setting $n=0$ in (9.220) yields

$$v_0(s) = v_0(s_0)\exp\left[-\frac{1}{2}\int_{s_0}^{s}\nabla^2\psi\, ds\right]. \tag{9.221}$$

The rays which are orthogonal to the surfaces $\psi = \text{const.}$ are straight lines, as seen from (9.216) which gives $(\mathbf{grad}\,\psi)\cdot(\mathbf{grad}\,\psi) = (d\psi/ds)^2 = 1$, where ds is the element of arc length along a ray, which proves this assertion.

Cylindrical Waves

We now consider waves whose wave fronts are concentric cylindrical surfaces. For the two-dimensional case (in the (x, y) plane) all the rays consist of a family of radiating lines $\theta = \theta_0$ emerging from the origin, where θ_0 is the parameter which determines the ray. The family of wavefronts consists of a family of concentric circles centered at the origin $r = r_0$, where each r_0 determines a wavefront. This means that $\psi = \pm r + \text{const.}$ Without loss of generality we may choose the constant to be zero. We have

$$\nabla^2\psi = \psi_{rr} + \left(\frac{1}{r}\right)\psi_{rr} = \frac{1}{r}.$$

We replace s by r in (9.220) and (9.221) where dr represents an element of arc length along the ray and the parameter θ determines the ray. Each v_n is then a function of r and θ. Then (9.221) becomes

$$v_0(r, \theta) = v_0(\theta)r^{-1/2} \qquad \text{where} \qquad v_0(\theta) = r_0^{1/2}(\theta)v_0(r_0(\theta), \theta), \quad (9.222)$$

where r^{-1} is the *curvature of the wave front*, $r = \text{const.}$ The factor $v_0(\theta)$ is $r_0^{1/2}v_0(r_0)$ evaluated on the ray $\theta = \text{const.}$ at $r = r_0(\theta)$. We have

$$\exp\left[-\frac{1}{2}\int_{s_0}^{s}\nabla^2\psi\,ds\right] = \exp\left[-\frac{1}{2}\int_{r_0}^{r}\frac{dr}{r}\right] = \left(\frac{r_0}{r}\right)^{1/2} = \left[\frac{G(r)}{G(r_0)}\right]^{1/2}, \quad (9.223)$$

where $G(r)$ is the Gaussian curvature or, in our two-dimensional example, the ordinary curvature r^{-1} of the wave front $\psi = \text{const.}$ at the point r on a ray. Then (9.220) can be written in the form

$$v_n(r, \;) = v_n(r_0)\left[\frac{G(r)}{G(r_0)}\right]^{1/2} - \tfrac{1}{2}[G(r)]^{1/2}\int_{r_0}^{r} G^{-1/2}(\bar{r})\nabla^2 v_{n-1}\,d\bar{r}. \quad (9.224)$$

If we now use (9.222) and (9.224) we find by induction that $v_n(r, \theta)$ can be written as the finite series

$$v_n(r, \theta) = \sum_{j=0}^{n} f_{jn}(\theta)r^{-(1/2)-j}. \quad (9.225)$$

Equation (9.225) tells us that $v_n(r, \theta)$ is separable into the product of functions of θ and $r^{-1/2-j}$ (summed over j from 0 to n). Inserting (9.225) into (9.224) yields the following recursive formulas for $f_{jn}(\theta)$:

$$f_{0n}(\theta) = r_0^{1/2}(\theta)v_n(r_0, \theta) - \sum_{j=1}^{n} r_0^{-j}f_{jn}, \qquad n > 1 \quad (9.226)$$

$$f_{jn} = \pm\tfrac{1}{2}j[(j-\tfrac{1}{2})^2 f_{j-1,n-1} + \ddot{f}_{j-1,n-1}], \qquad j \neq 0, \quad n > 1. \quad (9.227)$$

It is clear that r_0 is a function of θ. In (9.227) the $\pm$ signs are used for $\psi = \pm r$, respectively.

Putting the above results together, the asymptotic expansion (9.210) for a cylindrical wave becomes:

$$u(r, \theta) = r^{1/2} e^{+ikr} \sum_{n=0}^{\infty} (ik)^{-n} \sum_{j=0}^{n} f_{jn}(\theta) r^{-j}. \tag{9.228}$$

The coefficients $f_{jn}(\theta)$ can be calculated successively from (9.226) and (9.227). Suppose the asymptotic expansion of $u(r, \theta)$ is given on some ray $r = r(\theta)$. This means that the expansion coefficients $v_n(r_0, \theta)$ are given on the ray. Then (9.228) yields the asymptotic expansion of $\phi(r, \theta)$ for the cylindrical wave of the ray on which we prescribe data (the expansion of u). A variety of examples in the diffraction of plane, cylindrical, spherical waves, etc., is presented in [23].

In this section we attempted to give the reader a brief introduction to the fascinating field of asymptotic expansions vis-à-vis the relationship of the Hamilton–Jacobi theory to optics. We cannot go further into this area here. In the author's forthcoming book on wave propagation in electromagnetic media, this topic will be taken up in more detail.

Bibliography

Note: This bibliography represents the multidisciplined nature of this work. Thus, it contains references to various disciplines of applied mathematics (mathematical physics, theory of functions, tensor analysis, etc.) as well as the physical sciences of fluid and solid mechanics, and water waves. A selected sampling of the pertinent standard works in these fields is given rather than references to the numerous papers in these areas, which would only tend to confuse the nonspecialist reader. Therefore, this bibliography is not complete.

1. Ames, W.F., *Nonlinear Partial Differential Equations in Engineering*, Academic Press, 1966.
2. Bateman, H., *Partial Differential Equations of Mathematical Physics*, Cambridge University Press, 1959.
3. Bellman, R., *Perturbation Techniques in Mathematics, Physics, and Engineering*, Holt, Rinehart and Winston, 1964.
4. Birkhoff, G. and S. MacLane, *A Survey of Modern Algebra*, 4th ed., Macmillan, 1977.
5. Bliss, G.A., *Lectures on the Calculus of Variations*, University of Chicago Press, 1947.
6. Chester, Clive R., *Techniques in Partial Differential Equations*, McGraw-Hill, 1971.
7. Churchill, R.V., *Complex Variables and Applications*, 2nd ed., McGraw-Hill, 1960.
8. Churchill, R. V., *Fourier Series and Boundary Value Problems*, 2nd ed., McGraw-Hill, 1963.
9. Churchill, R.V., *Operational Mathematics*, 3rd ed., McGraw-Hill, 1972.
10. Courant, R. and K.O. Friedrichs, *Supersonic Flow and Shock Waves*, Interscience, 1948.
11. Courant, R. and D. Hilbert, *Methods of Mathematical Physics*, Vol. I—Linear Algebra, Eigenfunction Theory, and Theory of Vibrations, Vol. II—Wave Propagation in Solids and Fluids, Interscience, 1962.
12. Davis, J.L., *Finite Difference Methods in Dynamics of Continuous Media*, McGraw-Hill, 1986.
13. Davis, J.L., *Introduction to Dynamics of Continuous Media*, McGraw-Hill, 1987.
14. Ewing, W.M., W.S. Jardetsky, and F. Press, *Elastic Waves in Layered Media*, McGraw-Hill, 1957.

15. Feynman, R.P., R.B. Leighton, and M. Sands, *The Feynman Lectures on Physics*, Addison-Wesley, 1964.
16. Friedman, B., *Principles and Techniques of Applied Mathematics*, Wiley, 1956.
17. Goldstein, H., *Classical Mechanics*, Addison-Wesley, 1959.
18. Gurtin, M.E., *The Linear Theory of Elasticity, Handbuch der Physik*, Vol. 6a/2, Springer-Verlag, 1965.
19. Hadamard, J., *Lectures on Cauchy's Problem*, Dover, 1952.
20. Hildebrand, F.B., *Advanced Calculus for Applications*, 2nd ed., Prentice-Hall, 1976.
21. Jeffreys, H. and Bertha Jeffreys, *Methods of Mathematical Physics*, 2nd ed. Cambridge University Press, 1950.
22. Keller, J.B., "A Geometric Theory of Diffraction", in *Calculus of Variations and its Applications* (ed., L.M. Graves), *Proceedings of the Symposia on Applied Mathematics*, Vol. 8, pp. 27–52, American Mathematical Society, 1958.
23. Keller, J.B., R.M. Lewis, and B.D. Seckler, "Asymptotic Solution of Some Diffraction Problems", *Comm. Pure Appl. Math.*, **9** (1956), 207–265.
24. Kolsky, H., *Stress Waves in Solids*, Dover, 1963.
25. Lamb, H., *Hydrodynamics*, 1st American ed., Dover, 1945.
26. Landau, L.D. and E.M. Lifshitz, *Theory of Elasticity*, Pergamon Press, 1959.
27. Landau, L.D. and E.M. Lifshitz, *Fluid Mechanics*, Pergamon Press, 1959.
28. Love, A.E.H., *A Treatise on the Mathematical Theory of Elasticity*, Dover, 1944.
29. McLachlin, N.W., *Theory of Vibrations*, Dover, 1951.
30. Milne-Thomson, L.M., *Theoretical Hydrodynamics*, 2nd ed., Macmillan, 1950.
31. Morse, Philip M., *The Theory of Sound* (2 Vols.), 2nd ed., McGraw-Hill, 1949.
32. Rayleigh, J.W.S., *The Theory of Sound* (2 Vols.), 1st American ed., Dover, 1945.
33. Shapiro, A., *The Dynamics and Thermodynamics of Compressible Fluid Flow*, Vol. I, Ronald Press, 1958.
34. Sokolnikoff, I.S., *Mathematical Theory of Elasticity*, 2nd ed., McGraw-Hill, 1956.
35. Sokolnikoff, I.S., *Tensor Analysis*, 2nd ed., Wiley, 1964.
36. Sommerfeld, A., *Mechanics of Deformable Bodies*, Academic Press, 1950.
37. Stoker, James J., *Water Waves*, Interscience, 1957.
38. Timoshenko, S. and J.N. Goodier, *Theory of Elasticity*, McGraw-Hill, 1957.
39. Truesdell, C. and W. Noel, *The Nonlinear Field Theories of Mechanics, Handbuch der Physik*, Vol. III/3, Springer-Verlag, 1965.
40. Whittaker, E.T., *A Treatise on the Analytical Dynamics of Particles and Rigid Bodies*, 4th ed., Dover, 1944.
41. Whittaker, E.T. and G.N. Watson, *A Course of Modern Analysis*, American edition, Cambridge University Press, 1946.

Index